Costin Badica, Marcin Paprzycki (Eds.)

Advances in Intelligent and Distributed Computing

Studies in Computational Intelligence, Volume 78

Editor-in-chief
Prof. Janusz Kacprzyk
Systems Research Institute
Polish Academy of Sciences
ul. Newelska 6
01-447 Warsaw
Poland
E-mail: kacprzyk@ibspan.waw.pl

Further volumes of this series
can be found on our homepage:
springer.com

Vol. 57. Nadia Nedjah, Ajith Abraham and Luiza de Macedo Mourella (Eds.)
Computational Intelligence in Information Assurance and Security, 2007
ISBN 978-3-540-71077-6

Vol. 58. Jeng-Shyang Pan, Hsiang-Cheh Huang, Lakhmi C. Jain and Wai-Chi Fang (Eds.)
Intelligent Multimedia Data Hiding, 2007
ISBN 978-3-540-71168-1

Vol. 59. Andrzej P. Wierzbicki and Yoshiteru Nakamori (Eds.)
Creative Environments, 2007
ISBN 978-3-540-71466-8

Vol. 60. Vladimir G. Ivancevic and Tijana T. Ivacevic
Computational Mind: A Complex Dynamics Perspective, 2007
ISBN 978-3-540-71465-1

Vol. 61. Jacques Teller, John R. Lee and Catherine Roussey (Eds.)
Ontologies for Urban Development, 2007
ISBN 978-3-540-71975-5

Vol. 62. Lakhmi C. Jain, Raymond A. Tedman and Debra K. Tedman (Eds.)
Evolution of Teaching and Learning Paradigms in Intelligent Environment, 2007
ISBN 978-3-540-71973-1

Vol. 63. Wlodzislaw Duch and Jacek Mańdziuk (Eds.)
Challenges for Computational Intelligence, 2007
ISBN 978-3-540-71983-0

Vol. 64. Lorenzo Magnani and Ping Li (Eds.)
Model-Based Reasoning in Science, Technology, and Medicine, 2007
ISBN 978-3-540-71985-4

Vol. 65. S. Vaidya, L.C. Jain and H. Yoshida (Eds.)
Advanced Computational Intelligence Paradigms in Healthcare-2, 2007
ISBN 978-3-540-72374-5

Vol. 66. Lakhmi C. Jain, Vasile Palade and Dipti Srinivasan (Eds.)
Advances in Evolutionary Computing for System Design, 2007
ISBN 978-3-540-72376-9

Vol. 67. Vassilis G. Kaburlasos and Gerhard X. Ritter (Eds.)
Computational Intelligence Based on Lattice Theory, 2007
ISBN 978-3-540-72686-9

Vol. 68. Cipriano Galindo, Juan-Antonio Fernández-Madrigal and Javier Gonzalez
A Multi-Hierarchical Symbolic Model of the Environment for Improving Mobile Robot Operation, 2007
ISBN 978-3-540-72688-3

Vol. 69. Falko Dressler and Iacopo Carreras (Eds.)
Advances in Biologically Inspired Information Systems: Models, Methods, and Tools, 2007
ISBN 978-3-540-72692-0

Vol. 70. Javaan Singh Chahl, Lakhmi C. Jain, Akiko Mizutani and Mika Sato-Ilic (Eds.)
Innovations in Intelligent Machines-1, 2007
ISBN 978-3-540-72695-1

Vol. 71. Norio Baba, Lakhmi C. Jain and Hisashi Handa (Eds.)
Advanced Intelligent Paradigms in Computer Games, 2007
ISBN 978-3-540-72704-0

Vol. 72. Raymond S.T. Lee and Vincenzo Loia (Eds.)
Computation Intelligence for Agent-based Systems, 2007
ISBN 978-3-540-73175-7

Vol. 73. Petra Perner (Ed.)
Case-Based Reasoning on Images and Signals, 2008
ISBN 978-3-540-73178-8

Vol. 74. Robert Schaefer
Foundation of Global Genetic Optimization, 2007
ISBN 978-3-540-73191-7

Vol. 75. Crina Grosan, Ajith Abraham and Hisao Ishibuchi (Eds.)
Hybrid Evolutionary Algorithms, 2007
ISBN 978-3-540-73296-9

Vol. 76. Subhas Chandra Mukhopadhyay and Gourab Sen Gupta (Eds.)
Autonomous Robots and Agents, 2007
ISBN 978-3-540-73423-9

Vol. 77. Barbara Hammer and Pascal Hitzler (Eds.)
Perspectives of Neural-Symbolic Integration, 2007
ISBN 978-3-540-73953-1

Vol. 78. Costin Badica (Ed.)
Intelligent and Distributed Computing, 2008
ISBN 978-3-540-74929-5

Costin Badica
Marcin Paprzycki
(Eds.)

Advances in Intelligent and Distributed Computing

Proceedings of the 1st International Symposium on Intelligent and Distributed Computing IDC'2007, Craiova, Romania, October 2007

With 98 Figures and 19 Tables

Springer

Costin Badica
Software Engineering Department
Faculty of Automatics, Computers and Electronics
University of Craiova
Bvd. Decebal, Nr. 107
Craiova, RO-200440
Romania
E-mail: badica_costin@software.ucv.ro

Marcin Paprzycki
Systems Research Institute
Polish Academy of Sciences
ul. Newelska 6
01-447 Warsaw
Poland
E-mail: marcin.paprzycki@ibspan.waw.pl

Library of Congress Control Number: 2007936364

ISSN print edition: 1860-949X
ISSN electronic edition: 1860-9503
ISBN 978-3-642-09435-4
Springer Berlin Heidelberg New York

Springer is a part of Springer Science+Business Media
springer.com

Softcover reprint of the hardcover 1st edition 2008

Cover design: deblik, Berlin
Typesetting by the SPi using a Springer LaTeX macro package
Printed on acid-free paper SPIN: 12113828 89/SPi 5 4 3 2 1 0

Preface

Intelligent computing is becoming a mature field of information and communication technology, covering a hybrid palette of methods and techniques derived from classical artificial intelligence, computational intelligence, multi-agent systems etc. Intelligent computing is generally known to be computationally intensive from the point of view of resources required: time, memory, bandwidth. As such high resource consumption is one of its main drawbacks, distributed computing is expected to give an impetus to its practical utilization. Moreover, intelligent techniques have also proved their usefulness to the core distributed computing technologies like planning and scheduling, load balancing, replication, resource allocation and management, and its applications. Therefore, the emerging field of intelligent distributed computing can be expected both, to bring a fruitful cooperation and to pose new challenges of adaptation of both areas.

Intelligent and Distributed Computing 2007 – IDC 2007 was the first International Symposium aimed at bringing together researchers involved in intelligent and distributed computing to allow cross-fertilization and search for synergies of ideas and to enable advancement of research in these exciting sub-fields of computer science. IDC was started as an initiative of research groups from: (i) *Systems Research Institute, Polish Academy of Sciences in Warsaw, Poland* and (ii) *Software Engineering Department of the University of Craiova, Craiova, Romania*. IDC 2007 was held in Craiova, Romania during October 18-19, 2007.

This book represents the peer-reviewed proceedings of the IDC 2007. We received 52 submissions from 24 countries. Each submission was carefully reviewed by at least 3 members of the Program Committee. Acceptance and publication was judged based on the relevance to the symposium themes, clarity of presentation, originality and accuracy of results and proposed solutions. Finally 18 regular papers and 11 short papers were selected for presentation and were included in this volume, resulting in acceptance rates of 34,61 % for regular papers and 55.76 % for regular and short papers. The book contains also 5 invited papers authored by well-known researchers and practitioners in the field.

The 34 contributions in this book cover a broad area of topics related to intelligent and distributed computing, systems and applications, including: autonomic and adaptive computing; constraint satisfaction; cooperation and coordination; data mining and knowledge discovery; distributed problem solving and decision making; e-business, e-health and e-learning; genetic algorithms; image processing; information retrieval; intelligence in mobile and ubiquitous computing; knowledge-based systems; modeling with Petri nets; multi-agent systems; neural networks; ontologies and meta-data; planning; rough sets; rule-based systems; services and processes; software agents; uncertainty; XML technologies.

We would like to thank to Prof. Janusz Kacprzyk, editor of *Studies in Computational Intelligence* series and member of the Steering Committee for their kind support and encouragement in starting and organising the IDC 2007 event. We would like to thank to the Program Committee members for their work in promoting the even and refereeing submissions and also to all colleagues who submitted papers to the IDC 2007. We deeply appreciate the efforts of our invited speakers (in alphabetical order): Prof. Frances Brazier, Prof. Gabriel Ciobanu, Dr.Peter Noerr, Prof. Nikos Vlassis, and Prof. Jacek Zurada and thank them for their interesting lectures. Special thanks also go to Prof. Dan Popescu, Dean of the Faculty of Automation, Computers and Electronics, University of Craiova, to Prof. Dumitru Dan Burdescu, Head of the Software Engineering Department, University of Craiova and Director of the "Multimedia Applications Development" Research Centre and to all the members of the Software Engineering Department, University of Craiova for their help with organizing the IDC 2007 event. Last but not least we are grateful to our sponsors (in alphabetical order): *ANCS* (Mrs. Veronica Bubulete) *CS AEIC SA - Automation, Electronics, Informatics, Consulting* (Dr. Mircea Grosu), *MuseGlobal, Inc.* (Mr. Mihai Popa), *NetRom Software SRL* (Mr. Han in't Veld), *SyncRO Soft Ltd.* (Mr. Ştefan Vasile), and *TOP EDGE Engineering* (Dr. Dorian Dogaru).

Craiova, Warsaw
July 2007

Costin Bădică
Marcin Paprzycki

Organization

Organizers

Software Engineering Department, Faculty of Automation, Computers and Electronics, University of Craiova, Romania
"Multimedia Applications Development" Research Centre, Craiova, Romania

Honorary Chair

Mircea Petrescu, Romanian Academy and "Politehnica" University of Bucharest, Romania

Conference Chairs

Costin Bădică, Software Engineering Department, University of Craiova, Romania
Marcin Paprzycki, Systems Research Institute, Polish Academy of Sciences, Poland

Steering Committee

Costin Bădică, Software Engineering Department, University of Craiova, Romania
Janusz Kacprzyk, Systems Research Institute, Polish Academy of Sciences, Poland
Marcin Paprzycki, Systems Research Institute, Polish Academy of Sciences, Poland

Organizing Committee

Costin Bădică, Software Engineering Department, University of Craiova, Romania
Dumitru Dan Burdescu, Software Engineering Department, University of Craiova, Romania
Dan Popescu, Faculty of Automation, Computers and Electronics, University of Craiova, Romania

Invited Speakers

Frances Brazier, Department of Computer Science, Vrije Universiteit, Amsterdam, Netherlands
Gabriel Ciobanu, Romanian Academy and "A.I.Cuza" University of Iaşi, Romania
Peter Noerr, CTO, MuseGlobal, Inc., Salt Lake City, USA
Nikos Vlassis, Department of Production Engineering and Management, Technical University of Crete, Greece
Jacek Zurada, Department of Electrical and Computer Engineering, University of Louisville, Louisville, USA

Program Committee

Janos Abonyi, University of Veszprem, Hungary
Galia Angelova, Bulgarian Academy of Sciences, Bulgaria
Nick Bassiliades, Aristotle University of Thessaloniki, Greece
Giacomo Cabri, Universita' di Modena e Reggio Emilia, Italy
David Camacho, Universidad Autonoma de Madrid, Spain
Jen-Yao Chung, IBM T.J. Watson Research Center, USA
Gabriel Ciobanu, "A.I.Cuza" University of Iaşi, Romania
Valentin Cristea, "Politehnica" University of Bucharest, Romania
Luminiţa Dumitriu, "Dunarea de Jos" University of Galaţi, Romania
Vadim A. Ermolayev, Zaporozhye National University, Ukraine
Adina Magda Florea, "Politehnica" University of Bucharest, Romania
Chris Fox, University of Essex, UK
Maria Ganzha, Elblag University of Humanities and Economics, Poland
Adrian Giurca, Brandenburg University of Technology at Cottbus, Germany
Henry Hexmoor, Southern Illinois University at Carbondale, USA
De-Shuang Huang, Chinese Academy of Sciences, China
Axel Hunger, University of Duisburg-Essen, Germany
Mirjana Ivanović, University of Novi Sad, Serbia
Halina Kwasnicka, Wroclaw University of Technology, Poland
Ioan Alfred Leţia, Technical University of Cluj-Napoca, Romania
Heitor Silverio Lopes, Federal University of Technology - Parana, Brazil
Yannis Manolopoulos, Aristotle University of Thessaloniki, Greece
Urszula Markowska-Kaczmar, Wroclaw University of Technology, Poland
Beniamino Di Martino, Second University of Naples, Italy
Alexandros Nanopoulos, Aristotle University of Thessaloniki, Greece
Pavol Navrat, Slovak University of Technology in Bratislava, Slovakia
Viorel Negru, Western University of Timişoara, Romania
Ngoc-Thanh Nguyen, Wroclaw University of Technology, Poland
Mihaela Oprea, University Petroleum-Gas of Ploieşti, Romania
Doru Pănescu, "Gh.Asachi" University of Iaşi, Romania
George A. Papadopoulos, University of Cyprus, Cyprus

Elena Paslaru Bontas Simperl, Free University of Berlin, Germany
Dana Petcu, Western University of Timişoara, Romania
Jaroslav Pokorny, Charles University, Czech Republic
Shahram Rahimi, Southern Illinois University, USA
Omer Rana, Cardiff University, UK
Leszek Rutkowski, Czestochowa University of Technology, Poland
Yucel Saygin, Sabanci University, Turkey
Philippe Trigano, University of Technology of Compiegne, France
Laurenţiu Vasiliu, National University of Ireland, Ireland
Lucian Vinţan, Academy of Technical Sciences, Romania
Peter Vojtas, Charles University, Czech Republic
Gerd Wagner, Brandenburg University of Technology at Cottbus, Germany

Contents

Part I

Invited Papers

The world of autonomous distributed systems

Frances Brazier

IIDS group, CS Department,
Vrije Universiteit Amsterdam
de Boelelaan 1081a
1081 HV Amsterdam
The Netherlands
frances@cs.vu.nl

Today's world is dynamic, distributed and connected in time, space and tasks. Systems, both human and automated, interact continuously in highly dynamic environments. Some connections are relatively stable, others change very frequently, as systems and their needs change. Virtual organisations of autonomous systems are formed/emerge on the basis of characteristics such as availability, reliability, interests. Similarity, difference and/or other measures are the basis for evolving formations of organisations.

Different paradigms for distributed autonomous system development exist: p2p systems, the Grid, Autonomic Computing, Service Oriented Systems, Ambient Systems, to Multi-agent systems. Load balancing, crisis management, swarm applications, supply change management, energy management, data center management, traffic management, recommender systems, are examples of domains in which one or more of these approaches to distributed autonomous system design have been applied.

From the perspective of the users, as participants in such distributed systems, the precise technology is not of importance. The implications of the use of the technology is. This paper addresses a number of the issues shared by these paradigms and identifies a need for a framework for an understanding of the implications of the deployment of autonomous systems from the perspective of the human user..

1 Autonomous systems – the underlying technology

In most of the paradigms for distributed autonomous systems, autonomous systems have some implicit or explicit knowledge of the characteristics of their owner or the organisation they represent, of the tasks they pursue (or goals) , of their own 'reasoning' ability, of other systems' characteristics and roles in relation to their own. Autonomous systems also have some knowledge of trust

Frances Brazier: *The world of autonomous distributed systems*, Studies in Computational Intelligence (SCI) **78**, 3–7 (2008)
www.springerlink.com

relationships either implicitly or explicitly, of interaction/negotiation options, and of policies with respect to information sharing. Knowledge of their role in relation to other such systems is also often explicit.

The autonomous systems in virtual organisations, are often physically distributed and represent heterogeneous entities/institutes/organisations with different levels of accessibility, authorisation, and authentication. The can also change over time: systems come and go, as do connections. Most paradigms support both *uncoordinated* group formation , based solely on the individual system's initiative, and *coordinated* formation as the result of local management assigned to a coordinator within a virtual organisation.

In all of these paradigms interaction between systems can be *structured or unstructured, secure or not secure* .

Within the p2p paradigm, for example, interaction between peers, can be completely unstructured (eg flooding) or it can be structured (eg dht), it can be completely uncoordinated or coordinated (eg super peers), message passing can secure or not, depending on the design choices made. Characteristic of the p2p paradigm is scalability as a design criterion, thus the aim to limit the amount of information exchanged (note that this is not the same as the number of messages).

Within the Multi-Agent System paradigm, another example, the same variation with respect to structure, coordination and security is possible: interaction between agents can be structured (following eg FIPA interaction patterns) or not, can be coordinated (eg by a mediator agent) or uncoordinated, and interaction can be secure (eg JADE-S) or not.

As the paradigms can provide the same functionality there is no real reason for a user to need to know which paradigm is used. The user is interested in the options a technology provides to provide transparency.

2 User perspective – a need for transparency

Transparency is a necessary condition for user acceptance of autonomous systems: transparency of system use (eg the interface), transparency of task performance, but also transparency of responsibility and liability. Integrity of individual systems (both the underlying supportive middleware and the autonomous system's themselves) and integrity of interaction between autonomous systems, are important. Guaranteeing integrity comes at a cost. A user needs to understand the balance and the risks taken in relation to the technology chosen. The same thing holds for confidentiality of information. In most cases confidentiality is of importance and needs to be guaranteed.

For the Courts of Law, for example, for which a distributed multi-agent prototype system is currently being developed for distributed management of digital dossiers for criminal offenses, supported by the Public Prosecution, complete transparency and traceability is mandatory. The advantages of a distributed digital dossier in a physically distributed environment for the digitalisation of digital dossiers for criminal records, in which information is provided by heterogeneous entities/institutes/organisations with different levels of accessibility, authorisation, authentication, lies primarily in the timeliness of the data involved, consistency, correctness and efficiency. Transparency and complete traceability are very strong requirements for the virtual organization. Interaction within this well-defined trusted virtual organization is necessarily well-structured, coordinated and secure. Authorised users interact with the autonomous systems representing their own organizations. Figure 1 illustrates the interactions involved.

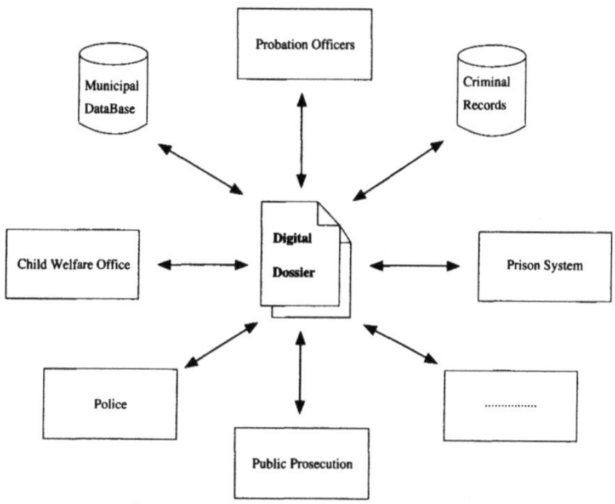

Fig. 1. Interaction with the Digital Dossier as used at a Court of Law

In other situations, such as, for example, mediated resource allocation, users interact with trusted mediators representing (virtual) organizations of resource providers. Users will want to be able to verify that all interaction between themselves and a mediator are confidential and securely logged for future reference if needed. A user does not, however, necessarily need to know about the underlying system on which offers made by a mediator are based. A user may, however, have the right to know why a request is, for example, is not honoured. To this purpose a mediator will need to be able to provide a rationale based on its own logs of interaction with the providers in its virtual organisation. Figures 2 and 3 illustrate two different virtual organisations: a well-structured, coordinated architecture, and an emergent, uncoordinated organisation, for which these logs will be needed.

An example of a virtual organisation of distributed autonomous systems for which different rules hold are open movie recommendation systems. Recommendation systems are based on user preferences and similarities. Users need to know that the value of recommendations depends on the trustworthiness of other users' data, and that the algorithms deployed do not guarantee

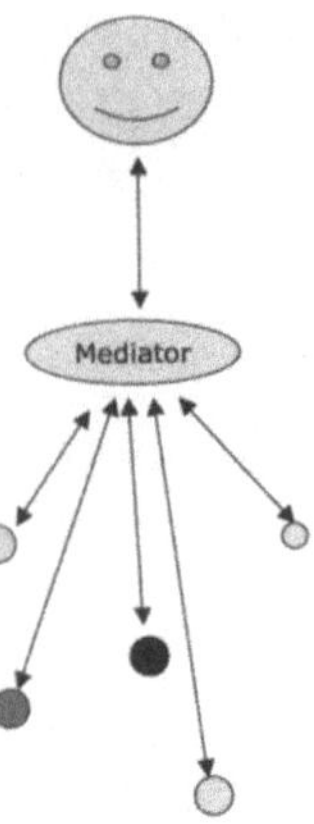

Fig. 2. A well-structured, coordinated virtual organisation

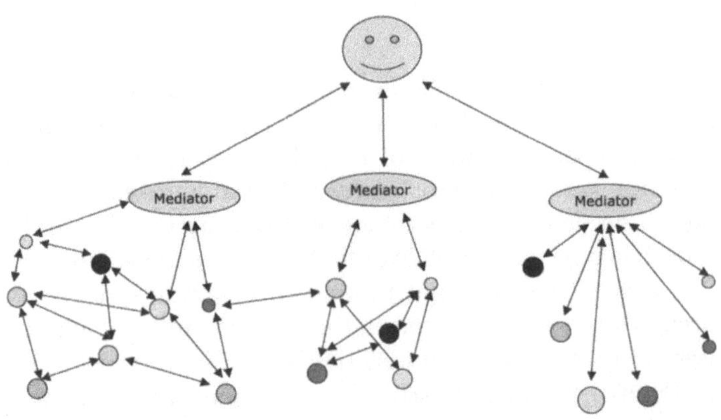

Fig. 3. An emergent, uncoordinated virtual organisation

successful recommendations. Insight in the risks and cost involved in this example, is transparent. Figure 4 depicts the user's role in a dynamic virtual organisation – as one of many.

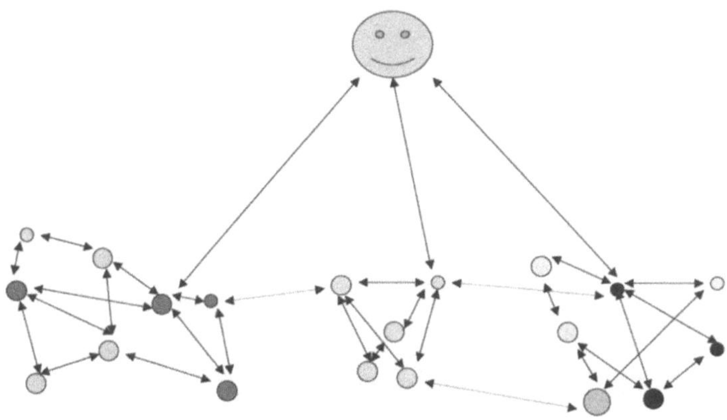

Fig. 4. The user's role in a dynamic virtual organisation

3 Discussion

For users to accept and use distributed autonomous systems, transparency is mandatory. Confidentiality, integrity, responsibilities and liabilities need to be transparent, as do the technological, legal and social implications of system failure. Our current research addresses these issues together with the development of technology to support secure and dedicated technology to support transparency in distributed autonomous systems. Technology to support coordinated and uncoordinated virtual organisations with structured and unstructured interaction patterns providing security mechanisms for confidentiality.

Acknowledgments

The author is grateful to the IIDS group and Martijn Warnier, in particular, for their contributions to this paper. The author is also grateful to Stichting NLnet, the NWO Token project and the BSIK-ICIS project for financial support.

1

Time and Space Coordination of Mobile Agents

Gabriel Ciobanu

"A.I.Cuza" University, Faculty of Computer Science
Blvd. Carol I no.11, 700506 Iaşi, Romania
Romanian, Academy, Institute of Computer Science
gabriel@info.uaic.ro

Summary. We study the evolution of agents able to move by using local knowledge and various migration means. We present two models of distributed systems with an explicit notion of location and some quantitative notions including time and capacity. We use these models for controlling mobility in open distributed systems by means of timers, bounded capacities and a simple migration primitive.

1.1 Introduction

The technology of agent systems, both hardware and software, is rather advanced. However design principles and techniques to define and verify their correct behaviour are at a more primitive stage. One approach in modelling the system behaviour has been the design of formal calculi in which the fundamental concepts underlying agent systems can be described and studied. In this paper we propose two formalisms for describing the behaviour of mobile agents in a distributed world. They are based on two existing formalisms called distributed π-calculus and ambient calculus, to which we add a network layer, a migration action, and other quantitative notions useful for coordination in time and space of the mobile agents.

A first approach could be given by defining a simple formal language for describing the systems in terms of their structure, namely how they are constructed from individual interacting agents. A semantic theory is defined in order to understand the behaviour of systems described in such a language, in terms of their ability to interact. Here a system consists of a finite number of agents which intercommunicate using a fixed set of named communication channels. This set of channels constitutes a connection topology through which all communication takes place; it includes both communication between agents, and between the system and its users.

The current agent systems are highly dynamic. Moreover, it is possible to create new communication links with other entities, and perhaps relinquish existing links. The π-calculus [10] and ambient calculus [4] are two formalisms seeking to address at least some dynamic aspects of such agents. The π-calculus includes the dynamic generation of communication channels and thus allows the underlying connection

Gabriel Ciobanu: *Time and Space Coordination of Mobile Agents*, Studies in Computational Intelligence (SCI) **78**, 9–23 (2008)
www.springerlink.com

topology to vary as systems evolve; it also allows private communication links to be established and maintained between agents. Ambient calculus is more oriented to dynamic aspects, working with ambients which represents units of movement.

Many concepts of the distributed systems are at most implicit (if not existing) in the π-calculus. Perhaps the most useful missing concept is that of domain (location), understood generally as a locus for computational activity. Thus one could view a distributed system as consisting of a collection of domains, each capable of hosting computational processes (agents), which in turn can migrate between domains. We use an extension of the π-calculus in which these domains have an explicit representation as locations. Distributed π-calculus is a formalism for agent systems in which dynamically created domains are hosts to resources which may be used by agents, and agents reside in domains, migrating eventually between domains for the purpose of using locally defined resources. Moreover, we use types to manage access control to resources in distributed systems. A domain may wish to restrict to certain agents the access to certain resources; we can think of resources having capabilities associated with them. Then domains may wish to distribute selectively to agents such capabilities on its local resources, and have agents manipulating these capabilities. Therefore in distributed π-calculus a system consists of a collection of domains, hosting agents, which can autonomously migrate between domains. These domains also host local channels on which agents communicate; but more generally these may be used to model local resources, accessible to agents currently located at the host domain. We describe the mobile agents by using a new primitive *go* $l.\ P$ which enables migration between domains. If an agent executing *go* $l.\ P$ is currently residing at k, it migrates to the domain l and continue there with the execution of P.

We define some quantitative ingredients over the systems described by the distributed π-calculus in order to coordinate the mobile agents in time and space. We add timers to distributed π-calculus. We assume a global notion of time, but we rather use a relative time of interactions given by timers. The global clock advances the time, and the interactions happen whenever the involved resources are available according to an interval of time given by timers of the interacting components (timers define timeouts for various resources, making them available only for a determined period of time). Using timers, we can control the concurrency of the components, and we can select between different choices in the system evolution. This provides a natural and flexible synchronization technique able to integrate and regulate dynamically the possible evolutions of the components. In a second formalism given by an extension of the mobile ambients, interaction in space relates to the ability to identify a specific domain in which a required and available resource exists. We use timers and timeout recovery processes over mobile ambients such that the resulting formalism provides a flexible coordination even in an open distributed environment where a number of possibly unknown entities are interacting. Several challenges related to these systems include finding the appropriate mechanisms to deliver adequate responses to time-critical demands, appropriate interaction methods, a decentralized control taking into account the time and space resources.

1.2 Timed Distributed π-calculus

Distributed π-calculus ($D\pi$) is an extension with types and locations of the π-calculus [10]. $D\pi$ is presented in [9], and provides a theoretical framework for describing communications between distributed processes with restricted resource access. In [6] we extend the distributed π-calculus by introducing timers over channel names in order to define timeouts for communications. The resulting formalism is called timed distributed π-calculus ($tD\pi$). Over this formalism we define a coordination of the whole system by assigning specific values to timers and defining a set of time constraints [7].

In $tD\pi$, waiting for a communication on a channel is no longer indefinite (like in $D\pi$); if no communication happens in a predefined interval of time, the waiting process goes to another state. This approach leads to a method of sharing the channels in time. The timer Δt of each channel makes the channel available for communication only for the period of time determined by the discrete value t. We consider timers for both input and output channels. The reason for adding timers to outputs comes from the fact that in distributed systems we have both multiple clients and multiple servers. This means that clients may switch from one server to another depending on the waiting time. To simplify our presentation we choose a simpler π-calculus and omit the syntax for *matching* or *summation*. A communication channel is considered a fixed resource at a location.

The syntax of *Input* and *Output* communication uses a pair of processes. For instance, an *Input* expression $a^{\Delta t}?(X : T).(P, Q)$ evolves to P whenever a communication is established during the interval of time given by Δt; otherwise it evolves to Q. The variable X is considered bound only in P and we should provide its type T; the type system is presented in [6].

Table 1: *Syntax of timed distributed π-calculus*

$u ::= x$	Variable Name	$P, Q ::= stop$	Termination
$\mid a^{\Delta t}$	Timed Channel	$\mid P \mid Q$	Composition
$l ::= x$	Variable Name	$\mid (\nu u : A)P$	Channel Restriction
$\mid k$	Location Name	$\mid go\ l.(P, Q)$	Movement
$v ::= bv$	Base Value	$\mid u!\langle v\rangle.(P, Q)$	Output
$\mid u \mid l$	Name	$\mid u?(X : T).(P, Q)$	Input
$\mid u@l$	Located Name	$\mid {}^{*}P$	Replication
$\mid (v_1,..,v_n)$	Tuple of Values	$M, N ::= M \mid N$	Composition
$X ::= x$	Variable	$\mid (\nu u@l : T)N$	Located Restriction
$\mid X@l$	Located Variable	$\mid l[[P]]_{\Gamma}$	Located Process
$\mid (X_1,..,X_n)$	Tuple of Variables		

Two channels are equal $a_1^{\Delta t_1} = a_2^{\Delta t_2}$ if and only if $a_1 = a_2$ and $t_1 = t_2$. Waiting indefinitely on a channel a is allowed by considering Δt as ∞. For example, an output process defined by the expression $a^{\infty}!\langle v\rangle.(P, Q)$ awaits forever to send the value v, simulating the behaviour of an output process in untimed π-calculus. In the expression below, two processes are running in parallel and can interact along the common channel a:

$$a^{\Delta t}!\langle v\rangle.(P,Q) \mid a^{\Delta t'}?(X:T).(P',Q') \longrightarrow P \mid P'\{^{v}/_{X}\}$$

We define type environment Γ as sets of *location types*. The purpose of the type environment associated with a specific process is to restrict the range of accessible resources the process can access. Formally, $\Gamma \subseteq \mathcal{L} \times K$ is a relation associating to a location name a location type. A location type is a set of *location capabilities* which may contain *channel types*, *move* capability (i.e., permission to migrate to that location), or *channel creation* capability (i.e., permission to create channels).

We extend the channel types of $D\pi$ with timers of the form Δt. Communication is now permitted on channels only in the interval of time given by the timer value t (i.e. until the timer of the channel type expires). These timers define the existence of the channel types inside the type environment. Timers decrease with each "tick" of an universal clock (we assume that we have an universal clock). Upon expiration, the channel types are discarded. Timers are created once with the channel types, and are activated when the types are added to the type environment. When the processes receive new channel names, types for the new channels become available. It means that the processes can communicate on the new channels according to the new types. For example, if a process receives through an input channel a located name $a@k$, then it gains the capability to move to location k, and to communicate on channel a.

We define a function ψ which affects only the set of capabilities. It decreases the timers of the channel types and removes the types with an expired timer. By removing channel types, it is possible to get location types with only *go* capability (we call them *empty locations*). A process can move to an empty location, but there it does not have the capability to perform any action, and consequently produces a *runtime error*. Thus ψ removes also the empty locations.

The passage of time is formalized by a *time-stepping function* ϕ_Δ defined over the set $\mathcal{P}_\Delta$ of tagged located processes. The possible communications are performed at every tick of the universal clock. Active channels are those that could be involved in these communications. ϕ_Δ affects the active channels which do not communicate at the tick of the universal clock (the channels involved in communication disappear together with their timers). Due to timers, the capabilities can be lost, which leads to "errors". We define ϕ_Δ to check the existence of the needed types and change the process accordingly. As ϕ_Δ decreases the channel timers, we extend it to take care of the type environments (by applying the cleanup function ψ). In the definition of ϕ_Δ we omit the channel type and the transmitted message in the input and output processes for brevity. For the *go k* syntax if the location type contains the capability *go*, then R is executed; if k is not defined in Γ, then Q is executed. If *go* is not present, the process is considered to do something against its permissions and an error is generated.

Well-typedness of processes is defined by a set of static rules (a detailed presentation of the static typing rules is given in [6]). These rules express the behaviour of a process with regard to its types. The subtyping relation <: is similar to the subtyping relation of $D\pi$, and $\Gamma(k)$ represents a type environment at location k. If a process tries to communicate on a channel for which it has no capability, it can still be well-typed if the alternative process Q is well-typed. Q is called the *safety process*.

Definition 1. *(Tagged time-stepping function)*
We define $\phi_\Delta : \mathcal{P}_\Delta \to \mathcal{P}_\Delta$, where Γ' is obtained by application of the cleanup function ψ. Note that we use a concise notation $a^{\Delta t}.(R, Q)$ to stand for both $a^{\Delta t}!\langle v\rangle.(R, Q)$ and $a^{\Delta t}?(X : T).(R, Q)$.

$$\phi_\Delta(l[[P]]_\Gamma) = \begin{cases} k[[R]]_{\Gamma'} & \text{if } P = go\,k.(R,Q) \text{ and } \Gamma(k) <: loc\{go\} \\ l[[Q]]_{\Gamma'} & \text{if } P = go\,k.(R,Q) \text{ and } k \notin dom(\Gamma) \\ l[[a^{\Delta(t-1)}.(R,Q)]]_{\Gamma'} & \text{if } P = a^{\Delta t}.(R,Q),\ t > 1 \text{ and } t \neq \infty \\ l[[Q]]_{\Gamma'} & \text{if } P = a^{\Delta t}.(R,Q),\ t \leq 1 \\ l[[Q]]_{\Gamma'} & \text{if } P = a^{\Delta t}.(R,Q),\ t > 1 \text{ and } \Gamma \not<: \Gamma(l,a) \\ \phi_\Delta(l[[R]]_\Gamma) \mid \phi_\Delta(l[[Q]]_\Gamma) & \text{if } P = R \mid Q \\ (\nu a@l{:}A)\phi_\Delta(l[[R]]_{\Gamma\{a@l:A\}}) & \text{if } P = (\nu a : A)R \\ l[[P]]_{\Gamma'} & \text{otherwise} \end{cases}$$

We write $\Gamma \vdash P$ and say that *process P is well-typed with respect to type environment Γ*; we also write $\Gamma \vdash_k P$ and say that *P is well-typed to run at location k.* To say that $P = a^{\Delta t}!\langle v\rangle.(R, Q)$ is well-typed to run at location k, with respect to type environment Γ, the following statements should hold: *(i)* $\Gamma \vdash_k v : T$ which means that v is a well-formed value at location k of type T; *(ii)* $\Gamma \vdash_k a : res\{w\langle T\rangle\}\Delta t$ which means that channel a exists at location k and may communicate values of type T for another t units of time; *(iii)* $\Gamma \vdash_k R$; $\Gamma \vdash_k Q$ which means that R and Q are well-typed at location k. For a tagged located process $k[[P]]_\Delta$, the well-typedness relation is denoted by $\Vdash$, and it is defined by using the well-typedness relation $\vdash_k$ for a process P running at location k.

Since the function ψ changes the capability set Γ by removing channel and location types, we are interested if the process is still well-typed under the new Γ'. The following lemma relates the typing environment of the processes with the passage of time.

Lemma 1. *(Well-typedness is preserved by the cleanup function)*
If $\Gamma \Vdash l[[P]]_\Delta$ then $\Gamma \Vdash \psi(l[[P]]_\Delta)$.

We consider the tagged located processes ranged over by N and M (e.g., N represents $l[[P]]_\Gamma$). We denote by $\nrightarrow$ the fact that rules (R_Γ-COM1) and (R_Γ-COM2) cannot be applied. Using these notations, we give the following reduction rules providing a dynamic semantics for $tD\pi$.

$$(R_\Gamma\text{-IDLE})\ \frac{l[[P]]_\Gamma \nrightarrow}{l[[P]]_\Gamma \to \phi_\Delta(l[[P]]_\Gamma)}$$

$$(R_\Gamma\text{-COM1})\ \frac{\Gamma(l,a) <: res\{r\langle T\rangle\}}{\begin{array}{c} l[[a^{\Delta t}!\langle v\rangle.(P,Q)]]_\Delta \mid l[[a^{\Delta t'}?(X:T).(P',Q')]]_\Gamma \to \\ \psi(l[[P]]_\Delta) \mid \psi(l[[P'\{^v/_X\}]]_{\Gamma\sqcap\{v@l:T\}}) \end{array}}$$

$$(R_\Gamma\text{-COM2})\ \frac{\Gamma(l,a) <: res\{ro\langle T\rangle\}}{\begin{array}{c} l[[a^{\Delta t}!\langle v\rangle.(P,Q)]]_\Delta \mid l[[a^{\Delta t'}?(X:T).(P',Q')]]_\Gamma \to \\ \psi(l[[P]]_\Delta) \mid \psi(l[[P'\{^v/_X\}]]_\Gamma) \end{array}}$$

$$(\mathrm{R}_\Gamma\text{-PAR})\ \frac{N \to N' \quad M \to M'}{N \mid M \to N' \mid M'} \qquad (\mathrm{R}_\Gamma\text{-RES})\ \frac{N \to N'}{(\nu a@l:T)N \to (\nu a@l:T)N'}$$

$$(\mathrm{R}_\Gamma\text{-CONG})\ \frac{N \equiv N' \quad N \to M \quad M \equiv M'}{N' \to M'}$$

We have several results, and here we mention two of them (more details are in [6]). These results claim that the passage of time does not interfere with the typing system, and that once well-typed, a process remains well-typed.

Proposition 1. *If* $\Gamma \Vdash l[[P]]_\Delta$ *then* $\Gamma \Vdash \phi_\Delta(l[[P]]_\Delta)$.

Theorem 1. *(Subject Reduction)*
For all tagged located processes

(a) If $N \equiv N'$ *then* $\Gamma \Vdash N$ *if and only if* $\Gamma \Vdash N'$.
(b) If $N \to N'$ *then* $\Gamma \Vdash N$ *if and only if* $\Gamma \Vdash N'$.

1.2.1 Interaction in Timed Distributed π-calculus

Throughout this paper we consider the following example which can motive and illustrate our approach. We assume the situation of a student finishing his lectures, moving out the university building and looking for an available vehicle in order to move to another location In front of the university there are a tram stop, a bus stop, and taxicabs. The student has the possibility to use any of the three types of vehicles: bus, tram and cab to reach the target location. The student can have a priority among them: the tram has the highest priority, followed by the bus, and finally the cab. Our scenario involves a tram, a bus and two cabs (a hired cab and an available cab). The trams and the buses are moving according to predetermined schedules which are known by the student.

In what follows we describe a system that contains two location and a bus. The other elements of the initial system (tram and cabs) could be described in a similar manner.

$$\begin{aligned}
in_{univ} &= bus^{\Delta t_2}_{univ}?(i).bus^{\Delta t_1}_{univ}!\langle stud_{univ}\rangle.in_{univ}\\
in_{camp} &= bus^{\Delta t_3}_{camp}?(i).bus^{\Delta t_4}_{camp}!\langle stud_{camp}\rangle.in_{camp}\\
bus &= go\ camp.bus^{\Delta t_6}_{camp}!\langle stud_1\rangle.bus^{\Delta t_7}_{camp}?(i).go\ univ.\\
&\quad\ bus^{\Delta t_8}_{univ}!\langle stud_2\rangle.bus^{\Delta t_5}_{univ}?(i).bus\\
system &= univ[[in_{univ}]] \mid bus \mid camp[[in_{camp}]]
\end{aligned}$$

where

$stud_1$ represents the number of students getting out of *bus* at *camp*,
$stud_2$ represents the number of students getting out of *bus* at *univ*,
$stud_{univ}$ represents the number of students getting into *bus* at *univ*, and
$stud_{camp}$ represents the number of students getting into *bus* in *camp*.

A possible evolution of the system is given by

$$system = camp[[in_{camp}]] \mid univ[[in_{univ} \mid bus]] =$$

$$= camp[[in_{camp}]] \mid univ[[bus^{\Delta t_1}_{univ}!\langle stud_{univ}\rangle.bus^{\Delta t_2}_{univ}?(i).in_{univ} \mid bus^{\Delta t_5}_{univ}?(i).$$

$$go\ camp.bus^{\Delta t_6}_{camp}!\langle stud_1\rangle.bus^{\Delta t_7}_{camp}?(i).go\ univ.bus^{\Delta t_8}_{univ}!\langle stud_2\rangle.bus]]$$

$$\rightarrow camp[[in_{camp}]] \mid univ[[bus^{\Delta t_2}_{univ}?(i).in_{univ} \mid go\ camp.$$

$$bus^{\Delta t_6}_{camp}!\langle stud_1\rangle.bus^{\Delta t_7}_{camp}?(i).go\ univ.bus^{\Delta t_8}_{univ}!\langle stud_2\rangle.bus]]$$

$$\rightarrow camp[[bus^{\Delta t_3}_{camp}?(i).bus^{\Delta t_4}_{camp}!\langle stud_{camp}\rangle.in_{camp} \mid bus^{\Delta t_6}_{camp}!\langle stud_1\rangle.$$

$$bus^{\Delta t_7}_{camp}?(i).go\ univ.bus^{\Delta t_8}_{univ}!\langle stud_2\rangle.bus]] \mid univ[[bus^{\Delta t_2}_{univ}?(i).in_{univ}]]$$

$$\rightarrow camp[[bus^{\Delta t_4}_{camp}!\langle stud_{camp}\rangle.in_{camp} \mid$$

$$bus^{\Delta t_7}_{camp}?(i).go\ univ.bus^{\Delta t_8}_{univ}!\langle stud_2\rangle.bus]] \mid univ[[bus^{\Delta t_2}_{univ}?(i).in_{univ}]]$$

$$\rightarrow camp[[in_{camp} \mid go\ univ.bus^{\Delta t_8}_{univ}!\langle stud_2\rangle.bus]] \mid univ[[bus^{\Delta t_2}_{univ}?(i).in_{univ}]]$$

$$\rightarrow camp[[in_{camp}]] \mid univ[[bus^{\Delta t_2}_{univ}?(i).in_{univ} \mid bus^{\Delta t_8}_{univ}!\langle stud_2\rangle.bus]]$$

$$\rightarrow camp[[in_{camp}]] \mid univ[[in_{univ} \mid bus]] \ = \ system$$

We use different (coloured) letters to emphasize the interacting elements.

We do not have a notion of *capacity* for the mobile process bus. In this description there is no way to remember the number of persons inside the bus, or the number of the persons who entered or exited the bus at each stop. Moreover, the notion of space used in $tD\pi$ is flat. A more realistic account of physical distribution is obtained using a hierarchical representation of space (locality) as in the ambient calculus [4].

1.3 Mobile Ambients with Quantitative Ingredients

Ambient calculus is a formalism for describing distributed and mobile computation in terms of ambients [4]. In contrast with other formalisms for mobile processes such as the π-calculus [10] where the computational model is based on the notion of *communication*, the ambient calculus is based on the notion of *movement*. An ambient is the unit of movement. The mobility of an ambient is controlled by the capabilities *in*, *out*, and *open*. In an ambient we have processes which may exchange messages. Several variants (e.g., [3, 13]) have been proposed by adding and/or removing features of the original calculus [4]. We extend mobile ambients by adding timers to ambients, capabilities, input and output actions. A timer Δt of each temporal resource makes the resource available only for a determined period of time t. We use timeout recovery processes acting when a timer expires. We define the available resources by a notion of capacity, and the resources consumed by an ambient when it moves into another ambient. A notion of domain defines the local computation resources making the ambient able to plan its movements and its navigation means. We think of a domain when we speak about proximity, local computation support or local knowledge. A movement changes the domain (local computation support, knowledge), and the ambient has a different domain when arriving at the destination. The ambients can adapt their behaviour according to the additional ingredients, depending mainly on timers, capacity and and local computation support (knowledge). We denote by $n^{\Delta t}_{(l,h,d)}[P]$ the fact that an ambient n has assigned a timer Δt, a *capacity* l representing the number of its free resources, a *weight* h counting the resources needed by a process to be executed, and a domain d. According to a global clock, the timer is decreasing its value; the timer Δt expires when $t = 0$. We use a pair $(n^{\Delta t}_{(l,h,d)}[P], Q)$ to denote a timed ambient, where Q represents a *safety process* used whenever the timer expires. If nothing happens in t units of time, the ambient $n^{\Delta t}_{(l,h,d)}$ is dissolved, process P running inside the ambient is reduced to $\mathbf{0}$, and safety process Q is executed. If $Q = \mathbf{0}$ we can simply write $n^{\Delta t}_{(l,h,d)}[P]$ instead of $(n^{\Delta t}_{(l,h,d)}[P], Q)$. The behaviour of an untimed mobile ambient is given by using ∞ instead of Δt, and 0 instead of h, l and d. We use a pair of processes for the input, output and movement processes. The process $open^{\Delta t} n.(P, Q)$ evolves to P whenever in time Δt the process becomes sibling to an ambient n; otherwise evolves to Q. The process $!\langle m\rangle^{\Delta t}.(P, Q)$ evolves to P whenever in time Δt the process becomes sibling to a process which is willing to capture the name m; otherwise evolves to Q.

We define various new technical ingredients. The domain of an ambient n represents the local knowledge given by the computation support (operating system, specific software); it includes information about the ports of movement, specific transport means and interaction links. The domain d is different for each location where the ambient can move using its movement capabilities.

The syntax of the coordinated mobile ambients (cMA) is defined in the following table:

Table 2: *Syntax of coordinated mobile ambients*

				$P, Q ::=$	processes
n, m		names			
x, y		variables		$\mathbf{0}$	inactivity
M	$::=$	capabilities		$M^{\Delta t}.(P, Q)$	movement
	in n		can enter n	$(n^{\Delta t}_{(l,h,d)}[P], Q)$	ambient
	out n		can exit n	$P \mid Q$	composition
	open n		can open n	$(\nu n)P$	restriction
				$!\langle m\rangle^{\Delta t}.(P, Q)$	output action
				$?(x)^{\Delta t}.(P, Q)$	input action
				$*P$	replication
				$P + Q$	choice

The coordinated mobile ambients use a discrete time. The passage of time is described by a (discrete) time-stepping function ϕ_Δ defined over the set $\mathcal{P}$ of cMA-processes. The possible actions are performed at every tick of a universal clock. ϕ_Δ affects the ambients and their capabilities which are not consumed. The consumed capabilities disappear together with their timers. If a capability or ambient has the timer equal to ∞, we use the equality $\infty - 1 = \infty$ when applying the time-stepping function ϕ_Δ. This function modifies a process accordingly with the global passage of time.

Definition 2. *We define the time-stepping function* $\phi_\Delta : \mathcal{P} \to \mathcal{P}$ *by*

$$\phi_\Delta(P) = \begin{cases} M^{\Delta(t-1)}.(R,Q) & \text{if } P = M^{\Delta t}.(R,Q),\ t > 0 \\ Q & \text{if } P = M^{\Delta t}.(R,Q),\ t = 0 \\ \phi_\Delta(R) \mid \phi_\Delta(Q) & \text{if } P = R \mid Q \\ (\nu n)\phi_\Delta(R) & \text{if } P = (\nu n)R \\ (n^{\Delta(t-1)}_{(l,h,d)}[\phi_\Delta(R)], Q) & \text{if } P = (n^{\Delta t}_{(l,h,d)}[R], Q),\ t > 0 \\ Q & \text{if } P = (n^{\Delta t}_{(l,h,d)}[R], Q),\ t = 0 \\ P & \text{if } P = *R \text{ or } P = \mathbf{0} \\ \phi_\Delta(R) + \phi_\Delta(Q) & \text{if } P = R + Q \end{cases}$$

The local communication given by the interaction between $!\langle m\rangle^{\Delta t}.(P,Q)$ and $?(x)^{\Delta t}.(R,Q)$ inside the same ambient does not consume time, so it is not included in the definition of ϕ_Δ. To see how this function is used, look at the reduction rules (Table 4).

The semantics of the timed mobile ambients is given by two relations: a structural congruence relation and a reduction relation. Structural congruence provides a way of re-arranging expressions such that the interacting parts can be brought together. The structural relation $\equiv_p$ over the timed mobile processes is the least relation satisfying the axioms and rules from the following Table:

Table 3: *Structural congruence*

(S-Refl)	$P \equiv_p P$ **(S-Sym)** $P \equiv_p Q$ implies $Q \equiv_p P$
(S-Trans)	$P \equiv_p R,\ R \equiv_p Q$ implies $P \equiv_p Q$
(S-Res)	$P \equiv_p Q$ implies $(\nu n)P \equiv_p (\nu n)Q$
(S-LPar)	$P \equiv_p Q$ implies $R \mid P \equiv_p R \mid Q$
(S-RPar)	$P \equiv_p Q$ implies $P \mid R \equiv_p Q \mid R$
(S-Repl)	$P \equiv_p Q$ implies $*P \equiv_p *Q$
(S-Amb)	$P \equiv_p Q$ and $R \equiv_p R'$ implies $(n^{\Delta t}_{(l,h,d)}[P], R) \equiv_p (n^{\Delta t}_{(l,h,d)}[Q], R')$
(S-Cap)	$P \equiv_p Q$ and $R \equiv_p R'$ implies $M^{\Delta t}.(P,R) \equiv_p M^{\Delta t}.(Q,R')$
(S-Par Com)	if $weight(P) = 0$ then $P \mid Q \equiv_p Q \mid P$
(S-Par Assoc)	$(P \mid Q) \mid R \equiv_p P \mid (Q \mid R)$
(S-Repl Par)	$*P \equiv_p P \mid *P$
(S-Res Res)	$(\nu n)(\nu m)P \equiv_p (\nu m)(\nu n)P$ if $n \neq m$
(S-Res LPar)	$(\nu n)(P \mid Q) \equiv_p P \mid (\nu n)Q$ if $(n) \notin fn(P)$
(S-Res RPar)	$(\nu n)(P \mid Q) \equiv_p (\nu n)P \mid Q$ if $(n) \notin fn(Q)$
(S-Res Amb)	$(\nu n)(m^{\Delta t}_{(l,h,d)}[P], Q) \equiv_p (m^{\Delta t}_{(l,h,d)}[(\nu n)P], Q)$ if $n \neq m$
(S-Zero Par)	$P \mid \mathbf{0} \equiv_p P$ **(S-Zero Res)** $(\nu n)\mathbf{0} \equiv_p \mathbf{0}$ **(S-Zero Repl)** $*\mathbf{0} \equiv_p \mathbf{0}$

Resources are preserved through reduction only in a *closed* system, namely a system which is surrounded by an ambient which cannot be opened, and any ambient can pass through it. In an *open* system, the resources are not preserved through reduction because a process may acquire new resources from the environment, or transfer resources to the environment. Some resources may become restricted, and so unavailable for any other process, e.g. $(\nu n)(n^{\Delta t}_{(l,h,d)}[P], Q)$.

For processes $M^{\Delta t}.(P, Q)$, $!\langle m\rangle^{\Delta t}.(P, Q)$ and $?(x)^{\Delta t}.(P, Q)$ the timers of P are activated only after the dissapearance of $M^{\Delta t}$, $!\langle m\rangle^{\Delta t}$, and $?(x)^{\Delta t}$, respectively. To preserve the timers of P, we define the time-preserving function $\psi : \mathcal{P} \to \mathcal{P}$ by $\phi_\Delta(\psi(P)) = P$. We denote by $\not\rightarrow$ the fact that rules (R-In), (R-Out), (R-Open) and (R-Com) cannot be applied. Note that the ambients can interact using these rules only if they are on the same domain (this expresses a certain topological requirement suggesting that they are close enough to interact), and the network support allows the application of these rules. The evolution is given by the reduction rules of Table 4. The function ϕ_Δ decreases the timers, and for the expired timers it discards the actions, capabilities and ambients. If one process evolves by one of the rules (R-In), (R-Out), (R-Open) and (R-Amb), while another one does not perform any reduction, then one of the rules (R-LPar), (R-RPar) should be applied. If more than one process evolve in parallel by applying one of the rules (R-In), (R-Out), (R-Open) and (R-Amb), then rule (R-Par) should be applied. When rules (R-In), (R-Out), (R-Open) and (R-Com) cannot be applied anymore, the rule (R-GTProgress) is applied to simulate the global passage of time, and so to permit the ambients to participate in other reductions in the next unit of time.

Table 4: *Reduction rules*

(R-GTProgress) $$\frac{P \not\rightarrow}{P \to \phi_\Delta(P)}$$

(R-In) $$\frac{h' \le l''}{\begin{array}{c}(n^{\Delta t'}_{(l',h',d)}[in^{\Delta t} m.(P, P') \mid Q], S') \mid (m^{\Delta t''}_{(l'',h'',d)}[R], S'') \to \\ (m^{\Delta t''}_{(l''-h',h'',d)}[(n^{\Delta t'}_{(l',h',d)}[\psi(P) \mid Q], S') \mid R], S'')\end{array}}$$

(R-Out) $$\frac{h'' \le l}{\begin{array}{c}(k^{\Delta t}_{(l,h,d)}[(m^{\Delta t'}_{(l',h',d')}[(n^{\Delta t''}_{(l'',h'',d')}[out^{\Delta t} m.(P, P') \mid Q], S'') \mid R], S')], S) \to \\ (k^{\Delta t}_{(l-h'',h,d)}[(n^{\Delta t''}_{(l'',h'',d')}[\psi(P) \mid Q], S'') \mid (m^{\Delta t'}_{(l'+h'',h',d')}[R], S')], S)\end{array}}$$

(R-Open) $$\frac{-}{\begin{array}{c}(m^{\Delta t'}_{(l',h',d')}[open^{\Delta t} n.\,(P, P') \mid (n^{\Delta t''}_{(l'',h'',d'')}[Q], S'')], S') \to \\ (m^{\Delta t'}_{(l'+l'',h',d')}[\psi(P) \mid Q], S')\end{array}}$$

(R-Com) $$\frac{-}{!\langle m\rangle^{\Delta t}.(P, Q) \mid ?(x)^{\Delta t'}.(P', Q') \to P \mid P'\{m/x\}}$$

(R-Amb) $$\frac{P \to Q}{(n^{\Delta t}_{(l,h,d)}[P], R) \to (n^{\Delta t}_{(l,h,d)}[\psi(Q)], R)}$$

(R-LPar) $$\frac{P \to Q}{R \mid P \to R \mid Q}$$ (R-RPar) $$\frac{P \to Q}{P \mid R \to Q \mid R}$$

(R-Par) $$\frac{P \to Q,\ P' \to Q'}{P \mid P' \to Q \mid Q'}$$ (R-Res) $$\frac{P \to Q}{(\nu n)P \to (\nu n)Q}$$

(R-Struct) $$\frac{P' \equiv_p P,\ P \to Q,\ Q \equiv_p Q'}{P' \to Q'}$$

We denote by $P \xrightarrow{\phi_\Delta}_t Q$ the fact that process P evolves to process Q after applying the rule (R-GTProgress) for $t \ge 0$ times, and with $t\phi_\Delta(R)$ the fact that function ϕ_Δ is applied t times to R. The passage of time cannot cause a nondeterministic behaviour.

Proposition 2. *If* $P \equiv_p Q$, $P \xrightarrow{\phi_\Delta}_t P'$ *and* $Q \xrightarrow{\phi_\Delta}_t Q'$ *then* $P' \equiv_p Q'$.

We can compare the behaviour of the agent systems by defining a bisimulation in cMA which requires processes to match their time passages. A binary relation $\mathcal{R}$ over processes is a strong simulation if whenever $(P, Q) \in \mathcal{R}$, if $P \xrightarrow{\phi_\Delta}_t P'$ then there exists Q' such that $Q \xrightarrow{\phi_\Delta}_{t'} Q'$, $t = t'$ and $(P', Q') \in \mathcal{R}$. A binary relation $\mathcal{R}$ is said to be a strong bisimulation if both $\mathcal{R}$ and $\mathcal{R}^{-1}$ are strong simulations. We say that P and Q are strongly bisimilar, written $P \sim_t Q$, if there exists a strong bisimulation $\mathcal{R}$ such that $P\mathcal{R}Q$. In this way we have a strong bisimulation for each $t \geq 0$. We say that P and Q are weakly bisimilar, written $P \approx_t Q$, if there exists a weak bisimulation $\mathcal{R}$ such that $P\mathcal{R}Q$.

Proposition 3.

i) *If* $P \equiv_p Q$ *then* $P \sim_t Q$*, for all* $t \geq 0$.
ii) $\sim_t$ *is an equivalence relation, and* $P \sim_t Q$ *implies* $P \approx_t Q$.
iii0 $\approx_t$ *is an equivalence relation.*

1.4 Evolution in Coordinated Mobile Ambients

We use the following example as a metaphor which motivates and illustrates our approach. We assume the situation of a process (student) having the goal of moving to a well-defined location. There are several transportation means with different properties: tram, bus, and taxi. Let us consider that the student can reach a tram stop, a bus stop, and two cabs. The student has the possibility to use any of the three types of transportation to reach the target location. The bus and the tram could move according to a predetermined schedule. Based on the local computation resources, the student can plan the movement to the target location. For an ambient A, $free\_resources(A)$ represents the (dynamically evolving) capacity l of A.

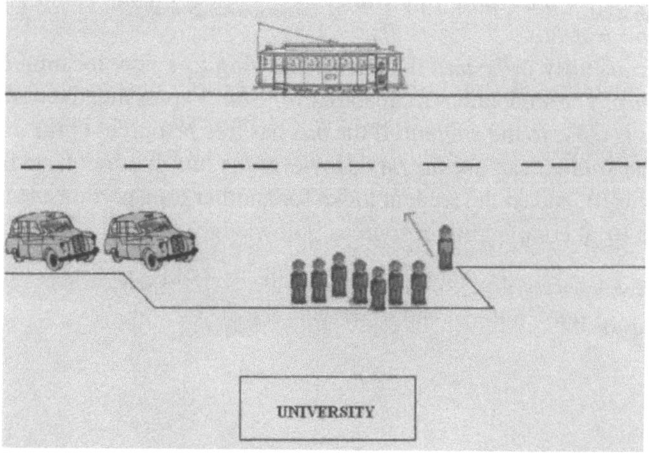

We encode the agents involved in this example; each of them is encoded as an ambient having the corresponding label. An ambient *student* has the possibility to use any of the three transport ambients *tram, bus* and *cab* to reach the destination. Since the transport means have different costs, it is possible to define a priority among them: the *tram* has the highest priority, followed by the *bus*, and finally by the *cab*. Based on the local computation resources, an agent can plan its movement from the current location to the target location where a specific goal should be fulfilled.

$$univ^{\infty}_{(l\_univ,h\_univ,d\_univ)}[student^{\infty}_{(1,1,d\_student)}[\ in^{\Delta t_{12}} tram\ ] \mid tram^{\infty}_{(l\_tram,h\_tram,d\_tram)}[\ out^{\Delta t_5} univ\ ]\ ] \text{ and } d\_student \neq d\_tram$$

Even if the student has a capability $in^{\Delta t_{12}} tram$ expressing the wish of using the tram, the *tram* is not close enough to the *student* ($d\_student \neq d\_tram$), and so it cannot be used by the student. After t_5 units of time the tram activates its capability $out^{\Delta t_5} univ$ and moves out of the domain *univ*.

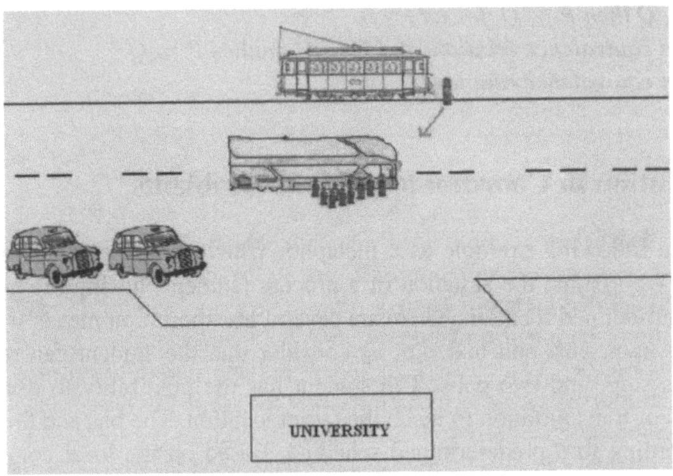

$$univ^{\infty}_{(l\_univ,h\_univ,d\_univ)}[student^{\infty}_{(1,1,d\_student)}[\ in^{\Delta t_{10}} bus\ ] \mid bus^{\infty}_{(l\_bus,h\_bus,d\_bus)}[\ out^{\Delta t_2} univ\ ]\ ] \mid tram^{\infty}_{(l\_tram,h\_tram,d\_tram)}[\ in^{\Delta 6} camp\ ]$$
and $d\_student = d\_bus$

By the capability $in^{\Delta 6} camp$ the tram is moving to a new location (domain) denoted by *camp*. The student has a capability $in^{\Delta t_{10}} bus$ expressing the wish of using the bus, and *bus* is close to the *student*. If the *bus* has free resources ($free\_resource(bus) > 0$), then the *student* can use the *bus*. However the *bus* does not have free resources (its capacity is 0), and so the student looks for another transport means in his domain by using the local computation resources (knowledge).

$$univ^{\infty}_{(l\_univ,h\_univ,d\_univ)}[student^{\infty}_{(1,1,d\_student)}[\ in^{\Delta t_{10}} bus\ ] \mid bus^{\infty}_{(0,h\_bus,d\_bus)}[\ out^{\Delta t_2} univ\ ]\ ] \text{ and } d\_student = d\_bus$$

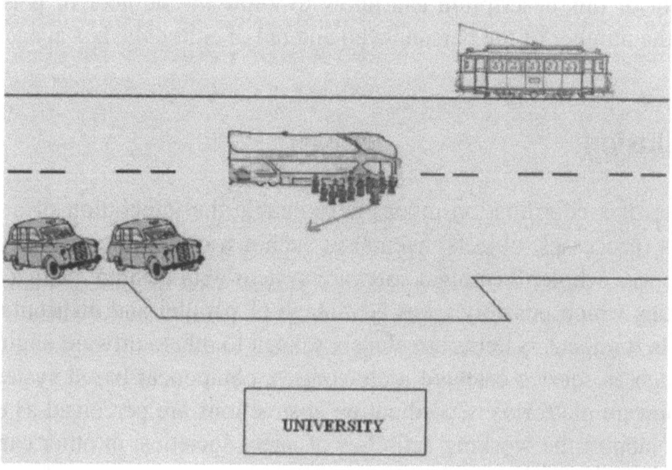

Then the *student* looks for a *cab*; if the *cab* is in the domain of the *student* and it has free resources $free\_resources(cab) > 0$, then the *student* enters the *cab*. One cab is not available (its capacity is 0). The other one is available with capacity 4; once hired, its capacity becomes 0 (it does not depend on the number of passengers).

$univ^{\infty}_{(l\_univ,h\_univ,d\_univ)}[student^{\infty}_{(1,1,d\_student)}[\ in^{\Delta t_{14}} cab\ ]\ |$
$cab^{\infty}_{(4,4,d\_cab1)}[\ out^{\Delta t_7} univ]\ |\ cab^{\infty}_{(0,4,d\_cab2)}[\ out^{\Delta t_7} univ]\ ]\ |$
$bus^{\infty}_{(l\_bus,h\_bus,d\_bus)}[\ in^{\Delta 6} camp\ ]$ and $d\_student = d\_cab1$

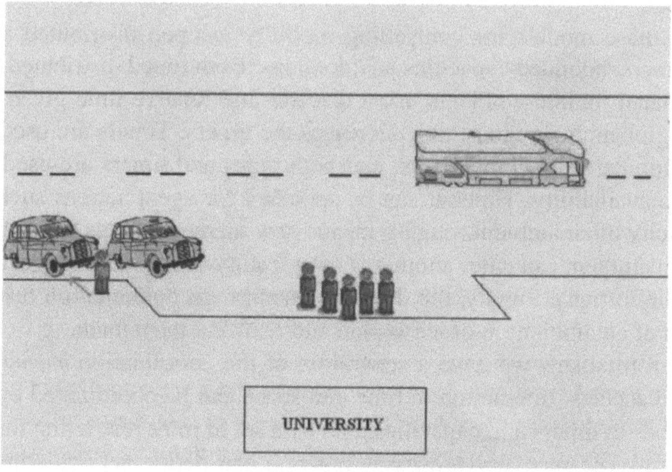

If none of the above steps are possible, the time-stepping function ϕ_{Δ} is applied, and then the evolution conditions are verified again.

This example describes the coordination in time and space given by the proposed formalism based on ambient calculus. The new ingredients used over mobile ambients are timers, timeout recovery processes, domains and capacity. A slight improvement of this description can allow to know the number of persons inside the bus, or the number of the persons who entered or exited the bus at each stop, etc.

1.5 Conclusion

The purpose of a coordination model is to enable the integration of a number of components (processes, objects, agents) in such a way that the resulting ensemble can execute as a whole, forming a software system with desired characteristics and functionalities which possibly takes advantage of parallel and distributed systems. Such models for agent systems are closely related to other software engineering approaches such as service-oriented architectures, component-based systems and related middleware platforms. Coordination abstractions are perceived as essential to design and support the working activities of agent societies; in other cases, service coordination, orchestration, and choreography can become essential aspects of the next generations of systems based on Web services.

The formal approaches presented in this paper are based on certain new ingredients: explicit domains, timers, timeout recovery processes and capacity. We consider first a flat representation of the space (timed distributed π-calculus), and then a hierarchical representation of the space by using ambient calculus extended with the new ingredients. Formal semantics and some basic results are presented. A non-monotonic behaviour of the systems are given by the timeout recovery processes (agents). We use an example to describe the coordination in time and space of a moving agent.

We use these models for controlling mobility in open distributed systems by means of timers, bounded capacities and domains. Both timed distributed π-calculus and coordinated mobile ambients use a discrete and relative time given by timers, based on a global clock whose tick decreases the timers. Timers are used to restrict the interaction between components, and both types and timers are used to control the resource availability. Timeout can be specified for agent actions such as in and out. A capacity of an ambient roughly means how many ambients can enter it (more precisely, an ambient can enter another if the capacity of the latter is greater than the weight of the former). Finally, the domains manage the computation resources and information of an ambient in order to plan and realize a movement.

In both formalisms we have a separation of the coordination aspects from the computation aspects. Interaction in time and space can be coordinated by assigning specific values to timers and capacities, and by a set of rules restricting the evolution of the system. The recovery processes (agents) can define actions when we have more than one interaction choice.

A related paper is [1] where more technical results are presented. Timed distributed π-calculus is presented in [6] as an extension of the distributed π-calculus by timers over channel names in order to define timeouts for communications. Over this formalism we define a coordination of the whole system by assigning specific values to timers and defining a set of time constraints [7].

Acknowledgements

The author thanks to Bogdan Aman for his collaboration on the concerning timed mobile ambients, and to Cristian Prisacariu for his collaboration on the timed distributed π-calculus. Many thanks to Oana Agrigoroaiei for useful comments.

References

1. B. Aman, G. Ciobanu. Timers and Proximities for Mobile Ambients. Lecture Notes in Computer Science vol.4649, Springer, 2007.
2. N. Busi, P. Ciancarini, R. Gorrieri, G. Zavattaro. Coordination Models: A Guided Tour. *Coordination of Internet Agents: Models, Technologies, and Applications*, Springer, 6–24, 2001.
3. M. Bugliesi, G. Castagna, S. Crafa. Boxed Ambients. *Theoretical Aspects of Computer Software*, Lecture Notes in Computer Science vol.2215, Springer, 38–63, 2001.
4. L. Cardelli, A. Gordon. Mobile Ambients. *Foundations of Software Science and Computation Structures*, Lecture Notes in Computer Science vol.1378, Springer, 140–155, 1998.
5. G. Ciobanu. Interaction in Time and Space. *Proceedings of Foundations of Interactive Computation*, 45–61, 2007.
6. G. Ciobanu, C. Prisacariu. Timers for Distributed Systems. *Electronic Notes in Theoretical Computer Science* vol.164(3), 81–99, 2006.
7. G. Ciobanu, C. Prisacariu. Coordination by Timers for Channel-Based Anonymous Communications. *Electronic Notes in Theoretical Computer Science* vol.175, 3-17, 2007.
8. M. Hennessy. *A Distributed pi-calculus.* Cambridge University Press, 2007.
9. M. Hennessy, J. Riely. Resource access control in systems of mobile agents, *Information and Computation* vol.173(1), 82–120, 2002.
10. R. Milner. *Communicating and Mobile Systems: the π-calculus.* Cambridge University Press, 1999.
11. G.A. Papadopoulos. Models and Technologies for the Coordination of Internet Agents: A Survey. *Coordination of Internet Agents: Models, Technologies, and Applications*, Springer, 25–56, 2001.
12. G.A. Papadopoulos, F. Arbab. Coordination Models and Languages. *Advances in Computers* vol.46, Academic Press, 329–400, 1998.
13. D. Teller, P. Zimmer, D. Hirschkoff. Using Ambients to Control Resources. *Concurrency Theory* , Lecture Notes in Computer Science vol.2421, Springer, 288-303, 2002.

The Difficulty of Search and an Interesting Solution

Peter Noerr

CTO MuseGlobal, Inc.

Summary. Fundamentally, the current searching paradigm does not work well. Current searching assigns keywords to documents; users think of keywords of their own and these keywords are then matched against those of the body of documents. When the same word appears there is a match and the document is returned to the user.

This relies on a number of assigned keywords on the document side, and usually a single search term from the user. Matching is a very hit or miss affair, involving problems of terminology, matching, indexing, relevance ranking and others which mean users are as often frustrated as happy with the results.

We will discuss a solution which enables a search system to create and maintain a richer, deeper context for both the document and the user, and employs novel matching and ranking technology to provide better answers.

Part of the technology requires a broad enough universe for searching which requires the use of federated search of disparate resources. This brings its own problems of syntactic and semantic normalisation as well as the intelligent utilisation of the results to provide better answers for the user.

1 The Search Process

There are two conventional measures of the performance of an Information Retrieval (or "Search") system - Recall and Precision. These are numerical measures of the basic user complaints:

"I didn't get anything useful"

"I got way too much stuff"

This paper attempts to look at a couple of reasons for these complaints and offer methods, and a working system, which attempt a novel approach to solving them.

Conventional inverted index search methods operate on a premise that is often not realised in the real world. A search as an activity is comprised of two distinct sets of operations which take place separated by a considerable period of time, and which are performed by different actors.

Peter Noerr: *The Difficulty of Search and an Interesting Solution*, Studies in Computational Intelligence (SCI) **78**, 25–34 (2008)
www.springerlink.com

Indexing

Documents are processed to assign them keywords. These may be manually assigned or computed from the content of the document and may include input from the context of the collection of documents being searched. The keyword are then processed by a search engine to create one or more indexes which may then be used to find the original documents.

Searching

The user sends keywords to the search engine, and these are matched by the search engine with the keywords in its inverted index(es) and a set of documents is produced as a result set.

No assumptions are made in the above about the existence or need for:

- different indexes (author, title, subject term, word),
- different search computations (threshold based),
- set operations on results (Boolean operators),
- search languages and their syntax and functionality,
- type of matching (exact character, wild characters, fuzzy matching),

Any of these search methodologies follow the same high level methods outlined above.

In fact even search engines not based in some fashion in inverted indexes (such as a full text scan or filter) follow the important precept above that there are terms from the documents, and there are terms from the users and the two must match for a record to be a 'hit'.

There exist search methodologies which do not rely on term matching and utilize other methods to characterize both the document and the query and to match them one at a time or via matrix operations. Although these systems utilize very different technology, the arguments and suggestion put forward below still hold as they treat the actual search process as a black box process and are only concerned with inputs and outputs. It should also be noted that these technologies are not currently in widespread use and are really excluded from this work.

Note that this paper is concerned with the search process. That is; the extraction of potentially relevant documents from a collection. It is not, per se, concerned with the important presentation aspects of a working user system. Conventional search systems display the documents to the user as an ordered list of results. The ordering of this list is the subject of much work on ranking so that the most relevant documents are presented first, thus saving user time. This issue of ranking will be touched on later, but is not the main focus of this paper. There are other forms of presentation of results as clusters or trees or relationship graphs and other exotic visual effects - collectively called "visualisation". These visualisations are not the subject of this paper and it will be assumed that conventional list presentations are used - though opportunities are created from this work for various forms of visualisation.

2 Three Search Mis-matches

2.1 Where to search

The most significant real-world problem, is that of looking (searching) in the right place. The user will obtain mediocre results at best if s/he looks in the wrong databases, at worst s/he will retrieve no results at all. In some cases the "right" place to search is well known and certain databases provide extensive core coverage of a subject whether it is INSPEC for physicists and engineers, or Biotechnology Abstracts for biotechnologists. However many subjects and even whole disciplines are not served by 'obvious' core databases. Equally the emerging cutting edges of many established disciplines are covered by the convergence of multiple bodies of knowledge and technologies. For these the user must look in multiple places to find comprehensive coverage and also to be sure of finding the unique.

Even if the user knows all the relevant databases and has access to them, a problem remains in that the user now has to sequentially apply the query to each of those search systems to get results from each one, and then is tasked with combining the results. This combination itself is no simple task as the documents are in different formats and at varying levels of completeness. And there are bound to be duplicates. All these mechanical, but extremely time consuming and error prone tasks must be accomplished to ensure even reasonably complete coverage of the query.

2.2 Extent of representation

In the matching process at the heart of all searching, the document targets are represented by a number of terms. The number is often limited only by the available information about the document, such as all the author's names. Often system limits are placed so that only the 20 "best" terms are used. The 20 is completely arbitrary and may indeed be expanded in the case of automatic indexing to all extracted terms after they are processed to remove duplicates and normalize their grammatical structure. It is very unusual for a document to be represented by as few as 2 or 3 terms to represent all of its content and the nuances therein.

However the user's search is typically comprised of between 2 and 3 terms.

If it is assumed that the extent of the user's query interest is as broad as that of the document's description, then there is a serious mis-match in the completeness of the search query when compared to the document description via its indexed terms.

2.3 Vocabulary

A third problem is the matter of vocabulary. If the language of the document and the query are different, then there will be no meeting possible and a search will fail because there are no matches found between the terms of the query and those of a perfectly suitable document. In the real world there are a number of reasons why this may happen:

- Natural language mis-match (English queries against a French document set)
- Terminology (technical terms (jargon) vs. layman's terms)
- Level of specificity ("physics" vs. "refraction of light")

As a subset of this problem is the matter of the search languages used. Where the search engine is capable of acting on very complex and precise Boolean queries, but the user persists in entering single term queries, the functional vocabulary of the search engine remains unused and wasted. Similarly, lack of knowledge of, or mis-use of, the types of available indexes for the search engine and its database mean the user is probably missing out on obtaining more precise result sets. This is most evident by the overwhelming use of "keyword" searching at the expense of more precise "author" or "subject" searching where the terms used are actually the author's name or a subject descriptor.

These three problems prevent the user from obtaining a small, precise, but broadly based set of results from a query. At the best this means more work for the searcher to gather the results and then to refine them and wade through the dross to find the gold. At worst it means the nuggets are missed completely and nothing but dross remains even after extensive, time consuming work.

3 Towards the light

To obtain a more practically useful search experience for the user requires the convergence of a number of technologies and methods of operation. Since this work is seeking a real world system it is important to note that, for a number of reasons, it is virtually impossible to obtain the "prefect score" of 100% Recall and 100% Precision in any practical case. Not least of these are the changing nature of the set of documents and where they are held, and also the changing nature of the user's perception of what is relevant.

What we attempt to describe below is a system which aims for "80%" in all things, in the belief that this is a major improvement in the technology of providing users with the information they are seeking. Obtaining the other "20%" is a task for tomorrow.

4 Three Search Mis-matches revisited

4.1 Where to Search

Computer systems are ideally suited to help with this problem. The work is repetitive, boring, involves large quantities of data and is time consuming.

There are essentially two possible methods of making sure the search covers all the suitable material:

- Aggregated, centralized search engines
- Federated search (or Metasearch)

In the aggregated search engines (of which the best known examples are the web search engines such as Google, Yahoo!, Ask, etc.) all possible relevant documents are aggregated in a single search engine and are indexed. There are variations on this where the documents are retained centrally or not, where the index is central or distributed and so on. But the essential characteristic is that a single search engine access can search all content in a consistent representation.

In federated search engines the documents are left where they are and are indexed by their own search engines. The federated search engine translates the user's query, sends it to the remote search engines, and aggregates the results.

In both cases the user enters a query into a single interface in a single syntax and receives a set of results from a wide variety of sources. If the variety is wide enough then the user has essentially all results available. Both technologies, on the face of it, are solutions to the user's dilemma of where to look. However the devil is in the details.

It is far beyond the scope of this paper to adequately compare all the operational pros and cons of the two technologies, but it is worth just noting the main differences where they impact the aspect of coverage.

Aggregated search engines rely on crawling or file imports to build their indexes. Crawling requires access and this is barred either technically or commercially to a large proportion of even the Web resources. This is the so-called "Deep Web". Aggregation is also a time consuming process and requires immense resources to process "everything" just in case it is needed. It is thus relatively slow and the Aggregators prioritise the sites and files they process, so much information is significantly out of date in their indexes. The largest problem however, is that they are omnivores and consume and index anything. Thus there is very little specificity in their content. They also rely on simple indexing from both a processing and user input standpoint, thus contributing to an explosion in the size of the results set for any query.

However they cover vast quantities of material, including from sources unknown to the user, and this material is included in the results.

In a nutshell they offer too many results from too small a part of what is available.

Federated search engines rely on the capabilities of the remote search engines, so their results may be variable. Although this is countered by translating the user query for each source, the variability of the sources is still there. A Federator must be set up for each remote source ahead of use, and the user must select those to be searched, This provides specificity of results, and a reduced result set size, but it cannot cover those sources the user (and the administrators doing the set up) do not know about. Since federated search engines work entirely in real time, they tend to retrieve small set of results for each source and combine those results. Thus, for reasonable response times, important results may be missed.

However, the Federated search engines are able to search the whole of the Deep Web and thus have a much bigger basic scope that the Aggregators. Also, since they work in real time, they are able to obtain the very latest results from the sources they search.

They provide the widest coverage, but may supply too few results in a reasonable time.

Since the Federated search engines can cover all sources (in a manner acceptable to the commercial interests of the content providers) and can have their quantity of results extended for only a time penalty, the suggested solution is to utilize federated search technology to provide results.

4.2 Extent of representation

The problem here is not with the database, as the documents in it are generally adequately represented as to their content. The problem is the representation of the user's query. The focus of the query is contained in the entered term(s), but the context and nuances of that query are lost.

Not only is there really no method in conventional search engines to represent the context of the query - other than to insert 10, 20 or more terms into the query, there is also no general method to utilize that context to improve the results.

To describe a user's context ("Topic of Interest" we call it) it is necessary to assume some long lived method of describing that interest and a method of applying it to the current search. If it is assumed the context can be adequately represented simply by a set of terms, most search engines have no way to apply this context to modify the results the core query returns. Conventional index search engines will treat all the terms as part of the search and AND them or OR them together as their basic logic requires. What is needed here is an additional method similar to 'filters' or 'limits' which allow the user to specify, for example, the language or material type of the results. Such a method has to be built into the basic search engine and it is conventionally not available.

Where experimental system do allow a user context (sometimes called a 'profile') then the conventional method of producing it is to either allow the user to edit it, or to build it by compiling a list of recent searches, or both.

We suggest a better method is to create the user context from the results obtained by the user, rather than the queries entered. Manual editing is still an essential requirement.

The advantages of using results rather queries are;

- Query terms tend to be repetitive - the user builds queries from the term s/he knows. So they have a narrow spread.

- Unknown terms will never occur in a query - limiting the chance for discovering new material
- The query is a single action with no ability to interact. The results are displayed and the user can interact with them and have the context change accordingly.
- The results are rich in terminology - this immediately helps with the vocabulary problem

Various technologies are needed to make any context aware search engine operate, and some are described below from our system.

4.3 Vocabulary

This problem is usually handled (when handled at all) by the use of dictionaries and thesauri to translate the user's terms into those of the search engine. This is a straight translation, conceptually one-for-one, merely ensuring that the query terms and document terms are from the same vocabulary.

Alternatives exist in the idea of "concept translation" where the query terms are matched against concepts in some concept database within the search engine, and are used as the translated concepts in the target "language". This of course may not be a natural language, rather a terminology "language".

The level of specificity is usually dealt with by use of a thesaurus to convert the query terms into those of the search engine(s). This interacts very strongly with the overall architecture as both the aggregating search engines and the federated search engines have their problems and advantages. Because the aggregators are omnivorous, they tend to have all levels of specificity present, and so any query will get some results. However there is generally little attempt to augment the descriptions or queries so all documents originally described at a particular level of specificity will be missed by a query at another one. Federators are able to translate the query to the required level for each source, thus getting optimum results, but they need to know the level of each source, and that is difficult information to obtain exactly.

These solutions are all well know and readily available in some systems, but are not very widespread. Since the vocabulary files are expensive to maintain this is a task which is not well handled in the real world, even though the answers are known in theory.

We suggested above that our proposed system has some spill-over benefit in this area by increasing the richness of the available vocabulary for handling the problems of jargon and level of specificity. In general it will have little impact on the natural language problems, but could do in very specific areas and with specific sets of data - namely databases with mixed languages documents.

5 An Interesting Solution

As a system to address the above problems in the way outlined, MuseGlobal has developed a number of technologies. Principally the Muse system is a federated search engine with extensive post-search processing capabilities.

The federation capabilities of Muse allow the user to select the source to be searched either singly for greatest precision or in cascading groups for greater coverage, speed and ease of use. Typically the groups are set up on a subject basis by an administrator and the user is only required to choose the particular groups of interest. There is little penalty in choosing more groups, so that is encouraged as long as the user stays away for choosing obviously inappropriate groups which will only serve to waste system resources, or retrieve results which are, at best, peripheral. Muse operates with a global library of sources which numbers over 4,000 and so most subject areas are covered by the available sources. If specific ones are needed then they are built on demand by MuseGlobal as part of the support for the system.

The post-search processing capabilities of Muse extend beyond those of the mechanical operations of removing duplicates and such like, to both syntactic and semantic mapping of results from each source on an individual basis into a normalized, multi-representational, internal form. This guarantees the results records are consistent and thus amenable to manipulation.

The above covers the broadness of scope of the system and allows a high recall. (Since the real world consists of an open ended set of documents, Recall as a mathematical measure cannot be calculated, but in terms of a count of relevant documents retrieved from "the universe" it provides a rankable number such that "high recall" or "higher recall" are meaningful phrases.) But it does nothing to address the issue of precision, or its cousin, relevance.

The Muse system includes a post-search processing module called Content Mining which enables the extraction and manipulation of Terms from retrieved documents. These Terms are multi-word phrases and may be thought of as either 'keywords' or 'concepts' depending on the parameters used for their extraction. The system is capable of extracting terms of different entity types (people, places, products, etc.) as well as simply subject terms. These extracted terms (the Muse function is called "distillation" to get the terms from the document) are represented either as lists or Term Clouds for both individual documents and groups of documents.

The Term Cloud is often referred to as a Topic, and is said to describe the Topic, or subject, of a particular document or result set. Terms within a Topic have both frequencies and weights. The frequency is the number of times the term has occurred within the document, and the weight is a modified form of the frequency taking into account the location of the terms within the document and weighting the occurrence frequencies accordingly. Both are simple open ended numerical values. When a Topic for a result set (or other group of documents) is created, functions exist to normalize the frequencies and weights if required. This would typically be done to compensate for the different size of documents.

Topics are attached to documents and results sets, but have an existence of their own. They are stored as Topic Maps, but do not, at this point, have relationships between the individual terms.

To enable the Topics to influence the search results for the user, a new form of relevance is introduced; Relevance by Topic. Conventionally Relevance scores each document in accordance with the occurrence of the search terms within the document. This gives a measure of how "close" the document is to the original aim of the searcher as expressed by the query. However we have argued earlier that the query is not a very good representation of the user's intent. So we introduce the idea of scoring and ranking the documents in accordance with how well they cluster around the centroid of the result set as represented by the Topic of that result set. By comparing the Topic of each document with that for the whole result set, we are able to obtain a measure of the "distance" of any document from the centre of the results set as expressed by the combined set of terms from all documents. Ranking documents by this method allows them to be displayed in a sequence showing how close they are to the consensus of what the documents are about, even if the query terms are not exactly in the centre.

This Relevance by Topic has the advantage that it is derived from the results and so can be affected by manipulating them, whereas conventional Relevance by Search - based on the search terms - is fixed. We also suggested that documents ranked towards the top of the list when ranked by Topic are more representative of the set of documents as a whole and thus indicate to the user more immediately whether this result set is worth pursuing, or if another search should be attempted.

Manipulation of the documents in the results set (for example by deletion of irrelevant ones) allows the Topic to be re-calculated and the ranking of the documents to change. Thus, as the user discards unwanted documents others similar to them will sink towards the bottom of the list, raising the more relevant (or, strictly, less irrelevant) to the top.

In this manner a set of records all entirely relevant can be obtained. The Topic derived from these records is thus a good descriptor of the Topic as a whole - as it is of interest to this particular user at this moment. If, after suitable editing, the Topic is saved, it becomes a Topic of Interest and represents one of the things the user is currently interested in. Consider by way of example the simple query for "water" issued by a sanitation engineer interested in pollution and by a boat manufacturer, interested in recreational uses. From the same set of results these two users would choose very different sets of results as relevant. The Topics derived from them will have overlap (both are about "water"), but will have many different terms incorporated.

The Topic of Interest includes terms from the documents, thus it is a vehicle for discovery of new terms and concepts unknown to the user. This is where the Topic helps with the vocabulary problems mentioned earlier.

Once a Topic of interest can be saved as an object in its own right then it becomes transferable and also can be used in various ways. The example system shows that the user is able to select a Topic of Interest at the point of submitting a query. The Muse system utilizes the context supplied by the Topic to inform the selection of records for, and the ranking of them within, the result set. Thus the water engineer who is also a boating enthusiast may select "pollution" or "sport" when entering new queries and get record selected and/or ranked for that Topic; differently for his different interests.

This system provides an easy to use method where:

- a context (Topic of Interest) for the user may be set
- a user may have multiple contexts
- the contexts are created automatically
- the contexts are derived from the results not the search
- the context can be used to refine results of future searches

This system improves the precision of results by utilizing Topics of Interest while ensuring a good recall with an underlying federated search engine.

A demonstration of the system and will be provided during the presentation, and further papers will disclose experimental results for this new type of context enabled search.

Distributed Decision Making for Robot Teams

Nikos Vlassis

Dept. of Production Engineering and Management, Technical University of Crete, Kounoupidiana 73100 Chania, Greece (vlassis@dpem.tuc.gr)

Summary. We review the problem of action coordination in a team of collaborative robots, viewed as a combinatorial optimization problem over a coordination graph. We outline a message-passing algorithm for action selection that approximately maximizes a payoff function that is additively decomposed over the graph. We discuss extensions to distributed stochastic optimal control problems, and outline some applications.

1 Introduction

In this article we review the problem of decision making of teams of robotic agents that must collaborate in order to solve a task. In order for solutions of this problem to scale to large teams of robots, the decision making of the robots must be done in a distributed fashion: each robot should choose its action based on local information available to it, without relying on a center to orchestrate the actions of each robot. Such distributed decision making problems are typical in engineering and related fields [Sandell et al., 1978].

A standard way to formalize the problem is to assume that the task of the robots is described by means of an objective function that the robots must optimize, and the optimization of this function must be done in a distributed fashion. Such distributed optimization problems have been extensively studied in the past [Bertsekas and Tsitsiklis, 1997]. When the action space of each robot is continuous and the objective function is a differentiable function of the actions, optimization of this function can be done by local gradient rules. Recent examples of such an approach in robotics include motion planning using a navigation function [Tanner and Kumar, 2005], and reconnaissance and information gathering [Mathews and Durrant-Whyte, 2007]. In these problems the objective function often has special structure (or it can be engineered to be so) such that the local decisions of the robots can guarantee the global coordination of the team.

In this paper we review a model in which the action space of each robot is discrete, and the objective function is additively decomposed into a number

Nikos Vlassis: *Distributed Decision Making for Robot Teams*, Studies in Computational Intelligence (SCI) **78**, 35–40 (2008)
www.springerlink.com

of terms, where each term involves only a small subset of the robots. This is the framework of *coordination graphs* [Guestrin et al., 2002] in which decision making becomes a combinatorial optimization problem. In the next sections we briefly review this model, we show how action selection can be carried out in a distributed manner, and we briefly review some applications.

2 Coordination graphs

A coordination graph is a decision-theoretic model of team decision making. Consider a team of n robots that are involved in a joint task, where each robot i can take action u_i from a discrete action set $\mathcal{U}_i$. Let $u = (u_1, \ldots, u_n)$ be the composite action of the team, taking values from the set $\mathcal{U} = \mathcal{U}_1 \times \ldots \times \mathcal{U}_n$. Further suppose that the joint task is defined through a global payoff function $Q : \mathcal{U} \mapsto \mathbb{R}$ that assigns to each composite action u a real number $Q(u)$, the shared payoff of the team for joint action u. The task involves computing the composite action $u^* = (u_1^*, \ldots, u_n^*)$ that maximizes the global payoff function:

$$u^* = \arg\max_u Q(u). \tag{1}$$

If we employed a centralized solution, the above maximization would only be practical in problems involving only few robots, since the number of composite actions scales exponentially with the number of robots. Here we are interested in distributed algorithms for optimizing (1), in particular algorithms that scale efficiently with the number of robots.

A coordination graph model assumes that the global payoff function $Q(u)$ can be written as a linear combination of k local payoff functions f_j, for $j = 1, \ldots, k$, each involving fewer robots. For example, for $n = 4$ robots and $k = 3$ local payoff functions each involving two robots, we have

$$Q(u) = f_1(u_1, u_2) + f_2(u_1, u_3) + f_3(u_3, u_4). \tag{2}$$

Here, for instance $f_2(u_1, u_3)$ involves only robots 1 and 3, with their actions u_1 and u_3. Such a decomposition can be graphically represented by a graph (hence the name) where each node represents a robot and each edge corresponds to a local payoff function. For example, the decomposition (2) can be represented by the graph of Fig. 1. Many practical problems can be modeled by such additively decomposable payoff functions. For example, in a robot soccer team, only nearby robots may need to coordinate their actions in order to improve global team performance (see Fig. 2, right). In these cases, global 'teamwork' is approximated by the sum of local 'teamworks'.

3 Coordination by message passing

Solving (1) when the global payoff function is additively decomposed as in (2) can be done with *nonserial dynamic programming* [Bertelé and Brioschi,

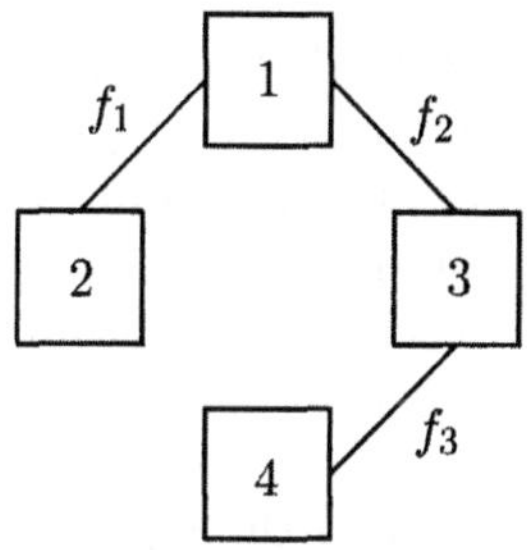

Fig. 1. A coordination graph for a team of 4 robots.

1972], an exact algorithm that admits distributed implementations and always computes the globally optimal u^* in finite time. However, depending on the structure of the coordination graph, the runtime of this algorithm can be exponential in the number of robots, and therefore it may not scale to large teams [Vlassis, 2007, ch. 4].

Here we describe an alternative algorithm, called *max-plus*, that is approximate but scales much better in large problems. Max-plus was originally developed for computing maximum a posteriori (MAP) solutions in Bayesian networks [Pearl, 1988], and is based on message-passing between neighboring robots on the graph. We briefly describe the algorithm here, and refer to [Vlassis, 2007] for more details.

Consider a payoff function that decomposes over a coordination graph as

$$Q(u) = \sum_{(i,j)} f_{ij}(u_i, u_j), \tag{3}$$

where $f_{ij} : \mathcal{U}_i \times \mathcal{U}_j \mapsto \mathbb{R}$ are local payoff functions that involve two neighboring robots on the graph, and the summation is over all (i, j) edges of the graph. In max-plus, at each time step each robot i sends a message μ_{ij} to a (randomly chosen) neighboring robot j, where $\mu_{ij} : \mathcal{U}_j \mapsto \mathbb{R}$ is a local payoff function defined as

$$\mu_{ij}(u_j) = \max_{u_i} \Big\{ f_{ij}(u_i, u_j) + \sum_{k \in \Gamma(i) \backslash j} \mu_{ki}(u_i) \Big\}, \tag{4}$$

where $\Gamma(i)\backslash j$ denotes all neighbors of robot i except robot j. Such messages are being exchanged until they converge to a fixed point, or until some termination condition is satisfied (e.g., some deadline expires). If we define local functions $Q_i : \mathcal{U}_i \mapsto \mathbb{R}$ for each robot i as

$$Q_i(u_i) = \sum_{j \in \Gamma(i)} \mu_{ji}(u_i), \tag{5}$$

and at each time step each robot locally computes

$$\tilde{u}_i^* = \arg\max_{u_i} Q_i(u_i), \tag{6}$$

then the profile of local optimizers $\tilde{u}^* = (\tilde{u}_1^*, \ldots, \tilde{u}_n^*)$ asymptotically approximates the optimal composite action u^* of (1), and in some cases it converges exactly to u^* in a finite number of steps [Wainwright et al., 2004].

The attractive feature of max-plus is that it decomposes a difficult global optimization problem (1) into a set of local optimization problems (6) that are much easier to solve. Although its convergence is not always guaranteed, max-plus is very easy to implement and it can compute good solutions fast [Kok and Vlassis, 2006]. Alternative algorithms for solving (1) that are provably complete are typically slower and more complicated to implement [Modi et al., 2005].

4 Distributed stochastic optimal control

In practical problems the objective function of the task may be more complicated. For instance, in an infinite-horizon stochastic control problem involving a team of robots, the objective function can be a value function of the form

$$V^\pi(x_0) = E\Big[\sum_{t=0}^{\infty} \gamma^t R(x_t, \pi(x_t))\Big], \tag{7}$$

where $x_t \in \mathcal{X}$ denotes the system state at time t and $R : \mathcal{X} \times \mathcal{U} \mapsto \mathbb{R}$ is a reward function. Here the task is to compute an optimal deterministic and stationary policy $\pi^* : \mathcal{X} \mapsto \mathcal{U}$ that maps each state x to a composite action $\pi^*(x)$ such that $V^{\pi^*}(x_0)$ is maximum for all states x_0 [Bertsekas, 2001]. In this case one can resort to standard dynamic programming for computing π^*, in which case the optimization step (1) will appear inside a Bellman backup operator. For example, the standard Q-learning rule for exploration tuple (x, u, r, x') reads

$$Q(x,u) := (1-\alpha)Q(x,u) + \alpha[r + \gamma \max_{u'} Q(x',u')], \tag{8}$$

and the maximization $\max_{u'} Q(x', u')$ is in the form (1) and hence it can be implemented by message-passing over a graph as above. This gives rise to a family of scalable reinforcement learning techniques for distributed stochastic optimal control [Kok and Vlassis, 2006].

5 Applications

Among other applications [Guestrin et al., 2002], coordination graphs have been used for coordinating a team of soccer robots (Fig. 2, right) [Kok et al., 2005], in robot formation control (Fig. 2, left) [Tsianos, 2005], and in multi-robot motion planning [Li et al., 2005]. In all these applications, the global payoff function of the task is decomposed as in (3), where the local payoff functions f_{ij} are defined by means of propositional payoff rules in the form

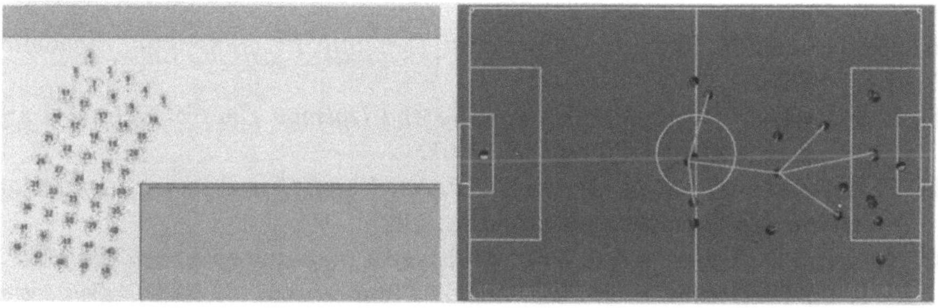

Fig. 2. Robot formation control using coordination graphs (left), and team coordination in robot soccer (right).

$$\text{rule}_{i,j}^m: \text{ if some condition holds for robots } i,j \text{ then } f_m$$

where f_m is a local payoff added to the global payoff Q. For instance, a local payoff rule between a goalkeeper i and some other teammate j in a robot soccer team could read

$$\text{rule: if has-role-defender}(j) \text{ and } u_i == \text{passTo}(j) \text{ then } 10$$

and similar rules can be defined for other situations during a game [Kok et al., 2005]. Action coordination can then be performed with nonserial dynamic programming or message-passing over the induced coordination graph. We refer to the above papers for more details.

6 Summary

We briefly examined the problem of distributed multi-robot decision making, cast as a combinatorial optimization problem over a coordination graph. We described an approximate action selection algorithm that is based on message-passing and which is able to mitigate the exponential complexity of the problem in several cases. Future research involves identifying good scheduling schemes for message-passing that guarantee high convergence rates and low approximation errors. Relationships to continuous-action coordination approaches, perhaps via relaxation techniques, are also interesting to study.

Acknowledgments

I would like to thank the organizing committee of IDC'2007, and in particular Costin Badica, for inviting me at IDC'2007.

References

U. Bertelé and F. Brioschi. *Nonserial Dynamic Programming.* Academic Press, 1972.

D. P. Bertsekas. *Dynamic Programming and Optimal Control*, volume I and II. Athena Scientific, 2nd edition, 2001.

D. P. Bertsekas and J. N. Tsitsiklis. *Parallel and Distributed Computation: Numerical Methods.* Athena Scientific, 1997.

C. Guestrin, D. Koller, and R. Parr. Multiagent planning with factored MDPs. In T. G. Dietterich, S. Becker, and Z. Ghahramani, editors, *Advances in Neural Information Processing Systems 14*, Cambridge, MA, 2002. The MIT Press.

J. R. Kok and N. Vlassis. Collaborative multiagent reinforcement learning by payoff propagation. *Journal of Machine Learning Research*, 7:1789–1828, 2006.

J. R. Kok, M. T. J. Spaan, and N. Vlassis. Non-communicative multi-robot coordination in dynamic environments. *Robotics and Autonomous Systems*, 50(2-3):99–114, 2005.

Y. Li, K. Gupta, and S. Payandeh. Motion planning of multiple agents in virtual environments using coordination graphs. In *Proc. Int. Conf. on Robotics and Automation*, Barcelona, Spain, 2005.

G. Mathews and H. Durrant-Whyte. Decentralised optimal control for reconnaissance. In *Proc. Int. Conf. on Information, Decision and Control*, Adelaide SA, Australia, 2007.

P. J. Modi, W.-M. Shen, M. Tambe, and M. Yokoo. ADOPT: Asynchronous distributed constraint optimization with quality guarantees. *Artificial Intelligence*, 161(1-2):149–180, 2005.

J. Pearl. *Probabilistic Reasoning in Intelligent Systems.* Morgan Kaufman, San Mateo, 1988.

N. R. Sandell, P. Varayia, M. Athans, and M. G. Safonov. Survey of decentralized control methods for large-scale systems. *IEEE Trans. Automatic Control*, AC-23(2):108–128, 1978.

H. Tanner and A. Kumar. Towards decentralization of multi-robot navigation functions. In *Proc. Int. Conf. on Robotics and Automation*, Barcelona, Spain, 2005.

K. Tsianos. Algorithms for optimal coordination of multiagent systems and applications. Master's thesis, National Technical University of Athens, Greece, June 2005.

N. Vlassis. *A Concise Introduction to Multiagent Systems and Distributed Artificial Intelligence.* Synthesis Lectures in Artificial Intelligence and Machine Learning. Morgan & Claypool Publishers, 2007.

M. J. Wainwright, T. S. Jaakkola, and A. S. Willsky. Tree consistency and bounds on the performance of the max-product algorithm and its generalizations. *Statistics and Computing*, 14:143–166, 2004.

Fully Coupled and Feedforward Neural Networks with Complex-Valued Neurons

Jacek M. Zurada[1] and Igor Aizenberg[2]

[1] Department of Electrical and Computer Engineering, University of Louisville
Louisville, Kentucky 20208, USA
jacek.zurada@louisville.edu

[2] Texas A&M University-Texarkana
P.O. Box 5518, 2600 N. Robison Rd. Texarkana, Texas 75505, USA
igor.aizenberg@tamut.edu

Summary. This paper discusses neural networks with complex-valued neurons with both discrete and continuous outputs. It reviews existing methods of their applications in fully coupled associative memories. Such memories are able to process multiple gray levels when applied for image de-noising. In addition, when complex-valued neurons are generalized to take a continuum of values, they can be used as substitutes for perceptron networks. Learning of such neurons is demonstrated and described in the context of traditional multilayer feedforward network learning. Such learning is derivative-free and it usually requires reduced network architecture. The notion of a universal binary neuron is also introduced. Selected examples and applications of such networks are also referenced.

1 Introduction

Neural networks consist of densely interconnected linear multiplier elements and nonlinear processing neurons. A particularly appealing feature of neural networks is their inherent ability to acquire knowledge by learning from data. Compared to biological neural networks, artificial neural networks are "neural" in the sense that they have been inspired by neuroscience, but not necessarily because they are true models of biological or cognitive phenomena. In fact, typical neural network architectures are more related to mathematical and/or statistical models, such as non-parametric pattern classifiers, clustering algorithms, nonlinear filters, and statistical regression models, rather than to neurobiological models [24].

Rooted in mathematics, computational physics and statistics, neural networks have emerged as a class of computing paradigm with versatile modeling capabilities. Thanks to their ability to learn from data, they offer substantial advantage over traditional engineering approaches to systems modeling. The traditional modeling in engineering has mostly relied on formal analytical techniques. Typically, a specialized algorithm is needed for each specialized task. As a result, many dedicated approaches have been developed for pattern classification and recognition, system

Jacek M. Zurada and Igor Aizenberg: *Fully Coupled and Feedforward Neural Networks with Complex-Valued Neurons*, Studies in Computational Intelligence (SCI) **78**, 41–50 (2008)
www.springerlink.com

identification and control, forecasting, image compression and for building expert systems.

In contrast to algorithmic approaches usually tailored to tasks at hand, neural networks offer a wide variety of versatile modeling techniques applicable to a large class of problems. Here, learning in data-rich environments leads to models of specific tasks. Through learning from specific data with rather general neural network architectures, neurocomputing techniques can yield problem-specific solutions [25].

Most neural networks are defined as consisting of 'two-state' neurons and evaluated accordingly. Interconnected neurons can in general be multiple-valued logic elements. The resulting networks retain many of the properties of binary networks, yet they become an extension to a multiple-valued computing paradigm. Attractor networks employing multilevel neurons are displaying much richer behavior than their original counterparts with bistable elements [25].

This paper focuses on novel links between conventional neural networks and multiple-valued logic concepts for both fully coupled (attractor) networks and perceptron type networks. Assumptions, basic concepts, advantages and disadvantages of such networks with multilevel neurons are reviewed.

2 Complex-Valued Neural Networks

Complex-valued neural networks become more and more popular. This is a natural process, because the use of complex-valued inputs/outputs, weights and activation functions makes it possible to increase the functionality of both a single neuron and a network, to make them more flexible, to improve their performance and to make their training faster. We will consider here two neurons that have been the first in the line of complex-valued neurons. There are a multi-valued neuron (MVN) and a universal binary neuron (UBN). Two main ideas behind these neurons are a model of the multiple-valued logic over the field of complex numbers and a complex-valued activation function, which is a function of the argument of the weighted sum. Both these ideas were introduced in 1971 in the seminal paper of N. Aizenberg et. al. [1]. This approach has opened a way for a significant increase of the single neuron functionality and for a simplification of its learning. The MVN has been introduced in [2] and the UBN has been introduced in [3]. Their observation along with detail derivation of their learning algorithms is presented in [4]. The main advantage of the UBN is its ability to learn any non-threshold (nonlinearly separable) Boolean function. Such classical nonlinearly separable problems as XOR and parity n becomes separable and can be easily learned using the single UBN [3, 4, 5].

The MVN is much more functional neuron than, e.g., traditionally used neuron with the sigmoid activation function. utilizes basic ideas of multiple-valued threshold logic over the field of complex numbers [4, 6]. The discrete-valued MVN has been comprehensively observed in [4]. A continuous-valued MVN and its learning are considered in [7, 8]. The most important properties of MVN are the complex-valued weights, inputs and output lying on the unit circle and the activation function, which maps the complex plane into the unit circle. The MVN learning algorithm is based on a simple linear error correction rule and therefore it is derivative free. Many applications of the MVN have been considering during the last years. The MVN has been used as a basic neuron in cellular neural networks [4], as a basic neuron of neural-based associative memories [4, 9, 10, 11, 12, 13, 14, 15, 16], as a basic neuron of neural-based pattern classifiers [12, 15, 16, 17, 18, 19, 20, 21] and as a generator of multi-valued nonlinear filter (MVF) [4, 22]. Recently proposed [7] and then developed [8], a multilayer neural network with multi-valued neurons (MLMVN) should be distinguished among other applications. Two very important advantages of the MLMVN are its derivative free backpropagation learning, which is based on the error-correction rule, and its much better adaptation to the highly nonlinear mappings that its predecessors show. It has shown much better performance than the multilayer perceptron (MLP) and different kernel-based networks (including SVM) solving such benchmark problems, as two spirals, parity n and Mackey-Glass time series prediction [7]. In [20], the MLMVN has been used for classification of gene expression microarray data showing better performance than other modern classifiers. In [21], the MLMVN has been used to learn the genetic code and it was shown that the genetic code can be learned completely using about 7 times smaller MLMVN than MLP. In [8], the MLMVN has been successfully used for solving the extremely highly nonlinear problem - blur and blur parameters identification, where it again outperformed other techniques including SVM.

We are going to observe here briefly the main details behind the MVN and UBN and the most impressive MVN's applications: the MLMVN and MVN-based associative memories.

3 Multi-Valued Neuron (MVN)

A single discrete-valued MVN performs a mapping between n inputs and a single output. This mapping is described either by a multiple-valued (k-valued) discrete function or continuous function of n variables $f(x_1, \ldots, x_n)$ with $n + 1$ complex-valued weights as parameters:

$$f(x_1, \ldots, x_n) = P(w_0 + w_1x_1 + \ldots + w_nx_n) \quad (1)$$

where $x_1, \ldots, x_n$ are the variables, on which the performed function depends and $w_0, w_1, \ldots, w_n$ are the weights. The values of the function and of the variables are complex. In the discrete case they are the kth roots of unity: $\varepsilon^j = \exp(\imath 2\pi j/k)$, $j \in \{0, \ldots, k-1\}$, $\imath$ is an imaginary unity and in the continuous case they all are lying on the unit circle. P is the activation function of the neuron. For the discrete case this function is [1]:

$$P(z) = \exp(\imath 2\pi j/k), \text{ if } 2\pi j/k \leq \arg z < 2\pi(j+1)/k \quad (2)$$

where $j = 0, 1, \ldots, k-1$ are values of the k-valued logic, $z = w_0 + w_1x_1 + \ldots + w_nx_n$ is the weighted sum , $\arg z$ is the argument of the complex number z. Equation (2) is illustrated in Fig.1. Function (2) divides a complex plane onto k equal sectors and

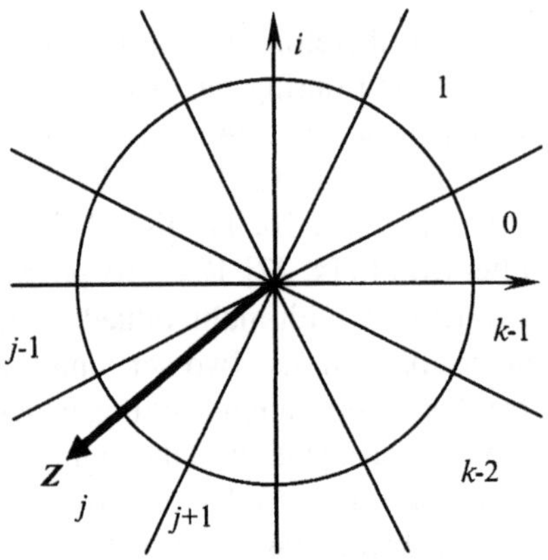

Fig.1. Geometrical interpretation of the MVN activation function

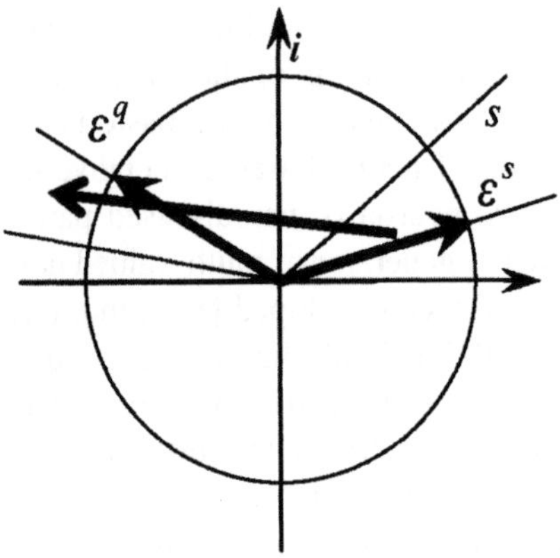

Fig.2. Geometrical interpretation of the MVN learning rule

maps the whole complex plane into a subset of points belonging to the unit circle. This is exactly a set of kth roots of unity.

For the continuous case $k \to \infty$ in (2) and then the angle value of the sector (see Fig.1) tends to zero. Hence, the function (2) is transformed in this case as follows [7]:

$$P(z) = \exp(\imath(\arg z)) = \varepsilon^{\imath\, Arg\, z} = \frac{z}{|z|} \tag{3}$$

where z is the weighted sum, $Arg\, z$ is a main value of its argument and $|z|$ is a modulo of the complex number z. The function (3) maps the complex plane into a whole unit circle, while the function (2) maps a complex plane just on a discrete subset of the points belonging to the unit circle.

The most efficient MVN learning algorithm is based on the error-correction rule [4]. This learning is reduced to the movement along the unit circle and it is derivative-free as well as classical perceptron learning. However, the perceptron is a binary neuron, while the MVN is either a multiple-valued or continuous-valued neuron. Let $\varepsilon^s = Y = P(z)$ be an actual output of the neuron. The error-correction learning rule is illustrated in Fig.2 and works as follows for the discrete-valued [4] and continuous-valued [7] MVN, respectively:

$$W_{r+1} = W_r + \frac{C_r}{(n+1)|z_r|}(\varepsilon^q - \varepsilon^s)\overline{X} \tag{4}$$

$$W_{r+1} = W_r + \frac{C_r}{(n+1)|z_r|}(\varepsilon^q - \varepsilon^{\imath\, Arg\, z})\overline{X} \tag{5}$$

where X is an input vector, n is the number of neuron's inputs, $\overline{X}$ is a vector with the components complex conjugated to the components of vector X, r is the index of iteration, W_r is the current weighted vector, W_{r+1} is a weighted vector after correction, $|z_r|$ is the absolute value of the current value of the weighted sum ($1/|z_r|$ is a variable self-adaptive part of the learning rate) and C_r is a constant part of the learning rate, which can always be taken equal to 1.

4 Multilayer Neural Network based on Multi-Valued Neurons (MLMVN)

The MLMVN has been considered in detail in [7] and then developed in [8]. It refers to the basic principles of the network with a feedforward dataflow through nodes proposed in [23]. The most important is that there is a full connection between the consecutive layers (the outputs of neurons from the preceding layer are connected with the corresponding inputs of neurons from the following layer). The network contains one input layer, $m-1$ hidden layers and one output layer. Let us use here the following notations. Let T_{km} be a desired output of the kth neuron from the mth (output) layer; Y_{km} be an actual output of the kth neuron from the mth (output) layer. Then the global error of the network taken from the kth neuron of the mth (output) layer is calculated as follows:

$$\delta^*_{km} = T_{km} - Y_{km} \tag{6}$$

Following the backpropagation learning algorithm for the MLMVN proposed in [7, 8], the errors of all the neurons from the network are determined by the global errors of the network (6). It is essential that the global error of the network consists not only of the output neurons errors, but of the local errors of the output neurons and hidden neurons. It means that in order to obtain the local errors for all neurons, the global error must be shared among these neurons.

Let w_i^{kj} be the weight corresponding to the ith input of the kj^{th} neuron (k^{th} neuron of the jth level), Y_{ij} be the actual output of the ith neuron from the jth layer ($j = 1, \ldots, m$), and N_j be the number of the neurons in the jth layer (it means that the neurons from the $j+1^{st}$ layer have exactly N_j inputs). Let $x_1, \ldots, x_n$ be the network inputs. Hence, the local errors are represented in the following way. The errors of the mth (output) layer neurons are:

$$\delta_{km} = \frac{1}{s_m}\delta^*_{km} \tag{7}$$

where km is a kth neuron of the mth layer; $s_m = N_{m-1} + 1$, i.e. the number of all neurons on the previous layer (layer $m-1$ which the error is backpropagated to) incremented by 1. The errors of the hidden layers neurons are computed as follows:

$$\delta_{kj} = \frac{1}{s_j}\sum_{i=1}^{N_{j+1}} \delta_{ij+1}(w_k^{ij+1})^{-1} \tag{8}$$

where kj specifies the kth neuron of the jth layer ($j = 1, \ldots, m-1$); $s_j = N_{j-1} + 1$, $j = 2, \ldots, m$, $s_1 = 1$ is the number of all neurons on the layer $j-1$ incremented by 1. Thus, the equations (7), (8) determine the error backpropagation for the MLMVN. It is worth to stress on its principal distinction from the classical error backpropagation: the MLMVN backpropagation does not depend on the derivative of the activation function. Moreover, both discrete-valued and continuous valued activation functions (2) and (3) that we use here are not differentiable at all as the functions of a complex variable.

A factor $1/s_j$ in (7), (8) ensures distribution of the particular neuron error among all the neurons on which this error depends. It should be mentioned that for the 1st hidden layer the parameter $s_1 = 1$ because there is no previous hidden layer, and there are no neurons the error may be shared with.

It is important that both error backpropagation and the learning process for the MLMVN are organized in the same manner independently of a particular modification of MVN (discrete-valued, continuous-valued, continuous inputs → discrete output and discrete inputs → continuous output).

After the error has been backpropagated, the weights of all neurons of the network can be adjusted using either the learning rule (4) or (5) (depending on the type of MVN and mapping, which we learn):

$$\begin{aligned} \widetilde{w}_i^{km} &= w_i^{km} + \frac{C_{km}}{(N_m+1)} \delta_{km} \overline{\overline{Y}}_{im-1}, \; i = 1, \ldots, n, \\ \widetilde{w}_0^{km} &= w_0^{km} + \frac{C_{km}}{(N_m+1)} \delta_{km}, \end{aligned} \tag{9}$$

for the neurons from the mth (output) layer (k^{th} neuron of the mth layer),

$$\begin{aligned} \widetilde{w}_i^{kj} &= w_i^{kj} + \frac{C_{kj}}{(N_j+1)|z_{kj}|} \delta_{kj} \overline{\overline{Y}}_{ij-1}, \; i = 1, \ldots, n, \\ \widetilde{w}_0^{kj} &= w_0^{kj} + \frac{C_{kj}}{(N_j+1)|z_{kj}|} \delta_{kj}, \end{aligned} \tag{10}$$

for the neurons from the 2nd till $m-1^{st}$ hidden layers (k^{th} neuron of the jth layer ($j = 2, \ldots, m-1$), and

$$\begin{aligned} \widetilde{w}_i^{k1} &= w_i^{k1} + \frac{C_{k1}}{(n+1)|z_{k1}|} \delta_{k1} \overline{x}_i, \; i = 1, \ldots, n, \\ \widetilde{w}_0^{k1} &= w_0^{k1} + \frac{C_{k1}}{(n+1)|z_{k1}|} \delta_{k1}, \end{aligned} \tag{11}$$

for the neurons of the 1st hidden layer, where n is the number of network inputs, N_j is the number of neurons in the jth layer. In general, the MLMVN training process should continue until the condition of the minimal mean square error will be satisfied:

$$E = \frac{1}{N} \sum_{s=1}^{N} \sum_{k} (\delta_{kms}^*)^2 (W) = \frac{1}{N} \sum_{s=1}^{N} E_s \leq \lambda, \tag{12}$$

where λ determines the precision of learning, E_s is the square error for the sth pattern $X_s = (x_1, \ldots, x_n)$. In particular, in the case when $\lambda = 0$ the equation (12) is transformed to $\forall k, \forall s \; \delta_{kms}^* = 0$, which means zero training error.

5 Universal Binary Neuron

A key point behind the universal binary neuron is a notion of P-realizable Boolean function.

The Boolean function $f(x_1, \ldots, x_n)$ is called a *P-realizable* function over the field of complex numbers $\mathbb{C}$, if the following correspondence holds for all the values of the variables x in the domain of the function f [3, 4]:

$$f(x_1, \ldots, x_n) = P_B(w_0 + w_1x_1 + \ldots + w_nx_n), \tag{13}$$

where $W = (w_0, w_1, \ldots, w_n)$, $w_j \in \mathbb{C}$, $j = 0, 1, \ldots, n$ is a complex-valued weighting vector and P_B is a binary predicate, which is defined as follows:

$$P_B(z) = (-1)^j, \text{ if } 2\pi j/m \le \arg z < 2\pi(j+1)/m, \ m = 2t, \ t \in \mathbb{N}, \tag{14}$$

where m is some even positive integer, j is a non-negative number $0 \le j \le m$. Definition of the predicate P_B (see (14)) is illustrated in Fig. 3. The predicate P_B separates the complex plane on $m = 2t$ equal sectors, and the function P_B is equal to 1 for the complex numbers in the even sectors $0, 2, 4, \ldots, m-2$, and it is equal to -1 for the numbers in the odd sectors $1, 3, 5, \ldots, m-1$.

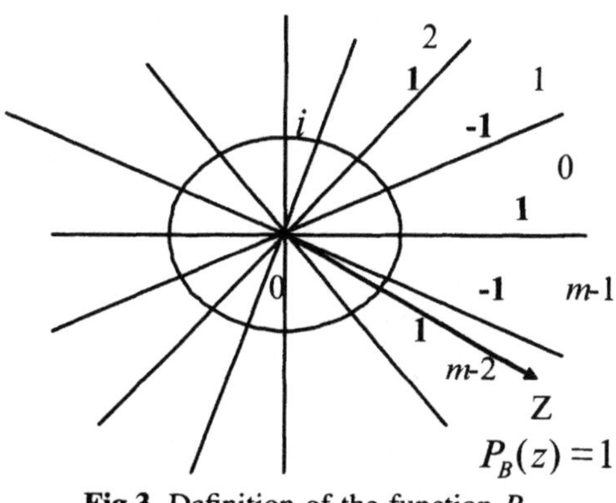

Fig.3. Definition of the function P_B (see (14))

Fig.4. Activation function (14) with $m = 4$

It is important to mention that we use here a Boolean alphabet $\{1, -1\}$ instead of a classical alphabet $\{0, 1\}$. If $y \in \{0, 1\}$ then $x \in \{-1, 1\}$ is obtained by a linear transformation $x = 1 - 2y$. Thus $0 \in \{0, 1\} \to 1 \in \{-1, 1\}$ and $1 \in \{0, 1\} \to -1 \in \{-1, 1\}$. The UBN over the field of complex numbers is a neural element [3, 4], which performs according to (13) with the activation function (14) for a given input/output mapping described by a Boolean function of n variables.

For $m = 2$ in (14) the functionality of the UBN coincides with the functionality of a traditional perceptron: it can implement only linearly separable Boolean functions. However, for $t > 1$ and $m > 2$ in (14) the functionality of the UBN with the activation function (14) is always higher than the functionality of the classical perceptron. For

Table 1. Solving the XOR problem on the single UBN using the weighting vector $(0, \imath, 1)$

#	x_1	x_2	$z = w_0 + w_1x_1 + w_2x_2$	$P_B(z)$	$XOR = x_1 \oplus x_2$
1)	**1**	**1**	$1 + \imath$	**1**	**1**
2)	**1**	**−1**	$-1 + \imath$	**−1**	**−1**
3)	**−1**	**1**	$1 - \imath$	**−1**	**−1**
4)	**−1**	**−1**	$-1 - \imath$	**1**	**1**

example, for $t = 2, m = 4$ the UBN has complete functionality for $n = 2$ [3, 4] (which means that all Boolean functions of two variables are P-realizable and can be learned by the single UBN). There are only two non-linearly separable Boolean functions of two variables: XOR and its negation. The XOR problem was considering always as a classical problem that can not be solved using a single neuron. However, this is not true, because the XOR problem can be solved using a single UBN. Let $t = 2$, $m = 4$ in (14). Thus, the function (14) separates the complex plane onto four equal sectors (quadrants): in the two sectors $P_B = 1$ and in the other two ones $P_B = 1$ (see Fig.4). Table (1) illustrates that the UBN solves the XOR problem using the weighting vector $(0, \imath, 1)$.

It is important to mention that all 256 Boolean functions of 3 variables are P-realizable and may be implemented on the single UBN with $m = 6$ in (14); all 65536 Boolean functions of 4 variables are P-realizable and may be implemented on the single UBN with $m = 8$ in (14); all $2^{2^5} = 2^{32}$ Boolean functions of 5 variables are P-realizable and may be implemented on the single UBN with $m = 10$ in (14) [4, 5]. Learning of the parity n problem up to $n = 14$ by the single UBN using a UBN learning algorithm is considered in [5].

The UBN learning can be reduced to the MVN learning [4]. Comparing the activation functions (2) and (14) (Fig.1 and Fig.3, respectively), we are coming to the following important conclusion. If the Boolean function $f(x_1, \ldots, x_n)$ is P-realizable with the weighting vector $(w_0, w_1, \ldots, w_n)$, and the predicate P_B is defined by the equality (14) then a partially defined (only on the Boolean sets of variables) m-valued function $\tilde{f}(x_1, \ldots, x_n)$ is the m-valued threshold function with the same weighting vector $(w_0, w_1, \ldots, w_n)$. But any P-realizable Boolean function can be implemented on a single UBN, and any m-valued threshold function can be implemented on a single MVN. This means that the UBN learning can be reduced to the discrete MVN learning according to rule (4). An incorrect output of the UBN for some input vector X from the learning set means that a weighted sum has fallen into an "incorrect" sector. Thus, the weights should be corrected to direct the weighted sum into one of the neighborhood sectors. A natural choice of the "correct" sector (left or right) is based on the closeness of the current value of the weighted sum. Let a current weighted sum z is laying in the sth sector, and this sector is "incorrect". Thus to apply the learning rule (4) for the UBN, we can choose

$$q = s - 1 \ (\mathrm{mod}\, m) \ if \ z \text{ is closer to } (s-1)^{st} \text{ sector} \quad (15)$$

$$q = s + 1 \ (\mathrm{mod}\, m) \ if \ z \text{ is closer to } (s+1)^{st} \text{ sector} \quad (16)$$

6 Summary

Both attractor networks and multilayer neural network based on multi-valued neurons have been shown to be a powerful tool for image reconstruction and for solving highly nonlinear problems of pattern recognition, classification, prediction, approximation and interpolation. When applied to these tasks, such networks have a number of unique properties and advantages in comparison with other neural networks models. These properties and advantages follow the unique features of a multi-valued neuron. In addition, attractor networks with complex-valued activation functions used as associative memory facilitate natural processing of gray-scale images and have the advantage of mathematical simplicity.

References

1. N. N. Aizenberg, Yu. L. Ivaskiv and D. A. Pospelov: About one generalization of the threshold function. In: *Doklady Akademii Nauk SSSR (The Reports of the Academy of Sciences of the USSR)*, vol.196, no.6, 1971, pp.1287-1290 (in Russian).
2. N. N. Aizenberg and I. N. Aizenberg: CNN Based on Multi-Valued Neuron as a Model of Associative Memory for Gray-Scale Images. In: *Proceedings of the Second IEEE Int. Workshop on Cellular Neural Networks and their Applications*, Technical University Munich, Germany, October 14-16, 1992, pp.36-41.
3. I. N. Aizenberg: The Universal Logical Element over the Field of Complex Numbers. In: *Kibernetika (Cybernetics and Systems Analysis)*, no.3, 1991, pp.116-121 (in Russian, journal is translated into English by Consultants Bureau, An Imprint of Springer Verlag New York LLC, Vol. 27, No 3, pp. 467-473).
4. I. Aizenberg, N. Aizenberg and J. Vandewalle: *Multi-valued and universal binary neurons: theory, learning, applications*, Kluwer Academic Publishers, Boston/Dordrecht/London, 2000.
5. I. Aizenberg: Solving the XOR and Parity n Problems Using a Single Universal Binary Neuron. In: *Soft Computing*, published Online First, June 2007, to appear in a regular hard copy: late 2007.
6. N. N. Aizenberg and Yu. L. Ivaskiv: *Multiple-Valued Threshold Logic*, Naukova Dumka Publisher House, Kiev, 1977 (in Russian).
7. I. Aizenberg and C. Moraga: Multilayer Feedforward Neural Network Based on Multi-Valued Neurons (MLMVN) and a Backpropagation Learning Algorithm. In: *Soft Computing*, vol.11, no.2, January 2007, pp. 169-183.
8. I. Aizenberg, D. Paliy, J. Zurada and J. Astola: Blur Identification by Multilayer Neural Network based on Multi-Valued Neurons. In: *IEEE Transactions on Neural Networks*, accepted, to appear early 2008.
9. S. Jankowski, A. Lozowski and J.M. Zurada: Complex-Valued Multistate Neural Associative Memory. In: *IEEE Trans. Neural Networks*, vol.7, 1996, pp.1491-1496.
10. H. Aoki and Y. Kosugi: An Image Storage System Using Complex-Valued Associative Memory. In: *Proc. of the 15th International Conference on Pattern Recognition*, Barcelona, 2000, IEEE Computer Society Press, vol.2, pp.626-629.
11. M. K. Muezzinoglu, C. Guzelis and J. M. Zurada: A New Design Method for the Complex-Valued Multistate Hopfield Associative Memory. In: *IEEE Trans. Neural Networks*, vol.14, No.4, 2003, pp.891-899.

12. H. Aoki, E. Watanabe, A. Nagata and Y. Kosugi: Rotation-Invariant Image Association for Endoscopic Positional Identification Using Complex-Valued Associative Memories. In: J. Mira, A. Prieto (Eds.): *Bio-inspired Applications of Connectionism*, Lecture Notes in Computer Science, vol.2085, Springer, 2001, pp.369-374.
13. D. L. Lee: Improving the capacity of complex-valued neural networks with a modified gradient descent learning rule. In: *IEEE Transactions on Neural Networks*, vol.12, no.2, Mar.2001, pp. 439-443.
14. D. L. Lee: Complex-valued Neural Associative Memories: Learning Algorithm and Network Stability. In: A. Hirose (Ed.): *Complex-Valued Neural Networks: Theories and Applications*, World Scientific, 2004.
15. H. Aoki: Applications of Complex-Valued Neural Networks for Image Processing. In: A. Hirose Ed.): *Complex-Valued Neural Networks: Theories and Applications*, World Scientific, 2004.
16. T. Miyajima and K. Yamanaka: Phasor models and their applications to communications. In: A. Hirose Ed.): *Complex-Valued Neural Networks: Theories and Applications*, World Scientific, 2004.
17. H. Aoki: A complex-valued neuron to transform gray level images to phase information. In: L. Wang J. C Rajapakse, K. Fukushima, S.-Y. Lee and Xin Yao (Eds.): *Proceedings of the 9th International Conference on Neural information Processing (ICONIP'2002)*, 2002, vol.3, pp.1084-1088.
18. I. Aizenberg, E. Myasnikova, M. Samsonova M. and J. Reinitz: Temporal Classification of Drosophila Segmentation Gene Expression Patterns by the Multi-Valued Neural Recognition Method. In: *Mathematical Biosciences*, vol.176 (1), 2002, pp.145-159.
19. I. Aizenberg, P. Ruusuvuori, O. Yli-Harja and J. Astola: Multilayer neural network based on multi-valued neurons (MLMVN) applied to classification of microarray gene expression data. In: *Proc. 4th TICSP Workshop on Computational Systems Biology (WCSB 2006)*, Tampere University of Technology, Finland, June, 2006, pp.27-30.
20. I. Aizenberg and J. M. Zurada: Solving Selected Classification Problems in Bioinformatics Using Multilayer Neural Network based on Multi-Valued Neurons (MLMVN). In: *Proceedings of 13th International Conference on Artificial Neural Networks (ICANN-2007)* (accepted), to appear: September 2007.
21. I. Aizenberg I. and C. Moraga: The Genetic Code as a Function of Multiple-Valued Logic Over the Field of Complex Numbers and its Learning using Multilayer Neural Network Based on Multi-Valued Neuron. In: *Multiple-Valued Logic and Soft Computing*, accepted, to appear: late 2007.
22. I. Aizenberg and C. Butakoff: Image Processing Using Cellular Neural Networks Based on Multi-Valued and Universal Binary Neurons. In: *Journal of VLSI Signal Processing Systems for Signal, Image and Video Technology*, vol.32, no.1-2, 2002, pp.169-188.
23. D. E. Rumelhart and J. L. McClelland: *Parallel distributed processing: explorations in the microstructure of cognition*. MIT Press, Cambridge, 1986.
24. J. M. Zurada: *Introduction to Artificial Neural Systems*, West Publishing Company, St. Paul, Minnesota, 1992.
25. J. M. Zurada: Neural Networks: Binary Monotonic and Multiple-Valued. In: *Proc. of the 30th IEEE International Symposium on Multiple-Valued Logic*, Portland, Oregon, May 23-25, 2000, pp.67-74

Part II

Regular Papers

Cluster-Based Route Discovery Protocol

Doina Bein[1], Ajoy K. Datta[2], and Shashirekha Yellenki[3]

[1] Department of Computer Science, University of Texas at Dallas, TX
siona@utdallas.edu
[2] School of Computer Science, University of Nevada Las Vegas, NV
datta@cs.unlv.edu
[3] School of Computer Science, University of Nevada Las Vegas, NV
rekha@cs.unlv.edu

Summary. In MANETS, node mobility induces structural changes for routing. We propose a route discovery algorithm for MANET based on link-cluster architecture. The algorithm selects the clusterheads and gateway nodes, and then builds routing tables for nodes both inside and outside the cluster. The algorithm attempts to minimize the number of clusterheads and gateway nodes to avoid storing redundant data. For intra-cluster routing, the shortest paths are maintained. For inter-cluster routing, we implement routing on-demand (the shortest paths are maintained only for the nodes that need to send packets). The proposed algorithm adapts to arbitrary movement of nodes, and joining and/or leaving of existent nodes.
Keywords: ad hoc network, cluster, location management, node mobility.

1 Introduction

A mobile ad-hoc network (MANET) is a self-configuring network of mobile hosts connected by wireless links, with an arbitrary topology. One mobile host may need to enlist the aid of other hosts in forwarding a packet to its destination due to the limited range of each mobile host's wireless transmissions. The mobility management of mobile nodes is important in mobile ad hoc networks. Minimal configuration and quick deployment make ad hoc networks suitable for emergency situations like natural or human-induced disasters, military conflicts, emergency medical situations, etc. The earliest MANETs, called "packet radio" networks, were sponsored by DARPA in the early 1970s. A major challenge faced in MANETs is locating the devices for communication, especially with high node mobility and sparse node density. Present solutions provided by the ad hoc routing protocols range from flooding [1] the entire network with route requests, to deploying a separate location management scheme [2] to maintain a device location database. Kawadia et al. [14] had given a general framework to support the implementation of ad-hoc routing protocols in Unix-like operating systems.

Doina Bein et al.: *Cluster-Based Route Discovery Protocol*, Studies in Computational Intelligence (SCI) **78**, 53–62 (2008)
www.springerlink.com

We present a protocol for routing in ad hoc networks that adapts fast to frequent host movement, yet requires little or no overhead during periods in which hosts move less frequently. Moreover, the protocol routes packets through a dynamically established and nearly optimal path between two wireless nodes. It also achieves higher reliability - if a node in a cluster fails, the data is still accessible via other cluster nodes. We consider a network with link-cluster architecture and discover an optimal route for the nodes to communicate with each other [3]. We use the concept of proactive protocols to route the packets within the cluster and the concept of reactive protocols to route the packets between the clusters. Such combination of proactive and reactive protocols used for routing the packets is called a hybrid protocol [4]. We also use the concept of location management when a node leaves a cluster to update the routing tables [5].

In Section 2 we present the architectural model and the variables used by the algorithm. The cluster-based route discovery algorithm is presented in Section 3, together with a proof of correctness in Section 4. We finish with concluding remarks in Section 5.

2 Preliminaries

Clustering is a scheme designed for large dynamic networks to build network control structure that increases network availability, reduces the delay in responding to changes in network state, and improves data security. Clustering is crucial for scalability as the performance can be improved by adding more nodes to the cluster. Link-cluster architecture [6, 7, 8] is a network control structure in which nodes are partitioned into clusters that are interconnected. The union of the members of all the clusters covers all the nodes in the network. In every cluster, nodes are classified in three ways: clusterhead, gateway, and ordinary node. A *clusterhead* schedules the transmissions and allocates resources within clusters. *Gateways* connect adjacent clusters. An *ordinary node* belongs to a single cluster (has a unique clusterhead). We will consider only *disjoint* clusters: a gateway node is a member of exactly one cluster and forms a link to a member of other cluster.

The unique identifier [9] for each node differentiates every single node in the network from the other. The node with the highest or lowest identifier becomes the clusterhead [10]. The change in clusterhead occurs only if two clusterheads move within the range of each other, where one of them relinquishes its role as clusterhead, or if an ordinary node moves out of range of all other nodes, in which case it becomes the clusterhead of its own cluster.

Cluster maintenance schemes are designed to minimize the number of changes in the set of existing clusters. They do not re-cluster after every movement, but instead make small adjustment to cluster membership as necessary. Nodes periodically select nodes that take on the role of a location server of their current location. All the gateway nodes and the clusterhead node which are present in the cluster region C_u of the clusterhead node u act as location servers for all the nodes in the cluster region C_u. When a node moves across two clusterhead regions, the node updates its home region C_u of the movement by a location update or by sending a leave message. A source node x from outside the cluster that needs to communicate with another node y in the cluster region C can use the clusterhead and gateway tables to identify the location of y and send a location query packet towards region C to obtain the current location of y. The first location server to receive the query for u responds with the current location of y to which data packets are routed.

3 Cluster-Based Route Discovery Algorithm

The algorithm uses a variable N_i^1 representing the one-hop neighborhood set of node i and a variable N_i^2 representing the two-hop neighborhood set of node i. Node i has a unique ID, $ID.i$. Variable $n.i$ is used to identify the neighbor of the shortest path to the clusterhead (for a non-clusterhead node). Every node in a network has a sequence table that keeps track of the messages already received by the node and makes the routing messages loop-free [11, 12]. Only gateways and clusterheads maintain the tables used for routing [3]. A clusterhead has another table that is used to route messages outside the cluster. This table has entries of all the destination and boundary gateway pairs. The gateway tables contain all the entries of the destination-clusterhead pairs of all the clusters they connect to. The routing table is updated whenever a new clusterhead is elected or some changes occur related to paths in the routing table. The ordinary nodes have only a variable indicating the neighbor on the shortest path towards their clusterhead.

The proposed protocol consists of three main steps: Clusterhead Election, Gateway Election, and Route Discovery that are implemented in three different modules.

The clusterhead selection protocol must satisfy three conditions: each non-clusterhead is within two hops from its clusterhead, and there are no adjacent clusterheads [13]. A node can act as a clusterhead as well as a gateway at the same time. A clusterhead will periodically do the following: checks the consistency of each variable, broadcasts CL_ANN messages to all its neighbors within its two hop distance, checks if any other clusterhead is in its range and if it finds one whose ID is bigger than itself then it gives up its clusterhead status by broadcasting CL_REJ messages. An ordinary node i periodically checks its clusterhead (alive or not) by sending a CL_REQ message to $n.i$. In case it has no clusterhead within its two hop distance, it sets its variables accordingly and waits for a CL_ANN message from a node within its two hops distance. It becomes a clusterhead if there is no clusterhead within its two hops distance. A CL_REQ message travels at most two hops from the sender. Once the CL_REQ message reaches the right destination but finds that its clusterhead moved from that location, the node in that particular location or the node which was supposed to be the one hop neighbor on the shortest path from the sender to the supposed-to-be clusterhead's location sends a *CL_CHG* message indicating that the previous clusterhead no longer exists in that location.

Algorithm 3.1 Clusterhead Selection Protocol

Actions of some node *i*

E.01 Timeout $\longrightarrow$
  if *i* is a clusterhead then sends *CL_ANN* to immediate neighbors
  else if *i* finds itself with faulty values or is "orphan" (has no CH),
    then elects itself as a clusterhead
    else *i* sends *CL_REQ* message to *n.i*

E.02 Receive *CL_ANN* from node *nb* $\longrightarrow$
  if *i* is an ordinary node and either the sender was its own CH or
    i has no current clusterhead, then updates its variable with respect to the sender as a clusterhead and forwards the message
  else if *i* is a clusterhead and the sender is a clusterhead with
    a lower ID and within 2 hops, then *i* accepts the sender as a clusterhead and sends *CL_REJ* to all neighbors

E.03 Receive *CL_REJ* from node *nb* $\longrightarrow$
  if *i* is a clusterhead, then drops the message
  else if the sender is *i*'s CH, then mark itself as "orphan" and forward it

E.04 Receive *CL_REQ* from node *nb* $\longrightarrow$
  if *i* is a clusterhead then
    if the sender belongs to its cluster, then send *CL_ANN* to sender
    else send *CL_CHG* to the sender
  else if the message is addressed to *i* then reply with *CL_CHG*
  else if the addressee is within two hops, then forward it to addressee
      else drop the message

E.05 Receive *CL_CHG* from node *nb* $\longrightarrow$
  if the message is regarding is *i*'s clusterhead, then *i* updates
    its variables accordingly and forwards the message to neighbors

E.06 Receive *CL_ACCEPT* from node *nb* $\longrightarrow$
  if *i* is a clusterhead and the addressee, then updates its routing table
    and sends the updated message to the bordering gateway nodes
  else if the message is not addressed to the node, then it forwards
    the message to its neighbors if the hop count < 2,
    but drops the message if the hop count ≥ 2

E.07 Receive *leave* from node *nb* $\longrightarrow$
  if *i* is a clusterhead and the addressee, then updates its routing table
    and sends the updated message to the bordering gateway nodes
  else if the message is not addressed to the node, then it forwards
    the message to its neighbors if the hop count < 2,
    but drops the message if the hop count ≥ 2

E.08 Receive *ctable_copy* from node *nb* $\longrightarrow$
if i is a clusterhead and the message is addressed to it, then the row contained in the message is copied into the routing table if the destination node is within 2 hop distance
E.09 Receive CL_CHG from node *nb* $\longrightarrow$
if i is gateway and the sender is clusterhead of one of its neighbors, then updates its GC_TABLE

The gateway selection protocol, a gateway node periodically checks if there is another gateway in two hop distance that connects the same clusters. If yes, it compares its own ID with it. If it has a smaller ID, then it relinquishes its role as a gateway by updating its $g.i$ variable and sending a GW_REJ message.

Algorithm 3.2 Gateway Selection Protocol

Actions of some node i

G.01 Timeout $\longrightarrow$ if i is a gateway and there is another gateway within 2 hops with a lower ID that connects at least the clusters, then sends GL_REJ to all neighbors
G.02 Receive GL_ANN from node *nb* $\longrightarrow$
if i is a clusterhead and the message is addressed to it, then updates its inter-cluster table
else if i is a gateway and there is another gateway within 2 hops with a lower ID that connects at least the clusters, then i sends GL_REJ to all neighbors
else it forwards the message to its neighbors if the hop count < 2, but drops the message if the hop count ≥ 2
G.03 Receive GW_REJ from node *nb* $\longrightarrow$
if i is a clusterhead and the message regards one of its bordering gateway node, it removes all such rows containing the sender's ID in the GW field of its tables
else it forwards the message to its neighbors if the hop count < 2, but drops the message if the hop count ≥ 2

In the route discovery module, we have intra-cluster and inter-cluster routing. For intra-routing (routing within the cluster), each clusterhead keeps data in its routing table about the nodes that belong to its own cluster. This data is collected in the clusterhead election module using CL_REQ messages. These messages are periodically sent by a non-clusterhead node to check the status of its own clusterhead and the path towards it. For inter-cluster routing (between the clusters), the clusterheads as well as the gateway nodes keep information of the gateway-destination and clusterhead-destination pairs, respectively to reach the temporary destination, which is a milestone in reaching the actual destination. This data is collected only when there is a need to communicate with the node and stored in the inter-cluster tables. These tables purge the routes that are unused for a long time and keep the entries updated.

The following steps are repeated until the route is found.

1. Sender checks with its clusterhead if its routing table has an entry for the destination node that it wants to communicate with. If the cluster-head has an entry, the sender gets the path from the clusterhead and uses it to communicate.

2. If the clusterhead's routing table does not have an entry, it checks with the clusterhead's gateway table. If it finds an entry, then it uses that route to communicate.

3. If the clusterhead's gateway table does not have an entry, then it checks with the gateway's cluster tables of all the bordering gateways for the route. If it finds the route, it uses that to communicate.

Algorithm 3.3 Route Discovery Module

Actions of some node i

A.01 Receive *Routedisc* from node *nb* $\longrightarrow$
    if the same message was received before, then drops it
    if i is a clusterhead
        if the message was addressed to i then sends back an *ack* message
        else if the destination node belongs to its cluster, it sends
            the *shortestpath* message to the sender
            else it updates its inter-cluster table and sends the
                updated message to the bordering gateway nodes
    else if i is a gateway
        if the message was addressed to i then sends back an *ack* message
        else if the destination node belongs to its inter-cluster table, it
            forwards it to all the clusterheads in its inter-cluster table
            else it updates its inter-cluster table and sends the
                updated message to the bordering gateway nodes
    else if i is an ordinary node
        if the message was addressed to i then it sends back an *ack* message
        else forwards the message to its neighbors
A.02 Receive *me_dest* from node *nb* $\longrightarrow$
    if i is the clusterhead of the destination and the sender does not belong
        to its inter-cluster routing table, it updates the table and sends the
        updated message to all its bordering clusterheads

---

else if i is a gateway
    if the clusterhead of the destination is at one hop distance,
        it forwards the message
    if the sender does not belong to the inter-cluster routing table,
        it updates the table and sends the updated message to
        all its bordering clusterheads
else if i is an ordinary node and the clusterhead of the destination is
    at one hop distance, it forwards the message

A.03 Receive *me_dest* from node *nb* $\longrightarrow$
    if i is a clusterhead or a gateway
        if the sender does not belong to its inter-cluster routing table,
            it updates its table and sends the updated message to all its
            bordering gateway nodes
        if it is not the destination, then it forwards the message to all
            the nodes in the specified in the field *route* of the message
    else if i is an ordinary node and it is not the destination, then it
        forwards the message to all nodes in the specified in the field *route*
        of the message

A.04 Receive *ack* from node *nb* $\longrightarrow$
    if i is a clusterhead, it updates its table and sends the updated message
        to the bordering gateways
    else if i is a gateway, it updates its table and sends the updated
        message to the bordering clusterheads
    else if i is an ordinary node, if the clusterhead of the destination is at
    one hop distance, it forwards the message to that particular neighbor

A.05 Receive *Ctable_update* from node *nb* $\longrightarrow$
    if i is a gateway
        if the message is from a neighboring clusterhead, it updates its
            inter-cluster routing table, else forwards it to its neighbors
    else if i is an ordinary node, not the addressee, but the addressee is
        a neighbor then it forwards the message to it

A.06 Receive *Gtable_update* from node *nb* $\longrightarrow$
    if i is a clusterhead
        if the message is from a gateway node that is present in its inter-
            cluster routing table, it updates its inter-cluster routing table
        else forwards it to its neighbors
    else if i is an ordinary node, not the addressee, but the addressee is
        a neighbor then it forwards the message to it

4 Proof of Correctness

Lemma 1. *The maximum number of hops between a clusterhead and a member of its own cluster is two.*

Proof. In clusterhead election module, Actions E.02 and E.06 ensure that any clusterhead announcement (CL_ANN) message or the clusterhead accept (CL_ACCEPT) message can travel at most a distance of two hops. For a node to be a member of a cluster it has to receive the clusterhead announcement message from a clusterhead and send the clusterhead accept message back to the clusterhead, which is possible only if the node is at a two-hop distance from its clusterhead.

Lemma 2. *No two clusterheads can be neighbors of each other.*

Proof. We prove this lemma by contradiction. Suppose there are two clusterheads that are neighbors. Action E.02 ensures that the clusterhead announcement (CL_ANN) message of one clusterhead reaches the other that is at one or two-hop distance from it (Lemma 1). When a cluster-head receives a clusterhead announcement message, it compares its own ID with the sender's ID. If its ID is less than that of the sender's ID, it relinquishes its role as a clusterhead and sends the clusterhead reject (CL_REJ) message to all its two-hop neighbors. Action E.03 ensures that the clusterhead reject message reaches all the two-hop neighbors. So, it no longer remains a clusterhead which contradicts our assumption that there can be two clusterheads that can be neighbors.

Lemma 3. *The minimum number of hops between two clusterheads is three.*

Proof. From Lemma 2, no two clusterheads can be neighbors of each other. Assume that the distance between two clusterheads is two hops. But that cancels one of the two clusterheads by comparing the IDs because the node between them becomes a gateway that acts as a common node for both clusters.

Lemma 4. *The maximum number of hops between the clusterheads of two neighboring clusters is five.*

Proof. Let us assume that the maximum distance between the two clusterheads is six. According to our module, Action E.02 makes sure that the clusterhead announcement message travels at most a distance of two hops. Then, there is at least one node that does not receive any clusterhead announcement message. This node waits for a timeout period (Action E.01) and at timeout, sets itself a clusterhead forming its own cluster.

Lemma 5. *If there exist only one link connecting two neighboring clusters then the eligible gateway nodes on both ends of the link will be selected as gateway nodes.*

Proof. We prove this lemma by contradiction. Suppose the nodes connecting the clusters are not gateway nodes. By the definition of a gateway, both nodes are eligible gateway nodes be-cause both of them have at least one neighbor that does not belong to its own cluster. In our module, we eliminate the eligible gateway nodes be-coming the gateway nodes only if they belong to the same cluster. So, both the nodes become the gateway nodes that contradict the assumption that they are not the gateway nodes.

Lemma 6. *If both the sender and destination are in the same cluster, a route discovery message is always acknowledged.*

Proof. When a node generates the route discovery (routedisc) message, it first sends it to its own clusterhead. Route discovery within a cluster means that the sender and destination belong to the same cluster. If the message reaches the destination before reaching the clusterhead, the destination node directly sends the acknowledgment (ack) message to the sender following the reverse path followed by the route discovery message. If the message reaches the clusterhead, all the clusterheads have en-tries for all the nodes in their intra-cluster table (routing table as named in our module) that belong to its own cluster. Once the clusterhead receives the message, it looks in its routing table, attaches the route from itself to the destination to the path followed by the route discovery message, and sends an acknowledgment message to the sender using a shortestpath message on the reverse path followed by the route discovery message.

Lemma 7. *If a node moves to another cluster, the route discovery algorithm will be able to find the node in finite time upon a request.*

Proof. When a node is in a cluster, it periodically acknowledges a clusterhead that it is still in the cluster. When the node moves out of the cluster, the clusterhead waits for a timeout interval, then removes all the rows with this node as destination from its intra- and inter-cluster routing tables, and updates the same to its boundary gateway nodes so that they can re-move the rows from their inter-cluster routing tables. If the node joins another cluster, it acknowledges the new clusterhead's CL_ANN message with a CH_ACCEPT message that it joined its cluster, and the new clusterhead updates its entry in its intra-cluster routing table. If the node itself becomes the clusterhead because it is not in two-hop distance from any clusterhead, then it broadcasts the message to all the clusterheads whose clusters it connects. The route is thus discovered.

5 Conclusion

We have presented a route discovery algorithm for MANET based on link-cluster architecture. The algorithm selects the clusterheads and gateway nodes, and then builds routing tables for nodes both inside and outside the cluster. The proposed protocol guarantees that in finite number of steps, the network is divided into clusters. The algorithm attempts to minimize the number of clusterheads and gateway nodes to avoid storing redundant data. For intra-cluster routing, the shortest paths are maintained. For inter-cluster routing, we implement routing on-demand (the shortest paths are maintained only for the nodes that need to send packets). The proposed algorithm adapts to arbitrary movement of nodes, and joining and/or leaving of existent nodes.

References

1. Y.B. Ko and N.H. Vaidya. Location-aided routing in mobile ad hoc networks. *Technical report 98-012, Texas A&M University*, 1998.
2. S. J. Philip, J. Ghosh, S. Khedekar, and C. Qiao. Scalability analysis of location management protocols for mobile ad hoc networks. *Wireless Communications and Networking Conference*, March 2004.
3. C. C. Chiang, H-K Wu, W. Liu, and M. Gerla. Routing in Clustered Multihop, Mobile Wireless Networks. *IEEE Singapore International Conference on Networks*, pages 197-211, 1997.
4. E. M. Royer and C. K. Toh. A Review of Current Routing Protocols for Ad hoc Mobile Networks. *IEEE Personal Communications*, 6(2), pages 46-55, April 1999.
5. Z. Kai, W. Neng, and L. Ai-Fang. A new AODV based clustering routing protocol. *International Conference on Wireless Communications, Networking, and Mobile Computing*, September 2005.
6. D. J. Baker and A. Ephremides. A Distributed Algorithm for Organizing Mobile Radio Telecommunication Networks. *Proceedings of the Second International Conference on Distributed Computer Systems*, pages 476-483, April 1981.
7. D. J. Baker and A. Ephremides. The Architectural Organization of a Mobile Radio Network via a Distributed Algorithm. *IEEE Transactions on Communications*, 29(11), pages 1694-1701, November 1981.
8. A. Ephremides, J. E. Wieselthier, and D. J. Baker. A Design Concept for Reliable Mobile Radio Networks with Frequency Hopping Signaling. *Proceedings of the IEEE*, 75(1), pages 56-73, January 1987.
9. S. R. Das, C. E. Perkins, and E. M. Royer. Performance Comparison of Two On-demand Routing Protocols for Ad Hoc Networks. *Proceedings of INFOCOM*, March 2000.
10. C. C. Chiang. Routing in Clustered Multihop, Mobile Wireless Networks. *Proceedings of the ICOIN*, 1996.
11. G. G. Chen, J. W. Branch, B. K. Szymanski. Self-selective routing for wireless ad hoc networks. *IEEE International Conference on Wireless And Mobile Computing, (WiMob'2005)*, August 2005.
12. C. E. Perkins and E. M. Royer. Ad-Hoc On-Demand Distance Vector Routing. *Second Annual IEEE Workshop on Mobile Computing Systems and Applications*, pages 99-100, February 1999.
13. T. Johansson, L. Carr-Motyckova. Bandwidth-constrained Clustering in Ad Hoc Networks. *The Third Annual Mediterranean Ad Hoc Networking Workshop*, pages 379-385, June 2004.
14. V. Kawadia, Y. Zhang, and B. Gupta. System Services for Implementing Ah-hoc Routing Protocols. *Procedeedings of the International Conference on Parallel Processing Workshops (ICPPW'02)*, pages 135-142, 2002.

Object-Oriented High Level Petri Nets as a Formal Language for Modelling and Design Multi-Agent Systems

Marius Brezovan[1] and Eugen Ganea[2]

[1] University of Craiova, Software Engineering Department
brezovan_marius@software.ucv.ro
[2] University of Craiova, Software Engineering Department
ganea_eugen@software.ucv.ro

Summary. Agent oriented systems are often complex systems and there is a need for developing modelling techniques that permit to manage such complex systems and to allow the process of system design to be sufficiently reliable and extensible.

In this paper we present a new class of Petri nets called Object Oriented High Level Petri Nets (OOHLPN) as a unified language used for modelling and design of multi-agent systems (MAS). The design of MAS is based on three fundamental perspectives: the functional, the interactional, and the organizational perspective. In order to allow the modelling and design of MAS from theses perspectives the proposed formalism has several properties: (a) it allow a connection with rule-based systems in order to incorporate the Belief-Desire-Intention (BDI) model, which is used to describe the mental states of agents, (b) it allows an asynchronous message passing between objects in order to describe the interactional perspective of MAS, (c) it allow the description of distributed and concurrent systems of objects in order to model the organizational perspective of MAS, and (d) it uses the inheritance and polymorphism concepts in order to allow an efficient method of MAS design. An example of using OOHLPN for MAS modelling is provided.

Key words: Multi-Agents Systems, Object-Oriented Petri Nets, Formal Specification, Distributed Object-Orientation, BDI Model

1 Introduction

Multi-agent systems (MAS) are adaptive and flexible systems, in which agents may be added or deleted at run-time, and the agent interactions may vary dynamically [20, 17]. Formal specification techniques can be used to support the design and implementation of MAS, in order to allow the developed systems to be robust, reliable, and efficient [12]. Also the use of formal specifications at the MAS design level can reduce the development cycle.

Marius Brezovan and Eugen Ganea: *Object-Oriented High Level Petri Nets as a Formal Language for Modelling and Design Multi-Agent Systems*, Studies in Computational Intelligence (SCI) **78**, 63–74 (2008)
www.springerlink.com

In this paper, a formal specification method is proposed, which can be used as a description language for modeling and design of MAS. The proposed formalism is called *Object-Oriented High Level Petri Nets* (OOHLPN) and it represents an extension of the standard High level Petri nets with the object-oriented concepts. Our purpose is to transform the high-level specification of these kinds of systems into an uniform object-oriented high level Petri net formalism while retaining the original system semantics.

Related work

The main difficulty of using Petri nets results from the contradiction between structural evolution of systems and the structural rigidity of Petri nets. This is the reason for introducing the object-oriented paradigm into the framework of Petri nets. Two formalisms proposed by Sibertin-Blanc [15] and Bastide [1] are Cooperative Objects, and their successor Cooperative Nets [16]. A Cooperative Object has a type, an identity and a state, a class of Cooperative Objects containing an interface and an implementation. Biberstein *et al.* [2, 5] proposed a formalism called CO-OPN that extends the use of algebraic data types to include the communication structures. An extended version of CO-OPN, called CO-OPN/2 [3] adds new features, as a synchronization mechanism, and the definition of sub-typing based on bisimulation. Valk [18, 19] proposed a Petri net model called Elementary Net Systems with two levels of activity, with synchronization constrained by a static labeling of transitions. This formalism has been augmented with object-oriented features resulting in a new formalism called Simple Object Systems.

Petri nets have been used in several approaches to model and specify MAS. Holvoet and Verbaeten have introduced the idea of enhancing agent oriented programming by using high-level nets to model agents, and extended this to a class of Petri nets called Generic Nets [8]. In some related work, Holvoet and Kielmann [9] have applied the CPN formalism to specifying agent systems based on their Objective-Linda coordination model. Moldt and Wienberg have developed an approach called AOCPN (Agent Oriented Colored Petri Nets) [13]. This system employed an object-oriented language, syntactically similar to C++, which maps onto CPN, extended by test arcs.

Organization

The paper is structured as follows. Section 2 presents the formal definition of the OOHLPN formalism. Section 3 presents an approach for modelling MAS by using OOHLPNs, and Section 4 presents an example of using OOHLPN for modelling multi-agent systems.

2 Formal Syntax of Object-Oriented High Level Petri Nets

Throughout this paper, we assume a universe U, which contains several disjoints sets including: OID which the set of all identifiers associated to objects,$SORT$ which is the set of all sorts, and $METH$ which is the set of public method names of object types. $SORT$, is partitioned in three disjoint sets, $SORT_D$, $SORT_{Ob}$ and $SORT_{Imp}$, corresponding to the names of (non object-oriented) data types, the names of object types, and the name of implementation modules for object types. The set $METH$ is considered as a sorted set, $(METH_{\hat{c}w,s})_{c \in SORT_{Ob}, w \in SORT^*, s \in SORT \cup \{\varepsilon\}}$, where $\hat{c}$ represents the reference type to the object containing the method, w is the sequence of sorts representing the input values, and s is the sort of the returned value.

OOHLPNs are constructed by using order-sorted signatures and order-sorted algebras in order to specify the notion of inheritance for hierarchies of implementation modules [7]. We use the notions of category and institutions in order to allow structured and composable signatures [4], and also structured and composables Petri nets. The reader is assumed to have some background in algebraic specifications, theory of categories and institutions, and algebraic theory of object-oriented programming.

Inheritance and subtyping are orthogonal notions, the contravariance rule captures the subtyping relation, whereas a covariant relation characterizes the specialization of code. In order to safety integrate these concepts, in OOHLPN formalism we separate the notion of a class into its two distinct elements: (a) an *interface*, which is used to define an object type, and (b) an *implementation module*, which is used to specify an object type implementation.

2.1 Interfaces and Types

OOHLPN safety integrate covariant and contravariant concepts by using *encapsulated multi-methods* and a multi-dispatching mechanism for messages [4]. An encapsulated multi-method is an overloaded method associated with an object type, which contains several methods (branches) having the same name. An *encapsulated multi-method* of an object type c with n branches is a pair (m, t), where $m \in \bigcap_{i=1}^{n} METH_{\hat{c}w_i, s_i}$ is the name of the multi-method, and $t = \{\hat{c}w_1 \to s_1, \ldots, \hat{c}w_n \to s_n\}$ represents its type. The set of all possible encapsulated multi-methods of c is denoted by $MMETH(c)$.

Definition 1. Let c be an object type. An *interface* defining c is a tuple, $intf = (c, super, Meth)$, where $super = s_1 \ldots s_n$ is a sequence of object types such that $\leq_{intf} = \{(c, s_1), \ldots, (c, s_n)\}$ is a partial order specifying the subtype relation associated to c, and $Meth \subseteq MMETH(c)$ is a finite set of encapsulated multi-methods of c.

In the OOHLPN formalism the weak form of substitutability principle is used because interfaces are defined at syntactic level.

Definition 2. Let $intf_i = (c_i, super_i, Meth_i)$, $i = 1, 2$, be two interfaces defining the object types c_1 and c_2. The object type c_2 is a *subtype* of c_1, iff the following relations hold: (a) $(c_2, c_1) \in \leq_{intf_2}$, (b) c_2 has at least all the messages of the supertype c_1, and (c) multi-methods of c_2 may have subtypes of the corresponding multi-methods of c_1 with the same name.

2.2 Modules and Implementations

The implementation module of an object type is realized in the OOHLPN formalism by using a class of Petri nets called *Extended High-Level Petri Net with Objects* (EHLPNO), which represent high level Petri nets enriched with some object orientation concepts, such as creating new objects inside transitions when they fire, and calling public methods of objects inside transitions.

Let t be a transition and x a variable of t associated to an object *ob*. A *method call* is a syntactical construction: $[b \leftarrow]\ x.m(a_1, \ldots, a_n)$, where m is the name of a method of *ob*, $a_1, \ldots, a_n$ are expressions containing input variables of t, and b is an (optional) variable of t. The set of all method calls is denoted by $MCALL$. The *object creation* can be specified by a syntactical construction: $v = new\ impl(e_1, \ldots, e_k)$, where v represents a variable associated to the newly created object, $e_1, \ldots, e_k$ are expressions containing input variables of t, and $impl$ is the name of the implementation module of the created object. The set of all object creation is denoted by $CREATE$.

Remark. There are no *destroy* methods in the OOHLPN formalism, because a garbage-collector mechanism is used in order to manage the references to objects.

In order to ensure the inheritance relation for implementation modules EHLPNOs are based on the concept of institutions with semi-composable signatures [6]. We use the institution, $\mathbf{NOS} = \langle \mathbf{Sign_{NOS}}, \mathbf{Sen_{NOS}}, \mathbf{Mod_{NOS}}, (\models_\Sigma)_{\Sigma \in |\mathbf{Sign_{NOS}}|} \rangle$, of regular order-sorted signatures with a Noetherian ordering on the sorts and signature morphisms preserving maximal elements of connected components [14]. The institution **NOS** has composable signatures, and the category of algebras $\mathbf{Mod_{NOS}}(\mathbf{Sig_{NOS}})$ is finitely cocomplete.

Definition 3. Given the institution **NOS**, a Boolean order-sorted signature, $Sig = (S, \leq, O)$, from $\mathbf{Sig_{NOS}}$, and $H = (S_H, \leq, O_H)$ an order-sorted *Sig*-algebra from the model functor $\mathbf{Mod_{NOS}}$, an *Extended High-Level Petri Net with Objects* is a tuple:

$$ehlpno = (NG, Sig, V, H, Type, AN, AC, M_0),$$

where:

(i) the net graph $NG = (P, T; F)$, the type function $Type : P \rightarrow S_H$, the net annotation $AN = (A, TC)$, the S-indexed set of variables V, and the initial marking $M_0 : P \rightarrow \bigcup_{p \in P} \mu Type(p)$ are defined as in the standard of High-Level Petri Nets [11];
(ii) $AC : T \rightarrow CREATE \cup MCALL \cup \{undef\}$ is the action annotation.

In order to distinguish between different types of transition annotations, from a graphical point of view, the guard functions are specified between parentheses (symbols '(' and ')'), while action annotations are specified between brackets (symbols '[' and ']').

The set of all Extended High-Level Petri Nets with Objects is denoted by $EHLPNO$. In order to define the inheritance relation for implementation modules we use compositional Petri nets. This requirement is accomplished by defining a cocomplete category of Extended High-Level Petri Nets with Objects, called *Categorical Extended High-Level Petri Nets with Objects* (CEHLPNO). It is out of the purpose of this paper to define the EHLPNO-morphisms and the category **EHLPNO**. The sets $EHLPNO$ and $CEHLPNO$ are in one-to-one correspondence to each other and it is not necessary to differentiate the terms $EHLPNO$ and $CEHLPNO$. The *plain union* of two EHLPNOs, N_1 and N_2, id denoted by $N_1 + N_2$, and it is the *amalgamated union* $N_1 +_{N_0} N_2$, where N_0 is the common subnet of N_1 and N_2. The plain union can be extended to several EHLPNOs.

Definition 4. Given the institution **NOS**, a Boolean order-sorted signature, Sig, from $\mathbf{Sig_{NOS}}$, and $intf$ an interface defining the object type c, an *implementation module* of the interface $intf$ is a tuple:

$$impl = (sm, ehlpno, Create, intf, inherit)$$

where: (a) $sm \in SORT_{Imp}$ is the name of the module, (b) $ehlpno$ is an Extended High-Level Petri Net with Objects, (c) $Create = (create, t_{create})$ represents the set of all *create* methods of c, and (d) $inherit \in \wp(SORT_{Imp})$ specify implementation inheritance relation of $impl$.

To each multi-method, m, of the interface $intf$, there is a corresponding subnet in $ehlpno$ having a pair of places, $(\#m, m\#)$, used to receive the input arguments of m, and to return the output information respectively.

In the OOHLPN formalism multiple inheritance relation is supported.

Definition 5. Let $impl_i = (sm_i, ehlpno_i, Create_i, intf_i, inherit_i)$, $i = 0, \ldots, k$ be $k+1$ implementation modules. The implementation module $impl_0$ *inherits* the modules $impl_1, \ldots, impl_k$, iff the following relations hold: (a) $inherit_0 = \{intf_1, \ldots, intf_k\}$, and (b) $ehlpno_0$ is the plain union of $ehlpno_1, \ldots, ehlpno_k$.

For each module $impl_i$ inherited inherited by $impl_0$ it is denoted that $impl_i \leq_I impl_0$.

2.3 Object-Oriented High Level Petri Nets

OOHLPN use the notions of subtype and inheritance hierarchy.

Definition 6. Let IT be a set of interfaces and IM a set of modules implementing the interfaces from IT. An *object-oriented system* associated to IT and IM is a triple:

$$OS = (IT, IM, Oid)$$

where $Oid : IM \to \wp(OID)$ is a function which associates a set of object identifiers to each implementation module, such that $Oid(impl_i) \cap Oid(impl_j) = \emptyset$ if $impl_i \neq impl_j$, or $Oid(impl_i) \subseteq Oid(impl_j)$ if $impl_i \leq_I impl_j$, assuming that $impl_i, impl_j \in IM$.

Definition 7. Let $OS = (IT, IM, Oid)$ be an object-oriented system as in the above definition. An *Object Oriented High Level Petri Net* associated to OS is a triple:

$$oohlpn = (OS, impl_0, oid_0)$$

where $impl_0 \in IM$, with the property that the object type associated to $impl_0$ is a root of the object type hierarchy of OS, and $oid_0 \in Oid(impl_0)$ is the object identifier associated to the initial object of *oohlpn*.

The initial implementation module $impl_0$ of an OOHLPN represents the higher abstraction level of a modelled system, and its initial object will exist at the beginning of the evolution of a dynamic system described by OOHLPN.

3 Modelling Multi-Agent Systems Using Object-Oriented High Level Petri Nets

In this Section we present an approach to the design and modelling of MAS based on our proposed class of Petri nets. In order to accurately describe MAS architecture we present MAS from three fundamental perspectives: the functional, the interactional, and the organizational perspective. From this point of view, Shoham's paradigm of Agent-Oriented Programming is transferred to a distributed environment by using OOHLPN.

3.1 Modelling Functional Perspective

The functional perspective of MAS is a description of the agents, which are responsible for interacting with the users and environments.

In our approach, an agent is described by the implementation module *Agent*, which is presented in Figure 1. *Agent* provides two methods, *GetMess* and *PutMess*, used for interacting with the environment or with other agents. Inside the agent, messages are managed by a special object, instance of an implementation module called *Communication*, which is presented in a later subsection. Messages are processed by an instance of the implementation module *TheoremProver*. The theorem prover can access also the knowledge base of the agent representing the agent's Beliefs, its list of plans representing the agent's Intentions, and its list of goals representing the agent's Desires..

Remark. In order to simplity notations in Figure 1 each arc with double arrows represents in fact two arcs in opposite direction having the same arc annotation.

3.2 Modelling Interactional Perspective

The interactional perspective describes the communication facility of MAS dealing with the interaction information among agents, and defining the rules that govern those interactions.

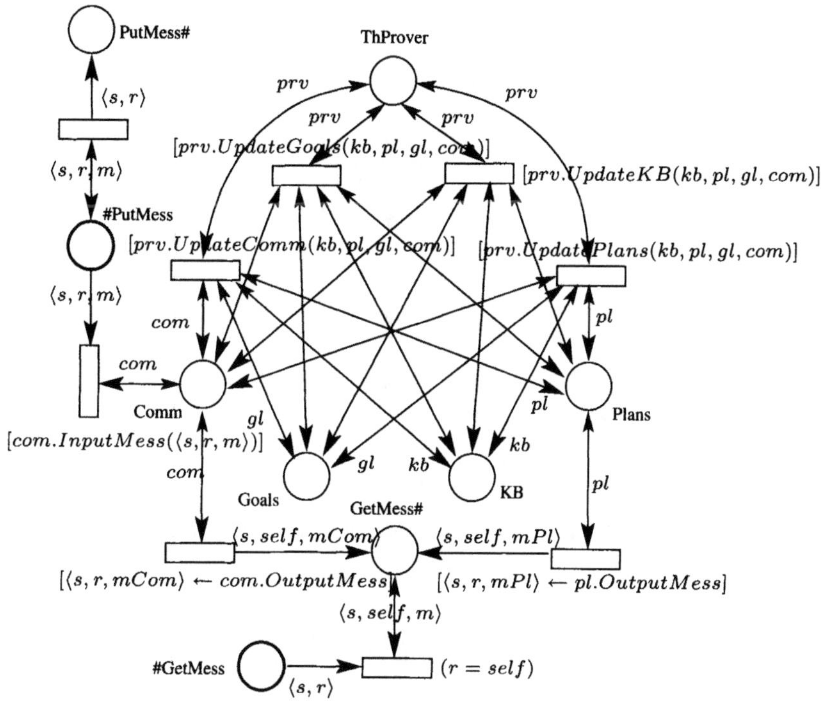

Fig. 1. The *Agent* implementation module

In our approach there are two levels that models the interactional perspective. Because agents have the capability to realize multiple conversations with other agents, we model each conversation by an distinct object, the management of these conversation objects being realized by using the implementation module *Communication*. All conversation objects are instances of some implementation modules that implements an object type, *ProtocolType*, which has two methods: *InputMess* and *OutputMess*, which allows objects to send and receive messages. For example, in Figures 2, 3, and 4 are presented the implementation modules *Pair−wise Negotiation*, *Seller*, and *Buyer* respectively.

All tokens representing messages are supposed to containing the reference of the sender object, the reference of the receiver object, and the message content.

One can observe that modules *Seller* and *Buyer* both inherit the module *Pair−wise Negotiation*.

3.3 Modelling Organizational Perspective

In our approach the organizational perspective represents an agent society which contains agents as objects. In Figure 5 is presented the implementation

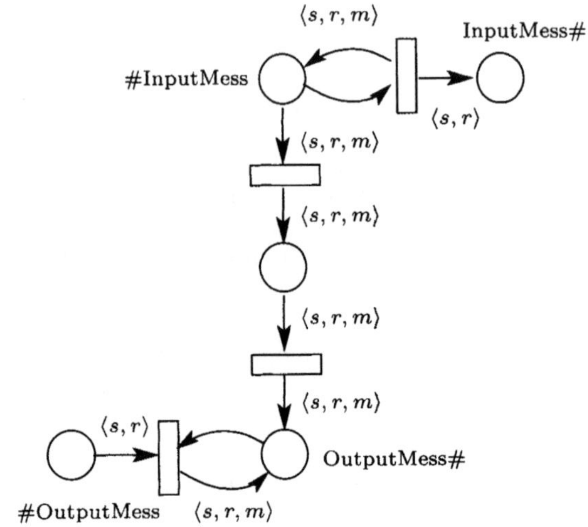

Fig. 2. The *Pair-wise Negotiation* implementation module

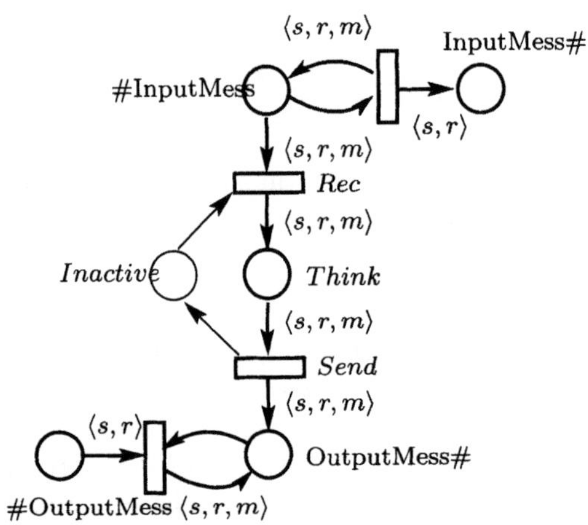

Fig. 3. The *Seller* implementation module

module *AgentSociety*, where the marking of the place *Agents* contains all alive agents as tokens.

The agents from an agent society can exchange messages with the environment, which is depicted in the figure by the place *Environment*. All messages exchanged by the agents and the environment are stored into the place *MsgBuf*, and the connected transitions to *MsgBuf* realize the service of exchanging messages. The transition *CreateAgent* allows the agents to create another agents.

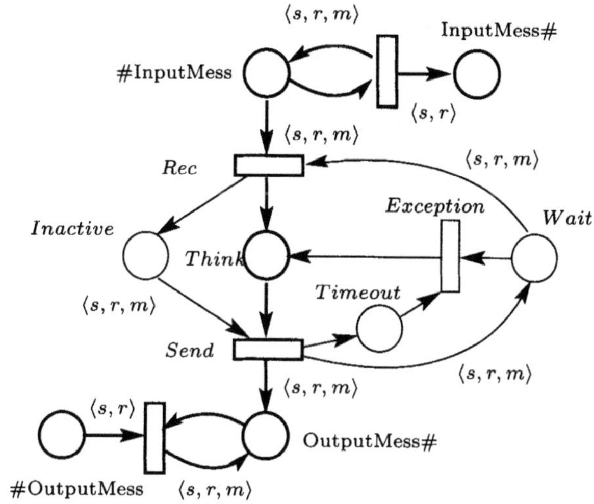

Fig. 4. The *Buyer* implementation module

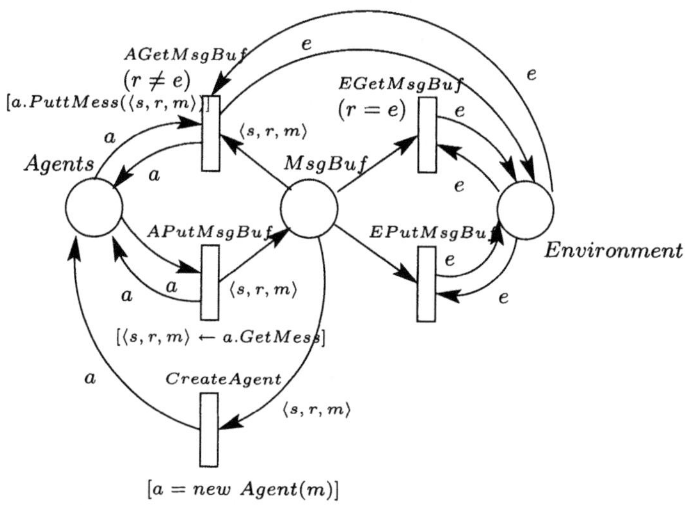

Fig. 5. The *AgentSociety* implementation module

4 The Packet-world

As an example of how our approach can be used, a packet-world [10] is modelled as a MAS. The packet-world consists of a number of different colored packets that are scattered over a rectangular grid. Agents that live in this virtual world have to collect those packets and bring them to their corresponding colored destination. In the packet-world agents can interact with the environment and in addition they have functionality to request information from each other. Instead of exploring the world to find a target an agent does not see, the agent can ask a visible colleague for the desired information.

The implementation modules corresponding to conversations of agents inherit an implementation module which is similar to the above presented module *Pair−wise Negotiation*:

- *EGetInfo* (ask the environment, in order obtain information concerning the actual position of the agent), which send a black token and receive the coordinates of the position;
- *EGetPacket* (ask the environment to update the packet-world after the agent picked a packet), *EPutPacket* (ask the environment to update the packet-world after the agent put down a packet), and *EDoStep* (ask the environment to update the packet-world after the agent performed a step), which send the direction and receive a black token;
- *AGetInfo* (ask another agent, in order to obtain information concerning a target), which send a color and receive a direction or unknown.

When two or more agents communicate each other, each agent contains its conversation objects as tokens. One of these tokens initiates the creation and the destruction of the all other cooperating objects. The initiator conversation object is created and destroyed by the theorem prover. The *Communication* implementation module is presented in Figure 6, and it realizes the management of conversation objects. The place *PObjBuff* is a container for all alive cooperating objects of an agent.

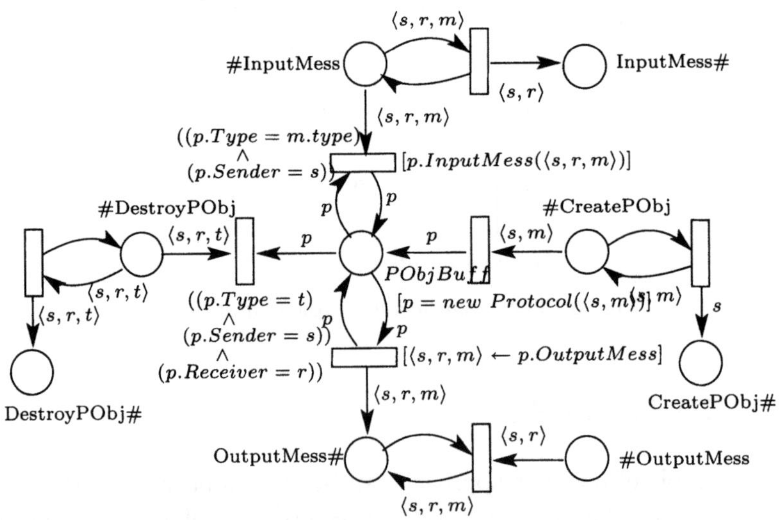

Fig. 6. The *Communication* implementation module

One implementation module not presented in this Section is the theorem prover. This is a problem connected to the rule-based systems and it is let to be solved in a future paper.

5 Conclusions

In this paper, we have proposed a new class of Petri nets, called Object Oriented High Level Petri Nets that can be used for the modelling and design of Multi-agent systems. The main features of the OOHLPN formalism are the following: (a) OOHLPNs preserve the property of Predicate/Transition nets to be viewed as the rule-based systems, which is inherited from the proposed standard of High-Level Petri nets, (b) unlike other approaches, OOHLPNs uses two distinct types of hierarchies: a hierarchy for object types and subtyping, and a different hierarchy for inheritance, and (c) the OOHLPN formalism uses the notions from theory of categories and institutions in order to achieve the concepts of inheritance and polymorphism.

We also proposed a method for modelling Multi-agent systems by using the OOHLPN formalism. Three fundamental perspectives of modelling MAS have been presented: the functional, the interactional, and the organizational perspective.

Future work will be carried out in two main directions: (a) to study thoroughly the connection between rule-based systems and OOHLPNs, and (b) to study thoroughly the connection between MAS and OOHLPNs and to develop a specific agent-oriented language based on OOHLPN formalism;

References

1. Bastide R, Sibertin-Blanc C, Palanque P (1993) Cooperative Objects: A Concurrent, Petri-Net Based, Object-Oriented Language. Proc. of the IEEE International Conference on Systems, Man and Cybernetics, 286-292
2. Biberstein O, Buchs D (1995) Structured Algebraic Nets with Object-Orientation. Workshop on Object-Oriented Programming and Models of Concurrency 95, 131-145
3. Biberstein O, Buchs D, Guelfi N (2001) Object-Oriented Nets with Algebraic Specifications: The CO-OPN/2 formalism. Lecture Notes in Computer Science 2001: 70-127
4. Bruce K, Cardelli L, Castagna G, The Hopkins Objects Group, Leavens G, Pierce B (1996) On Binary Methods. Theory and Practice of Object Systems, 1(3): 221-242
5. Buchs D, Guelfi N (1991) CO-OPN: a Concurrent Object Oriented Petri Net Approach. Proc. of the International Conference on Application and Theory of Petri Nets, 432-454
6. Goguen J, Burstall R (1999) Institutions: Abstract Model Theory for Specification and Programming. Journal of the Association for Computing Machinery, 39(1): 95-146

7. Goguen J, Meseguer J (1992) Order-Sorted Algebra I: Equational Deduction for Multiple Inheritance, Overloading, Exceptions and Partial Operations. Theoretical Computer Science, 105(2) : 217-273
8. Holvoet T, Verbaeten P (1996) Synchronization specications for agents with net-based behavior descriptions. Proc. of CESA '96 IMACS Conference, Symposium on Discrete Events and Manufacturing Systems, Lille, France, 613-618
9. Holvoet T, Keilmann T (1997) Behavior specication of active objects in open generative communication environments. In: Hesham El-Rewini and Yale N. Patt (eds) Proc. of the HICSS-30 Conference, Track on Coordination Models, Languages and Systems, 349-358. IEEE Computer Society Press, 7-10
10. Huhns S (1999) Multi-agent Systems and Societies of Agents. In: G. Weiss, Multiagent Systems, MIT Press
11. ISO/IEC 15909 Committee (2002) High-level Petri Nets - Concepts, Definition and Graphical Notation. Final Draft. International Standard ISO/IEC 15909, Version 4.7.3
12. Luck M, dIverno M (1995) A formal framework for agency and autonomy. Proc. of the First Intl. Conf. Multi-Agent Systems, 254-269
13. Moldt D, Wienberg F (1997) Multi-agent-systems based on coloured petri nets. Proc. of the 18th International Conference on Application and Theory of Petri Nets, Toulouse, France. Lecture Notes in Computer Science, 1248: 82-101
14. Mossakowski T (1996) Representations, hierarchies and graphs of institutions. Ph.D.thesis, Universitat Bremen
15. Sibertin-Blanc C (1994) Cooperative nets. Lecture Notes in Computer Science 815: 471-490
16. Sibertin-Blanc C (2001) CoOperative Objects: Principles, Use and Implementation. Lecture Notes in Computer Science 2001: 216-246
17. Xu H, Shatz S M (2003) A framework for model-based design of agent-oriented software. IEEE Transactions on Software Engineering, 29(1): 15-30
18. Valk R (1995) Petri Nets as dynamical objects. Proc. of the 16th International Conference on Application and Theory of Petri Nets, 26-30
19. Valk R (1998) Petri Nets as Token Objects - An Introduction to Elementary Object Nets. Lecture Notes in Computer Science. 1420: 1-25
20. Zambonelli F, Omicini A (2004) Challenges and research directions in agent-oriented software engineering. Autonomous Agents and Multi-Agent Sytems, 9(3): 253-283

Considerations on the Design and Implementation of an Agent-Based Auction Service

Adriana Dobriceanu, Laurenţiu Bîscu, Costin Bădică, and Elvira Popescu

Software Engineering Department, University of Craiova, Romania
adriana@yahoo.com, laurbiscu@yahoo.com,
{badica_costin,popescu_elvira}@software.ucv.ro

Summary. In this note we present some considerations on the architecture, design and implementation of an agent-based service for automating auctions. We discuss conceptual architecture, types of agents, agent communication and design and implementation issues. Some initial experimental results obtained using our proof-of-concept implementation are also included.

1 Introduction

Recently, interest for e-commerce automation has also been concentrated on the development of global agent-based e-commerce environments that enable dynamic trading between business partners ([7]). In particular, negotiation is a necessary and important activity to allow engagement of business parties (either individuals or institutions) in non-trivial business relationships. In this context, a considerable effort has been spent on the study, design and implementation of automated negotiations in an agent-based setting [6].

Probably auctions are the most popular example of price negotiations in the consumer-to-consumer sector, with `eBay.com` as market leader. Auctions spread also into the business-to-business sector (e.g. `Dovebid.com` or `Liquidation.com`). Auctions are also used by retailers for clearing shelves of excess and out-of-date inventory (see for example `CompUSA.com`, [12]).

As electronic auctions spread a lot with the advent of the Web, there are really many providers of auction services. Therefore, it may be quite difficult for a buyer or a seller to find the most appropriate auction to meet his or her needs. Usually this activity requires the use of a Web browser and of a general purpose search engine to look for a sought-after product or appropriate auction service according to a given set customer requirements including product and/or auction features.

An important advance over this manual approach is the provision of special Web applications known as *auction directories*.

An *auction directory* is a Web application that provides an up-to-date list of the most popular auction sites. Examples are `auction-lynx.com` or `reallyfirst.com`. Sites listed by an auction directory are usually organized either alphabetically, by the

Adriana Dobriceanu et al.: *Considerations on the Design and Implementation of an Agent-Based Auction Service*, Studies in Computational Intelligence (SCI) **78**, 75–83 (2008)
www.springerlink.com

category of the products sold or the rank of their popularity – facility known as an *auction top*. These facilities allow a human user to easily navigate the auction directory in order to find an appropriate auction that matches his or her needs. Additionally, more advanced auction directories (eg. `internetauctionlist.com`) provide a search facility for auctions taking into account various auction characteristics like: auction name (or keyword present in the auction name), category of auctioned product, location (country and/or state), auction starting date, a.o. Even more advanced auction directories provide facilities for real-time querying and monitoring of active auctions (eg. `auctionlotwatch.co.uk`).

In this paper we take this idea of an auction service from the human-driven Web world to the software agents world by proposing an agent-based auction service. This work builds on the existing experience with rule-based automated negotiation in agent systems [6]. However, the main difference is that now an auction is seen as a separate service, rather than being entirely incorporated into an e-shop infrastructure as in [7]. The service will have to be rented by a shop that would like to sell its products through an auction, for example for clearing its shelves on 'sales' time.

An important advantage of this approach is that it is applicable to other e-commerce scenarios, rather than only when a shop is selling its products to its customers. For example, let us consider Romanian regulations for awarding of procurement contracts for public institutions [4] – e-procurement scenario. This is a complex procedure that at a certain stage uses a reverse sealed-bid auction. The contractor may employ a sealed-bid auction service for achieving this task. Actually, this really happens when the contractor is using the Web service `e-licitatie.ro`, but the process is heavily manual. A significant improvement would be the automation of this process by means of software agents, and it is exactly here where our proposal is useful.

The paper is organized as follows. In section 2 we discuss the architecture of an agent-based auction service. We cover service infrastructure and communication protocol between the auction participants and the auction service. In section 3 we briefly introduce our proof-of-concept implementation and we discuss some initial experimental results. Last section concludes and points to future work.

2 Conceptual Architecture of an Auction Service

We understand automated negotiations as a process by which a group of software agents communicate with each other to reach a mutually acceptable agreement on some matter [13]. In this paper we focus our attention on *auctions* – a particular form of negotiation that spread during the last years with the advent of the Internet and the Web. Auctions are negotiations where resource allocations and prices are determined by bids exchanged between participants according to a given set of rules [14].

In automated negotiations (including auctions) it is important to distinguish between *negotiation protocols* (or *mechanisms*) and *negotiation strategies*. The protocol comprises public "rules of encounter" between negotiation participants by

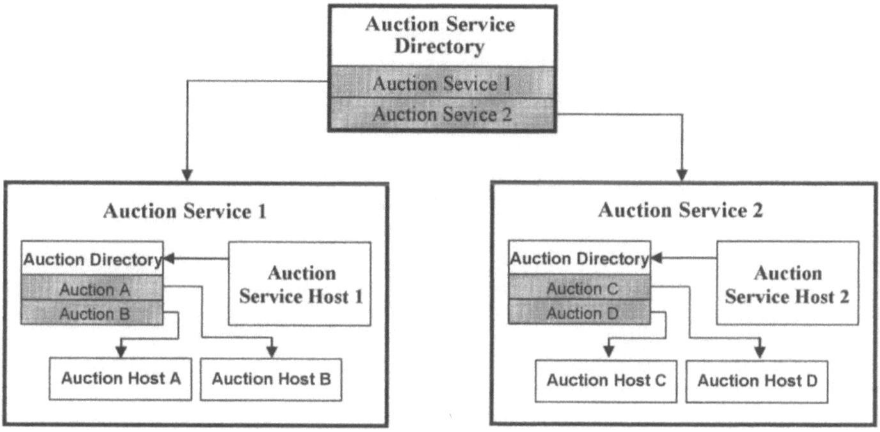

Fig. 1. Auction services

specifying the requirements that enable them to interact and negotiate. The strategy defines the private behavior of participants aiming at achieving their desired outcome. This behavior must be consistent with the protocol and is chosen to optimize participant welfare ([15]).

We consider an e-commerce environment that is populated with one or more auction services, with each service capable of providing one or more types of auctions. Trading partners (represented by buyer and seller agents that "live" in this environment) are capable of discovering appropriate auction services and use them in order to achieve their goals of buying and/or selling goods.

Auction services (*AS*) are registered with a global directory – *Auction Service Directory* or *ASD*. For each service both the name and the description of the service must be registered with the *ASD*. Then, a customer agent can look for a particular *AS* by querying the *ASD*. Also, when a new *AS* is created and added to this environment, it should be registered with the *ASD*, and when an *AS* goes off-line, it should be deregistered from the *ASD*.

An *AS* implements a specific type of auction and it has two important functions: i) management of active auctions including activities like auction creation and auction termination; ii) coordination of auction participants, including activities like bid submission and auction clearing, for each auction active within the service. Each *AS* includes a collection of agents that help the *AS* to achieve its functions.

An *AS* contains a local directory of active auctions known as the *Auction Directory* or *AD*. An active auction, known also as *auction instance* (*AI*) is a process that coordinates behaviors of participants that are registered with the auction.

An *AS* is managed by a specialized agent known as *Auction Service Host* or *ASH*. *ASH* serves as *AS* entry point and it is also in charge with creation, deletion and management of *AI*s.

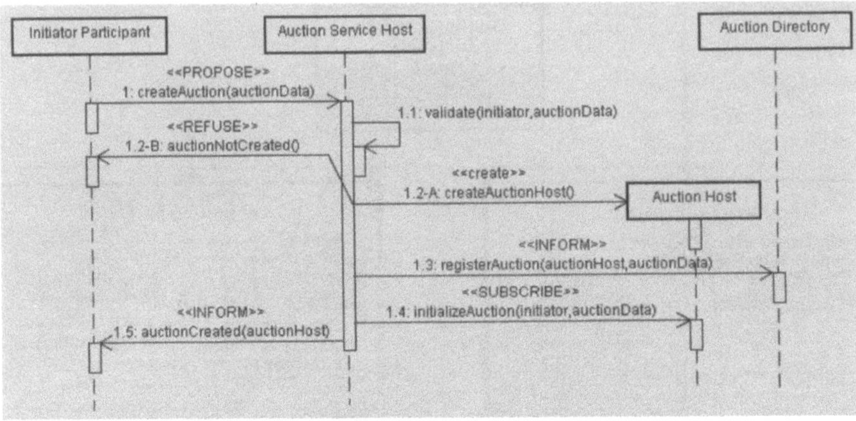

Fig. 2. Auction initiation

We have conceptualized an *AI* using the negotiation model inspired from [9] and [6] (an alternative model that uses statecharts to represent negotiation protocols is presented in [10]). An *AI* is managed by a specialized mediator agent known as the *Auction Host* (*AH*). For each *AI* active inside an *AS*, there is an associated entry in *AD* containing the identifier of the corresponding *AH* together with an appropriate description of the *AI*. Creation and deletion of *AH*s together with their registration and respectively deregistration with the *AD* are handled by the *ASH* agent.

Figure 1 shows a system with two auction services *Auction Service 1* and *Auction Service 2*. Each service currently runs two *AI*s, i.e. *Auction Service 1* runs *Auction A* and *Auction B*, while service *Auction Service 2* runs *Auction C* and *Auction D*.

An agent, usually a buyer or seller agent, registered to participate in an auction hosted by an *AS* is called *Auction Participant* or *AP*. Additionally, an auction is initiated by a special *AP* called *Auction Initiator Participant – AIP*.

Sequence diagram shown in figure 2 illustrates a scenario for initiating an auction. After deciding what *AS* to use (this part is not shown on the sequence diagram), an *AIP* issues a request for initiating an auction to the chosen *AS* – message *createAuction()*. The request is handled by the corresponding *ASH* agent. Assuming that the request was successful, a corresponding *AI* is created within the *AS*. This involves the creation of an *AH* agent – message *createAuctionHost()*, with an associated entry in the *AD* – message *registerAuction()*. The auction is then initialized using message *initializeAuction()* and finally the id of the *AH* is returned to the *AIP* – message *auctionCreated()*.

Sequence diagram shown in figure 3 illustrates a scenario with a participant registering to an auction. After deciding what *AS* to use, the *AP* queries the *AS* for all the active auctions – message *queryAS()* to the *ASH*. The *ASH* delegates the query to the *AD* – message *queryAD()*. The *AD* returns a list of active auctions to the *ASH*. Then *ASH()* queries separately each of the active auctions to get their current public status (i.e. the currently highest bid for English auctions, or number of unsold items in a Dutch auction) – message *requestAuctionInfo()* to *AH*. Each *AH* responds to the *ASH* with the auction current public status and the *ASH* returns this data to the participant. The *AP* analyzes all data collected about current public status of active auctions within a given time period in order to decide what auction to choose. For example, in an English auction, the participant (buyer in this case) will definitely not choose to register to auctions for which their current highest bid is higher than its own private reserve price (maximum price the buyer would agree to pay for the auctioned product). After the participant decides what auction to choose – message *chooseAuction()*, it will register to that chosen auction. Assuming the registration is allowed, *ASH* registers participant to the auction – message *addParticipant()* and then it requests the current public status of the auction from the *AH*. This value is returned to the *ASH* which in turn passes it to the *AP*, also confirming its successful registration with the auction. *AP* is now ready to start bidding using message *makeBid()*.

Note that as we are using the negotiation model from [9], an *AP* is allowed to dynamically join an auction. This basically means that as soon as an *AP* successfully registers to an auction, it gets the current public status of the auction and it may start bidding immediately. With respect to this, our model is different from [7] where buyers are collected and released in batches to a given auction. While [7] solution seems to match better with the overall goal of the e-commerce multi-agent system discussed there in order to optimize flow of commodities, we believe that the model of dynamically joining a given auction fits better with the auction service model that we are adhering to in this paper. Actually, this is the model employed by current Web auction houses (eg. `eBay.com`).

Note also that two consequences of borrowing the [9] negotiation model are: i) *AP*s communicate with the *AH* using a *generic negotiation protocol*. Negotiation is seen as the process of exchanging proposals (or bids) via a common space that is governed by an authoritative entity – *AH* (having the role of negotiation host from [9]). Status information describing negotiation state and intermediary information is automatically forwarded by the host to all entitled participants according to the information revealing policy of that particular auction (see [9, 5] for details; ii) *Negotiation rules* are used for enforcing the negotiation mechanism. Rules are organized into a taxonomy: rules for participants admission to negotiations, rules for checking the validity of proposals, rules for protocol enforcement, rules for updating the negotiation status and informing participants, rules for agreement formation and rules for controlling the negotiation termination (see [6, 8] for details).

3 Initial Implementation and Experiments

We have realized a proof-of-concept implementation of an English auction service following the design outlined in this paper using JADE 3.4 agent platform [3]. For this purpose we have reused parts of the implementation discussed in [6] including negotiation infrastructure and negotiation rules for describing an English auction, and we have extended it with functions for management of multiple active auctions

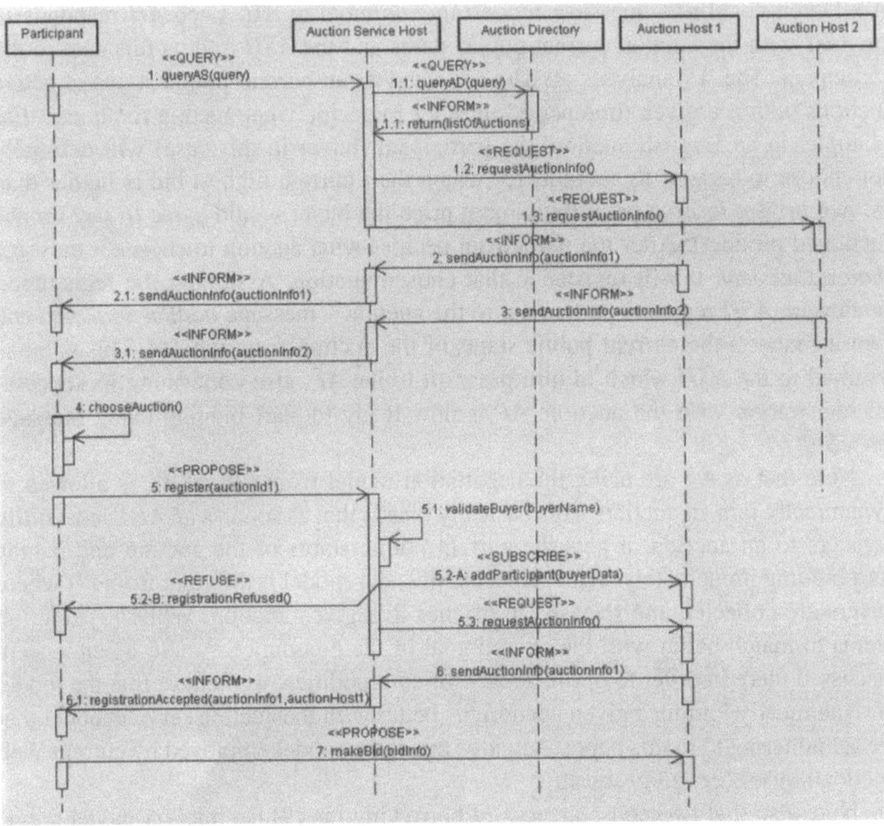

Fig. 3. Participant registration

within the auction service. We have implemented the following types of agents: *ASH*, *AH*, *AP* (in the role of buyer) and *AIP* (in the role of seller). Currently, *AD* is implemented as a list data structure internal to the *ASH*, rather than as a separate agent. Following [6], each *AH* incorporates a Jess rule ([2]) engine that is initialized with rules for describing English auctions [8].

We have performed some experiments using this implementation to asses the performance of the service as a function of the number of auction participants. For this purpose we have measured:

i) the *latency* of the service i.e. the time elapsed since a participant submitted a bid until it receives a response from the service;
ii) *number of bids* as a function of the number of participants and
iii) *throughput* as the ratio indicating the number of bids processed by the service per unit of time.

Experiments were carried out for 2, 3, . . . , 10, 15 and 25 identical participants – buyers in this case. We have chosen a very simple strategy that actually simulates a

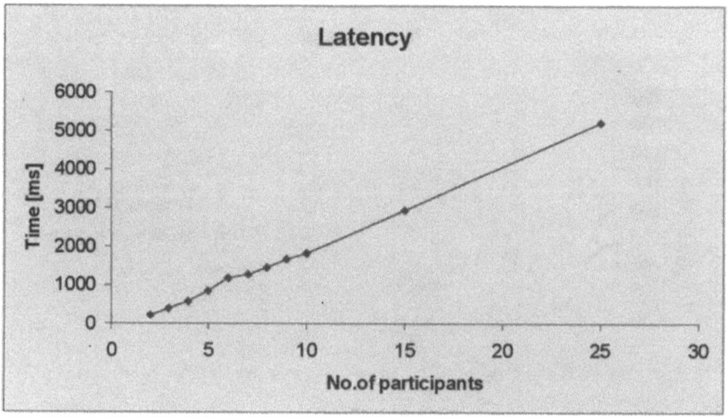

Fig. 4. Latency

sniping scenario [1]: each participant agent submits a first bid immediately after it is granted admission to the auction and subsequently, whenever it gets a notification that another participant issued a bid that was accepted by the host. Note that this sniping scenario is a situation when the service is heavily overloaded, as participants bid as fast as they can with the hope to overbid their opponents.

The value of the bid is determined by incrementing the currently highest bid with a fixed bid increment. Additionally, each participant has its own valuation of the negotiated product. If the current value that the buyer decided to bid exceeds her private valuation (known as *reserve price*) then the proposal submission is canceled (i.e. product became "too expensive"). In the experiments we have chosen a bid increment of 6 and a reserve price of 90. The auction starting price was 0. The experiments were carried out on a single machine.

Figure 4 illustrates the latency of the service as a function of the number of participants. The graphic shows a liner increase of the latency with the number of participants.

Figure 5 illustrates the total number of bids submitted by participants as a function of the number of participants. Note that because of the sniping scenario many bids were rejected.

Figure 6 illustrates the service throughput measured under the sniping conditions as the average number of bids submitted and processed by an *AH* per unit of time. Note that the throughput decreases as the number of participants is increasing.

4 Concluding Remarks

In this note we have presented our initial results concerning implementation and experiments of an agent-based auction service. The work discussed here can be extended in several directions that we would like to pursue in the near future. They

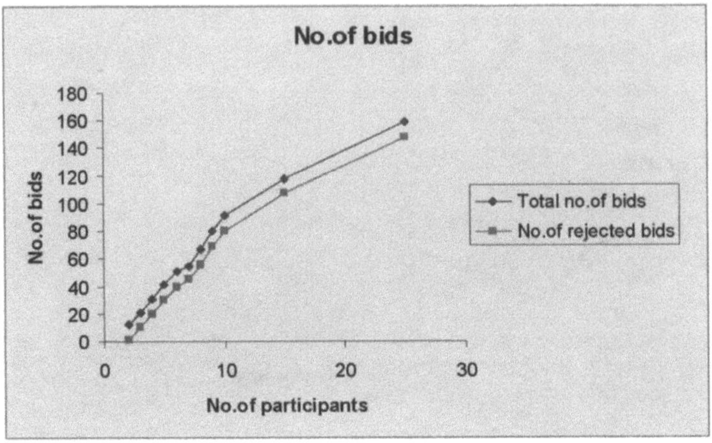

Fig. 5. Number of bids

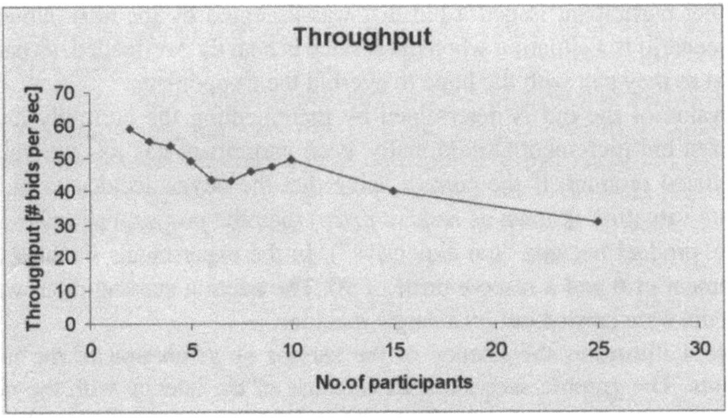

Fig. 6. Throughput

concern: i) enhancing the representation of auction services and auction instances; ii) conducting more experiments in order to asses the performance of our auction service, taking into account that several auctions can be performed in parallel; iii) improving the architecture of the auction service using for example ideas from [11].

References

1. *Auction sniping.* `http://en.wikipedia.org/wiki/Auction_sniping`
2. *Jess, Sandia Lab.* `http://herzberg.ca.sandia.gov/jess/`
3. *JADE: Java Agent Development Framework.* `http://jade.cselt.it`

4. Government Decision no. 925/2006 for approving the application norms of the provisions referring to the award of procurement contracts from the Government's Emergency Ordinance no. 34/2006. `http://www.anrmap.ro/`
5. Bădică, C., Bădică, A.: Formal modeling of agent-based English auctions using finite state process algebra. In N. Nguyen, A. Grzech, R. Howlett, and L. Jain (eds.), *Agent and Multi-Agent Systems: Technologies and Applications. Proc.KES-AMSTA-2007*. LNCS 4496, 248-257, Springer, 2007.
6. Bădică, C., Ganzha, M., Paprzycki, M.: Implementing Rule-Based Automated Price Negotiation in an Agent System. In: *Journal of Universal Computer Science*, vol. 13, no. 2, 244–266, 2007.
7. Bădică, C., Ganzha, M., Paprzycki, M.: Developing a Model Agent-based E-Commerce System. In: *E-Service Intelligence Methodologies, Technologies and Applications*, Studies in Computational Intelligence 37, 555–578, Springer, 2007.
8. Bădică, C., Giurca, A., Wagner, G.: Using Rules and R2ML for Modeling Negotiation Mechanisms in E-Commerce Agent Systems. In: *Preproc.* 2nd *Int. Conf. Trends in Enterprise Application Architecture, TEAA'2006*, 71–85, 2006.
9. Bartolini, C., Preist, C., Jennings, N.R.: A Software Framework for Automated Negotiation. In: *Proc. SELMAS'2004, LNCS 3390*, Springer Verlag, 213–235, 2005.
10. Benyoucef, M., Rinderle, S.: Modeling e-Negotiation Processes for a Service Oriented Architecture. In: *Group Decision and Negotiation*, vol.15, no.5, 449-467, Springer, 2006.
11. Hillston, J., Kloul, L.: Performance investigation of an on-line auction system. In: *Concurrency and Computation: Practice and Experience*, 13(1): 23- 41, 2001.
12. Laudon, K.C., Traver, C.G.: *E-commerce. business. technology. society* (2nd ed.). Pearson Addison-Wesley, 2004.
13. Lomuscio, A.R., Wooldridge, M., Jennings, N.R.: A classification scheme for negotiation in electronic commerce. In: F. Dignum, C. Sierra (Eds.): *Agent Mediated Electronic Commerce: The European AgentLink Perspective*. LNCS 1991, Springer Verlag, 19–33, 2002.
14. McAfee, R.P., McMillan, J.: Auctions and bidding. In: *Journal of Economic Literature*, 1987, 25(2):699–738.
15. Wooldridge, M.: *An Introduction to MultiAgent Systems*, John Wiley & Sons, 2002.

Automatic Streaming Processing of XSLT Transformations Based on Tree Transducers

Jana Dvořáková

Department of Computer Science,
Faculty of Mathematics, Physics and Informatics,
Comenius University, Bratislava, Slovakia
dvorakova@dcs.fmph.uniba.sk

Summary. Streaming processing of XML transformations is practically needed especially if we have large XML documents or XML data streams as the transformation input. In this paper, we present the design of an automatic streaming processor of transformations specified in XSLT language. Unlike other similar systems, our processor guarantees bounds on the resource usage for the processing of a particular type of transformation. This feature is achieved by employing tree transducers as the underlying formal base. The processor includes a set of streaming algorithms, each of them is associated with a tree transducer with specific resource usage (memory, number of passes), and thus captures different transformation subclass. The input XSLT stylesheet is analyzed in order to identify the transformation subclass to which it belongs. Then the lowest resource-consuming streaming algorithm capturing this subclass is applied.

1 Introduction

A typical XML transformation processor (e.g., processors for the popular transformation languages XSLT and XQuery) is tree-based, i.e., it reads the whole input document into memory and then performs particular transformation steps according to the specification. References to any part of the input document are processed in a straightforward way by traversing the in-memory representation and the extracted parts are combined to form a required output fragment. The output document may be constructed either in the memory or sequentially as a data stream.

This approach has been sufficient for most of XML documents. However, now there appear more and more XML documents that require specific handling, such as large XML documents and XML data streams. A natural solution is to employ streaming processing, when the input document is read as a stream, possibly in several passes; and the output document is generated as a stream in one pass. Only a part of the input document may be accessed at a time, and thus advanced techniques must be used to process references to

Jana Dvořáková: *Automatic Streaming Processing of XSLT Transformations Based on Tree Transducers*, Studies in Computational Intelligence (SCI) **78**, 85–94 (2008)
www.springerlink.com

the input document and connect the extracted parts to the proper position within the output document.

Currently, the most frequently used XML transformation languages are XSLT [10] and XQuery [9], both general (Turing-complete) languages intended for tree-based processing. There is great interest in the identification of XSLT and XQuery transformations that allow efficient streaming processing. The key issue in the design of a streaming processor is to find the way of handling the non-streaming constructs of the languages. The goal of this work is to propose a system that performs automatic streaming processing of XSLT transformations, so that the upper bounds on the resource usage for particular transformation subclasses are guaranteed. This feature is achieved by employing tree transducers as the base of the processor. More precisely, the processor represents a demonstration implementation of the formal framework for XML transformations introduced in [4]. In this paper, the framework is simplified and customized in order to facilitate the implementation. It contains a general model – an abstract model of general, tree-based transformation languages, and a set of streaming models that differ in the kind of memory used and number of passes over the input allowed. Each streaming model can simulate some restricted general model. The framework contains a simulation algorithm for each such pair *streaming model* → *restricted general model*. The framework is abstract, and thus can be used to develop automatic streaming processors for other general transformation languages as well.

The implementation of the framework for XSLT language includes the implementation of streaming models and two modules: (1) an analyzer that associates the input XSLT stylesheet with the lowest resource-consuming streaming model that is able to process it, and (2) the translator that automatically converts the XSLT stylesheet into the streaming model chosen according to the associated simulation algorithm. The processor based on the framework is easily extensible since new transducers and algorithms may be specified and implemented, as well as optimizable since the current algorithms may be replaced by the more efficient ones. Although there are some XML transformations such that their streaming processing is always high resource-consuming (e.g., complete reordering of element children), most of the practical transformations can be processed with reasonable bounds on the resource usage.

Related Work. Much of the previous work is devoted to streaming evaluation of XPath expressions, e.g. [1, 2]. Besides, several automatic streaming processors for XSLT and XQuery have been implemented. XSLT processor based on SPM (Streaming Processing Model) [3] is a simple one-head, one-pass transformer without an additional memory. The conversion of XSLT stylesheet to a streaming algorithm is well-described, however, only a small subset of XSLT is captured. Existing processors for XQuery language [5, 6] are equipped with memory buffers. Therefore they are able to handle large subsets of XQuery, but the algorithms employed are not provided with a complexity analysis, and therefore the resource requirements for processing a particular type of transformation are not known. Empirical studies presented in [3, 5, 6] show that the streaming processors introduced tend to be less resource-consuming than the tree-based processors, but the results hold only for ad-hoc transformations chosen for the experiments. To our best knowledge, our work is the first attempt to design an automatic streaming processor for a large subset of a general XML transformation language that guarantees specific resource usage for a given transformation class.

2 Complexity of Streaming Processing

In this section, we specify the relevant complexity measures for the streaming algorithms for XML transformations.

The basic constructs of the XML document are elements, element attributes, and text values. It may be represented as a tree that is obtained by a natural one-to-one mapping between elements and internal nodes of the tree. Text values appear in the leaves. Reading a document in document order then exactly corresponds to the preorder traversal of the constructed tree.

The tree-based processing of XML transformations is flexible in the sense that the input document is stored in the memory as a tree and can be traversed in any direction. On the contrary, during the streaming processing the elements of the input document become available stepwise in document order and similarly the output elements are generated in document order. The actual context is restricted to a single input node. Clearly, one-pass streaming processor without an additional memory is able to perform only simple transformations, such as renaming elements and attributes, changing attribute values, filtering. It must be extended to perform more complex restructuring. The common extensions are (1) allowing more passes over the input document, (2) adding an additional memory for storing temporary data. The extensions can be combined[1]. We obtain the corresponding complexity measures for streaming processing of XML transformations:

1. the number of the passes over the input tree, and
2. the memory size.

It is reasonable to consider the complexity of the streaming processing in relation to the tree-based processing. As mentioned in Sect. 1, all XML transformations can be expressed in both XSLT and XQuery, and processed by their tree-based processors. Various transformation subclasses can be then characterized by putting restrictions on these general transformation languages, typically by excluding certain constructs.

When designing streaming algorithms, we have a choice regarding three settings – the type of the memory used (none, stack, buffers for storing XML

[1] More passes over the input tree are not possible for XML data streams that must be processed "on the fly".

fragments), and the values of the two complexity measures mentioned. Streaming algorithms with different settings may capture different transformation subclasses. Since the transformation subclasses are characterized as some subsets of the general transformation language considered, the key issue in the algorithms is to realize a streaming simulation of the non-streaming constructs included in the restricted language.

We use tree transducers to design the streaming algorithms formally and to model transformation subclasses. They are included in the formal framework for streaming XML transformations that we describe in the next section.

3 Formal Framework

The framework is intended as a formal base for automatic streaming processors of the general transformation languages. It does not cover all XML transformations. In order to keep the models employed simple and comprehensible, we restrict it to model mainly the transformations that capture the relevant problems of streaming processing. In Sect. 4, we mention how some of the restrictions on the transformation set can be overcome in the implementation.

The framework consists of the following formal models:

1. a basic general model for tree-based processing of XML transformations and its restrictions,
2. a basic streaming model for streaming processing of XML transformations and its extensions.

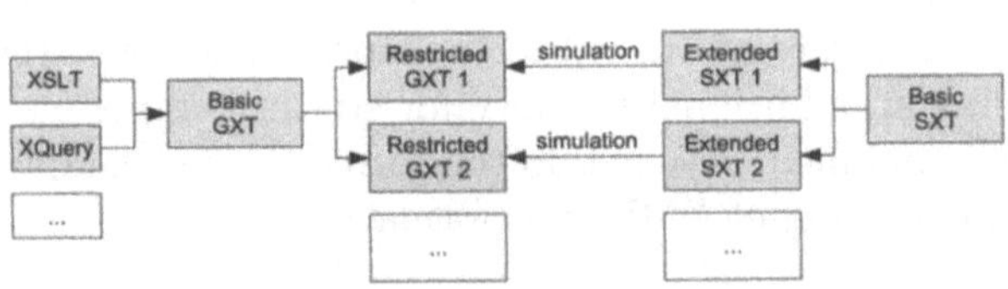

Fig. 1. A schema of the formal framework

Both models are based on tree transducers, models for tree transformations [7] originated in the formal language theory. We introduce two novel models – a *general XML transducer (GXT)* used as the general model, and a *streaming XML transducer (SXT)* used as the streaming model. They are defined in common terms in order to facilitate development of the simulation algorithms.

The overall schema of the framework is shown in Fig. 1. The basic SXT represents a simple one-pass streaming model without an additional memory. Following the ideas from Sect. 2, it can be extended by memory for storing temporary data and by allowing more passes over the input document. The basic GXT represents the most powerful general model. As already mentioned, it does not capture all XML transformations, but only a subset significant in the case of streaming processing.

For each extended SXT, the transformation subclass captured is identified by imposing various restrictions on the basic GXT. The inclusion is proved by providing an algorithm for simulating this restricted GXT by the given extended SXT.

XML Document Abstraction. In what follows, we do not consider element attributes and data values[2]. Let Σ be an alphabet of element names. The set of *XML trees* over Σ is denoted by $\mathcal{T}_\Sigma$, the empty XML tree is denoted by ε. An *indexed XML tree* may in addition have some leaves labeled by symbols from a given set X. A set of XML trees over Σ indexed by X is denoted by $\mathcal{T}_\Sigma(X)$. In the *rightmost indexed XML tree*, the element of the indexing set occurs only as the rightmost leaf. The set of rightmost indexed XML trees is denoted by $\mathcal{T}_\Sigma(X)_r$.

Selecting Expressions. We use simple *selecting expressions* derived from XPath expressions [8] to locate the nodes within the XML tree. The selecting expression is a *path* consisting of a sequence of *steps*. It can be either absolute (starting with /), or relative. The *step* consists of two components – *axis* and *predicate*. They are specified as outlined below. Comparing to the XPath language, the set of expressions is restricted and the syntax of some constructs is simplified – we explain the meaning in parentheses. The semantics of the selecting expressions follows the semantics of the equivalent XPath expressions.

step :	*axis* [*predicate*]	
axis :	$\times$ (self), $\downarrow$ (child), $\downarrow*$ (descendant), $\uparrow$ (parent), $\uparrow*$ (ancestor), $\leftarrow$ (left sibling), $\overset{*}{\leftarrow}$ (preceding), $\rightarrow$ (right sibling), $\overset{*}{\rightarrow}$ (following)	
predicate :	$*$	(select all elements)
	name	(select the elements named *name*)
	i	(select the element on i-th position within siblings)
	step	(select the elements having context specified by *step*)

The names of the elements are taken from an alphabet Σ. We denote the set of selecting expressions over Σ by $\mathcal{S}_\Sigma$.

General XML Transducer (Fig. 2a). The input heads of GXT traverse the input tree in any direction and the output is generated from the root to the leaves. At the beginning of a transformation, the transducer has only one input head, which aims at the root of the input tree, and one output head, which aims at the root position of the empty output tree. During a single transformation step, the whole input tree is available as a context. One or more new computation branches can be spawned and the corresponding input control is moved to the input nodes specified by selecting expressions. At the same time, the output heads may generate a new part of the output.

[2] We refer the reader to [4] for the definition of the extended framework including both element attributes and data values.

Formally, the GXT is a tuple $T = (Q, \Sigma, \Delta, q_0, R)$, where Q is a finite set of states, Σ is an input alphabet, Δ is an output alphabet, $q_0 \in Q$ is an initial state, and R is a set of rules of the form

$$Q \times \Sigma \rightarrow \mathcal{T}_\Delta(Q \times \mathcal{S}_\Sigma) .$$

For each $q \in Q$ and $\sigma \in \Sigma$, there is exactly one rhs such that $(q, \sigma) \rightarrow rhs \in Q$. The right-hand side of a rule contains an XML tree over the output alphabet indexed by *recursive calls* – pairs of the form (q, exp), where q is a state and exp is a selecting expression that returns a sequence of input nodes to be processed recursively. A simple example of a GXT transformation follows.

Example 1. Let $T = (Q, \Sigma, \Sigma, q_0, R)$ be a GXT where $Q = \{q_0\}$, $\Sigma = \{\alpha, \beta, \gamma\}$. and R consists of the rules

$$(q_0, \alpha) \rightarrow \varepsilon \, , \tag{1}$$
$$(q_0, \beta) \rightarrow \alpha((q_0, \downarrow[*])) \, , \tag{2}$$
$$(q_0, \gamma) \rightarrow \gamma((q_0, \downarrow[2]), (q_0, \downarrow[1])) \, . \tag{3}$$

The transducer processes the input trees over alphabet Σ. The subtrees at nodes named α are completely removed (rule 1), the nodes named β are renamed and get a new name α (rule 2), and at last, when encountering a node named γ, the first two children are processed in reversed order (rule 3).

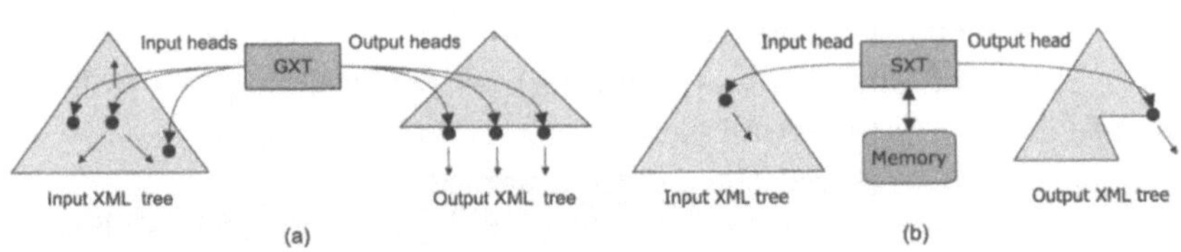

Fig. 2. The processing model of the transducers: (**a**) the GXT; (**b**) the SXT

Streaming XML Transducer (Fig. 2b). The SXT has a single input head that traverses the input tree in preorder, and a single output head that generates the output tree in preorder. Each node is visited twice during a single pass – once when moving top–down, and once when moving bottom–up. Thus, we recognize two types of SXT states (1) the states indicating the first visit of nodes and (2) the states indicating the second visit of nodes. During a single transformation step, the input head either moves one step in preorder or stays at the current position. At the same time, an output action is performed, depending on the type of rule applied. When applying a *generating rule*, a new part of the output is connected to the current position of the output head, and then the output head moves to the position under the rightmost leaf of the new part. When applying a *closing rule*, no output is generated, only the output head is moved one step upwards in preorder within the output tree.

Formally, the streaming XML transducer (SXT) is a tuple $T = (Q, \Sigma, \Delta, q_0, R)$, where $Q = Q_1 \cup Q_2$, $Q_1 \cap Q_2 = \emptyset$ is a finite set of states, Σ, Δ are as above, $q_0 \in Q_1$ is the initial state, and $R = R_g \cup R_c$, $R_g \cap R_c = \emptyset$ is a finite set of rules of the form:

$$R_g : Q \times \Sigma \times Pos \rightarrow \mathcal{T}_\Delta(Q \times \mathcal{S}_\Sigma)_r \ , \quad R_c : Q \times \Sigma \times Pos \rightarrow Q \times \mathcal{S}_\Sigma \ ,$$

where $Pos = \{leaf, no\text{-}leaf\} \times \{last, no\text{-}last\}$[3]. For each $q \in Q$ and $\sigma \in \Sigma$ there is at most one rhs such that for each $pos \in Pos$ there is a rule $(q, \sigma, pos) \rightarrow rhs \in R$. Furthermore, for each $(q, \sigma, pos) \rightarrow rhs \in R$, $rec(rhs) = (q', exp)$[4], one of the following preorder conditions holds:

1. *moving downwards*: $q \in Q_1$, and
   - $pos[1] = no\text{-}leaf$, $q' \in Q_1$, $exp = \downarrow[1]$, or
   - $pos[1] = leaf$, $q' \in Q_2$, $exp = \times[*]$,
2. *moving upwards*: $q \in Q_2$, and
   - $pos[2] = no\text{-}last$, $q' \in Q_1$, $exp = \rightarrow[1]$, or
   - $pos[2] = last$, $q' \in Q_2$, $exp = \uparrow[*]$,
3. *no input move*: q, q' are of the same kind, $exp = \times$.

The left-hand side of a rule consists of a state, an element name and a node position. The position is used to determine the preorder move within the input tree and it consists of two predicates – the first one indicating a leaf node, and the second one indicating a last node among the siblings. The right-hand side is an XML tree rightmost indexed by a recursive call.

In [4], it is demonstrated how to design streaming algorithms within the original framework. Namely, there are identified restrictions that must be put on the GXT in order to make possible the simulation by the one-pass SXT using stack size proportional to the height of the input tree. As the result, the simulation algorithm for the *local and order-preserving* transformations is presented. Since in this paper we focus on the implementation of the framework, we do not mention particular simulation algorithms.

4 Design of XSLT Streaming Processor

We describe an automatic streaming processor for XSLT transformations based on the framework introduced. The models within the framework are abstract, and thus the framework provides means to develop efficient streaming algorithms for XML transformation subclasses at abstract level, and to

[3] If $pos \in Pos$ is a node position, we refer to its first component by $pos[1]$ and to its second component by $pos[2]$.

[4] If rhs is a particular right-hand side, we refer to its recursive call by $rec(rhs)$.

adapt them to an arbitrary general transformation language. We first focus on the general issues regarding framework implementation, and then we deal with an adaptation for the XSLT transformation language in more detail.

As mentioned in the previous section, the formal framework is restricted in several ways. Some of the restrictions can be easily overcome in the implementation, while others require more complex handling.

- *Restrictions on the XML document.* Attributes and data values are associated with elements. They can be easily added to the implementation – if such construct needs to be processed, it is accessed using the same path like the parent element. On the other hand, if the construct needs to be generated in the output, its content can be retrieved using selecting expressions similar to those used for recursive processing of elements.
- *Restrictions on the selecting expressions.* The simple selecting expressions used capture the typical problems that arise during the streaming location of the nodes in XML document (context references in predicates, backward axis). Other constructs must be handled separately – however, the techniques used for constructs included in our restricted set may be often exploited. Moreover, there has been already carried on a research on the streaming processing of large subsets of XPath language [1, 2].
- *Restrictions on the general transformation language.* A part of the restrictions in GXT results from the restrictions on selecting expressions, and others are caused by excluding certain general transformation constructs, such as loops, variables, functions. However, the GXT models transformations that reorder the nodes within an XML tree with respect to the document order, which is the important issue in streaming processing of XML transformations.

Let us now describe the design of the prototype XSLT streaming processor. The GXT represents an abstract model for general transformation languages. Since our intention is to adapt the framework for the XSLT language, it does not need to be implemented directly. Instead, we are looking for a correspondence between restricted GXTs and XSLT subsets. The basic GXT can be easily converted into an XSLT stylesheet:

(1) The initial XSLT template is created. Its purpose is to set the initial mode that equals to the initial state q_0 of the GXT.

```
<xsl:template match="/">
   <xsl:apply-templates select="." mode="q0">
</xsl:template>
```

(2) Each rule $(q, name) \rightarrow rhs$ of GXT is translated into an XSLT template:

```
<xsl:template match="name" mode="q">
   ... template body ...
</xsl:template>
```

The template body depends on the rhs. If it is a single recursive call (q', exp), it is mapped to the `xsl:apply-templates` instruction, where the `select` expression contains the translation of the expression *exp* into a corresponding XPath expression *exp'*:

```
<xsl:apply-templates select="exp'" mode="q'">
```

If *rhs* contains an indexed XML tree with some element nodes, all of them must be generated as new elements within the template body. A single element named *name* is generated using the `xsl:element` instruction:

```
<xsl:element name="name">
   ... element content ...
</xsl:element>
```

The element content contains instructions for generating child elements and child recursive calls, if exist. Recursive calls are translated using the `xsl:apply-templates` instruction as mentioned above. In a similar way, each restricted

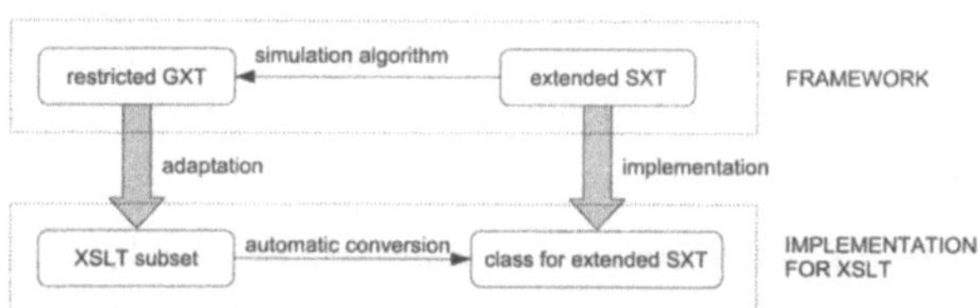

Fig. 3. An implementation of the framework for XSLT language

GXT can be translated to the corresponding XSLT subset. According to the principle of the formal framework, a restricted GXT (GXT_r) can be simulated by some extended SXT (SXT_e) such that the simulation algorithm is known. Then XSLT stylesheets from the XSLT subset associated with GXT_r can be converted to SXT_e using the simulation algorithm. The conversion can be performed automatically since the simulation algorithm exactly determines how to convert constructs of the given XSLT subset into the rules of SXT_e. The resulting SXT_e is constructed explicitly as an object and its method $transform()$ performs streaming processing of the transformation specified by the stylesheet. The relation between the framework and the implementation for XSLT is shown in Fig. 3.

To sum up, the streaming processor works in three steps:

1. *Analysis.* The analyzer examines the constructs in the input XSLT stylesheet (both XPath constructs and XSLT constructs themselves).It checks whether there is specified an XSLT subset that allows all the constructs encountered. If there are more such subsets, the smallest one is chosen.
2. *Translation.* The translator creates an object for the extended SXT associated with the XSLT subset chosen. The creation is automatic, following the simulation algorithm provided for the XSLT subset.
3. *Processing.* The method $transform()$ of the new SXT object is run on the input XML document. The streaming transformation performed is equivalent to the one specified by the input XSLT stylesheet.

5 Conclusion

We have presented a design of an automatic streaming processor for XSLT transformations. Comparing to other similar processors, the contribution of our approach is that the resource usage for streaming processing of particular types of XSLT transformations is known. Our processor includes several streaming algorithms, and it automatically chooses the most efficient one for a given XSLT stylesheet. The process of choice has a solid formal base – a framework consisting of tree transducers that serve as models both for the streaming algorithms and for the transformation types.

We have already implemented tree transducers included within the framework, a major part of the analyzer, and the translator for processing the local and order-preserving XSLT transformations. In the future work, we plan to include algorithms for the local and non-order-preserving transformations to obtain a processor for a large subset of practically needed XML transformations. We intend to carry out performance tests and comparison to other implementations subsequently.

Acknowledgments. This work was supported in part by the grants VEGA 1/3106/06 and GUK 358/2007.

References

1. Bry F, Coskun F, Durmaz S, et al. (2005) The XML Stream Query Processor SPEX. In: ICDE 2005 1120–1121. IEEE Computer Society, Washington
2. Chen Y, Davidson S B, Zheng Y (2006) An Efficient XPath Query Processor for XML Streams. In: ICDE 2006 79. IEEE Computer Society, Washington
3. Guo Z, Li M, Wang X, Zhou A (2004) Scalable XSLT Evaluation. In: APWEB 2004, LNCS 3007/2004:190–200. Springer Berlin / Heidelberg
4. Dvořáková J, Rovan B (2007) A Transducer-Based Framework for Streaming XML Transformations. In SOFSEM (2) 50–60. Institute of Computer Science AS CR, Prague
5. Florescu D, Hillery C, Kossmann D, et al. (2003) The BEA/XQRL Streaming XQuery Processor. In: VLDB Journal 13/3 294–315. Springer-Verlag New York
6. Ludscher B, Mukhopadhyay P, Papakonstantinou Y. (2002) A Transducer-Based XML Query Processor. In: VLDB 2002 227–238. Morgan Kaufmann
7. Thatcher J W (1973) Tree Automata: An Informal Survey. Currents in the Theory of Computing 4:143–172. Prentice-Hall, Englewood Cliffs, NJ
8. W3C (1999) XML Path Language (XPath), version 1.0, W3C Recommendation. `http://www.w3.org/TR/xpath`
9. W3C (2007) XQuery 1.0: An XML Query Language, W3C Recommendation. `http://www.w3.org/TR/xquery`
10. W3C (1999) XSL Transformations (XSLT) Version 1.0, W3C Recommendation. `http://www.w3.org/TR/xslt`

Knowledge Processing for Web Search – An Integrated Model

Peter Gurský[1], Tomáš Horváth[1], Jozef Jirásek[1], Stanislav Krajči[1], Róbert Novotný[1], Veronika Vaneková[1], and Peter Vojtáš[2]

[1] Institute of Computer Science, P. J. Šafárik University, Košice, Slovakia
[2] Department of Software Engineering, Charles University, Prague, Czech Republic

Summary. We propose a model of a middleware system enabling personalized web search for users with different preferences. We integrate both inductive and deductive tasks to find user preferences and consequently best objects. The model is based on modeling preferences by fuzzy sets and fuzzy logic. We present the model- theoretic semantic for fuzzy description logic f-$\mathcal{EL}$ which is the motivation of creating a model for fuzzy RDF. Our model was experimentally implemented and integration was tested.

Key words: middleware, fuzzy DL, fuzzy RDF, relevant objects, user preferences

1 Introduction and Motivation

One of the main roles of semantic web is to enable automatic access to web resources and services to be easily used by middleware engines or agents. Our research leads to the model of a middleware system permitting users to search objects of one domain but from heterogeneous sources. In this paper we present different approaches to solve the main task of such a system – return to users the best objects relative to their preferences.

Let us consider the following example: Imagine a user looking for a hotel which has *good price*, *good distance from an airport* and has *good equipment in rooms*.

The meaning of goodness (preference) can be different for each user (or group of users). For the price of hotels, one user could prefer cheap hotels (student), second prefers expensive hotels (manager) and the other one prefers middle price (professor). The natural meaning of this ordering is that the user determines the relations "better" or "worst" between two values of a given property. For each such property, user has a notion about the ordering of objects based on real value of property from an attribute domain. We call these orderings of particular attribute domains the *user local preferences*. The *user global preference* will be modeled by fuzzy aggregation operators.

Peter Gurský et al.: *Knowledge Processing for Web Search – An Integrated Model*, Studies in Computational Intelligence (SCI) **78**, 95–104 (2008)
www.springerlink.com

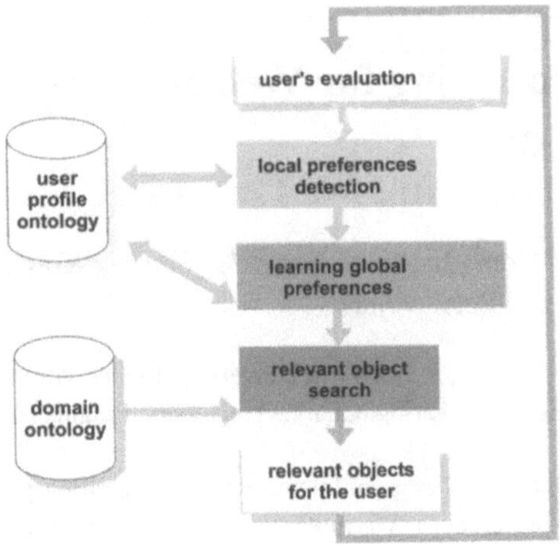

Fig. 1. Flow diagram of user profile dependent part

2 Web Search

In our system different users can retrieve different answers for the same question. Searching is based on user preferences that can be created during current and previous sessions as well.

For the first contact with a new user we can assign him to a user type. We can consider several user types for a particular domain. The suitable user type identification at the time of first contact requires huge testing with real users and is the aim of our future research. We assume that it can be done by a form with easy questions like gender, job position, age of children etc.

To identify suitable objects for user we need to have his local and global preferences. We can guess them by user type profile but we need to allow user to refine the profile to correspond his real preferences. For many users, it is hard to specify their local or global preferences exactly (by functions, equations, etc.). Instead, they rather express their local preferences in natural language (high salary, small distance, etc.), similarly global preferences. If we give to users a form to explain local preferences for all of our properties as well as for the global preference, it could be easily boring for them. Thus we have to use different way to obtain local and global preferences. We follow the idea that everybody can easily say, for a particular object, how suitable it is. Thus we will require from user to evaluate the objects in scale from the worst to the best.

Whole user dependent searching process is illustrated on Figure 1. The objects to evaluation can be a representative set of samples, suitable objects for some user type profile or any objects retrieved during a session. After evaluation of some objects, our system can learn user preferences and give to user suitable objects from the whole set of objects. After new objects are given to user, he can evaluate them and start the whole process again.

Learning process has two main phases – learning of local preferences and learning of global preferences.

2.1 Detecting Local Preferences

To learn local preferences, we give to user a representative sample of hotels (see Table 1) with several attributes, e.g. distance from the aiport, price of the accommodation and equipment of rooms. The user classifies hotels into categories *poor*, *good* and *excellent*.

Table 1. Representative sample of hotels evaluated by the user

Hotel	Distance	Price	Equipment	Evaluation
Apple	100 m	$ 99	nothing	poor
Danube	1300 m	$ 120	tv	good
Cherry	500 m	$ 99	internet	good
Iris	1100 m	$ 35	internet, TV	excellent
Lemon	500 m	$ 149	nothing	poor
Linden	1200 m	$ 60	internet, TV	excellent
Oak	500 m	$149	internet, TV	good
Pear	500 m	$ 99	tv	good
Poplar	100 m	$ 99	internet, TV	good
Rhine	500 m	$ 99	nothing	poor
Rose	500 m	$ 99	internet, TV	excellent
Spruce	300 m	$ 40	Internet	good
Themse	100 m	$ 149	internet, TV	poor
Tulip	800 m	$ 45	internet, TV	excellent

We distinguish four basic types of local preferences, according to which attribute values are preferable for a given user. We call these basic types *higher-best*, *lower-best*, *middle-best* and *marginal-best*.

The local preferences can be detected by statistical methods, e.g. regression or QUIN [3], etc. Using the linear regression we can detect just two basic types (*higher-best* and *lower-best*). In contrast to QUIN, the regression is resistant against statistically irrelevant values. Thus for the purposes of detecting the mentioned four basic types, the polynomial regression is the most appropriate approach. In case of the Table 1, we checked that the higher distance (*higher-best* type of ordering) and the lower price (*lower-best* type of ordering) are appropriate for the user.

2.2 Learning Global Preferences

We learn user's global preferences by the method of ordinal classification with monotonicity constraints [9] based on Inductive Logic Programming (ILP) system ALEPH [13].

The user's global preferences are computed by using his local preferences which can be well represented in ILP. For illustration, consider that we have discretized

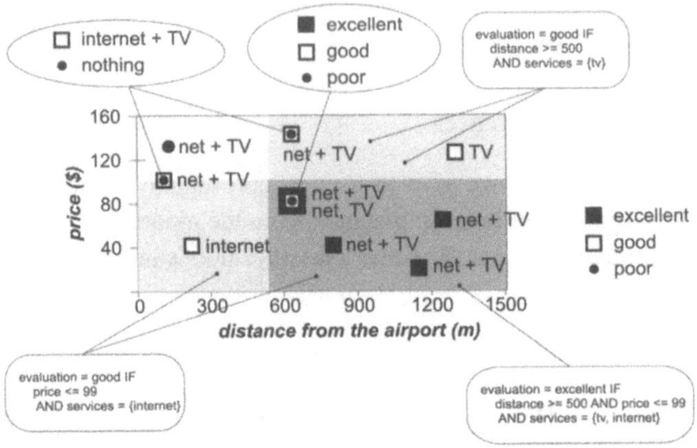

Fig. 2. The results of our approach in the case of our illustrative example

values of distance to three classes: *near*, *middle* and *far*. The different orderings of these classes for different users can be:

near ≤ middle ≤ far	(non-decreasing)
near ≤ far ≤ middle	(user prefers *near* values to *far* values, *middle* is least preferred)
far ≤ middle ≤ near	(non-increasing)
far ≤ near ≤ middle	(*far* values are preferred to *near* values, *middle* is least preferred)
middle ≤ near ≤ far	(the middle values are "better" than *near* or *far*)
middle ≤ far ≤ near	(the middle values are "better" than *far* or *near*)

An usual aggregation function can be easily simulated by monotone classification rules in the sense of many valued logic (on Figure 2 the computed user's global preferences from our illustrative data are presented):

evaluation = excellent IF distance ≥ 500 AND price ≤ 99
AND services = {TV, internet}
evaluation = good IF (distance ≥ 500 AND services = {TV})
OR (price ≤ 99 AND services = {internet})

We can see that the ordered meaning of the classification is preserved in our results (this is proved in [8] as the *igap-consistency* of our approach): hotels classified in the grade "excellent" by the user also fulfills requirements for "good" and "poor" hotels, and "good" hotels fulfills requirements for "poor" ones (e. g. the hotel with 800 m distance from airport and $45 price equipped with internet and TV is "at least as appropriate as" the hotel 300 m far from the airport and $40 price equipped just with internet) according to the local preferences (more far and more cheap is the better).

From our results we can obtain also additional information about crisp attributes (equipment). The meaning of any classification rule is as follows: If the attributes of object x fulfill expressions on the right side of the rule (body) then the overall value of x is at least the same as on the left side of the rule (head). We can, of course, assign the explicit values to vague concepts like excellent or good. During the simulation of computation of aggregation function we can simply test the validity of requirements of the rules from the strongest rule to weaker ones. When we find the rule that holds, we can say that the overall value of the object is the value on the left side of the rule. Since we test the sorted rules, we always rank the object with the highest possible value.

2.3 Fuzzy RDF Based on Fuzzy Description Logic

In this section we analyze the model of fuzzy RDF/OWL based on a model of fuzzy description logic. The terms like cheap, expensive or near represent fuzzy sets as in [16]. Our model of fuzzy RDF includes such fuzzy sets stored as RDF triples.

One important feature of our model is that we can prepare fuzzy RDF independently from user global preference. This is because we can adapt to user by adjusting his aggregation function @. This makes the processing of data more effective, because we do not need to order data for every user query. We already have the data ordered and we just combine the relevancies from fuzzy RDF into one result.

We introduce the model of building fuzzy RDF based on fuzzy description logic f-$\mathcal{EL}$ proposed in [15]. This logic removes some features of both classical and fuzzy description logic (like negation, universal restriction and fuzzy roles), but it adds aggregation operator @. We lose the ability to describe fuzziness in roles (i. e. our data from domain ontology are crisp; we do not consider uncertainty in values. User preferences are represented as fuzzy concepts and they are the source of fuzziness in results.), but we gain combination of particular user preferences to a global score by his @. Advantage of this description logic is lower complexity of querying. Expressivity is lower than that of full fuzzy description logics DL (DL) but still sufficient for our task and embedability into web languages and tools (see [14]).

Basic building boxes of description logic are concepts and roles. Here, roles express properties of resources (hotels in our case). Although the basic model of expressing RDF triples ⟨*subject*, *predicate*, object⟩ corresponds to oriented graphs, we use here the language of logic: *predicate*(*subject*, *object*).

The alphabet consists of sets $\mathcal{N}_C$ of concepts names, $\mathcal{N}_R$ role names and $\mathcal{N}_I$ instance names. The roles in f-$\mathcal{EL}$ are crisp and concepts are fuzzy. Our language of description logic further contains constructor $\exists$ and a finite set of aggregation functions symbols $@_U$ for each user and/or for each group of users. Concept descriptions in f-$\mathcal{EL}$ are formed according to the following syntax rules

$$C \to \top \mid A \mid @(C_1, \ldots C_n) \mid \exists r.C$$

In order to give this syntax a meaning, we have to define interpretations of our language. In f-$\mathcal{EL}$ we have interpretations parameterized by a (possibly partially, usually linearly) ordered set of truth values with aggregations. For a preference structure – a set of truth values $P = [0, 1]$, a P-interpretation is a pair $\mathcal{I} = \langle \Delta^{\mathcal{I}}, \bullet^{\mathcal{I}} \rangle$ with nonempty domain $\Delta^{\mathcal{I}}$ and fuzzy interpretation of language elements:

- $A^{\mathcal{I}} : \Delta^{\mathcal{I}} \to P$, for $A \in \mathcal{N}_C$

- $r^{\mathcal{I}} \subseteq \Delta^{\mathcal{I}} \times \Delta^{\mathcal{I}}$, for $r \in N_R$
- $(\exists r.C)^{\mathcal{I}} = \sup\{C^{\mathcal{I}}(y) : (x,y) \in r^{\mathcal{I}}\}$
- $(@(C_1,\ldots,C_n))^{\mathcal{I}}(x) = @^{\bullet}(C_1^{\mathcal{I}}(x),\ldots,C_n^{\mathcal{I}}(x))$

Example 1. Suppose that we have $N_R = \{price,\ distanceFromAirport\}$ where *price* and *distanceFromAirport* are RDF predicates from domain ontology.

$N_C = \{cheap_U, close_U\}$ where $cheap_U$ and $close_U$ are fuzzy functions explicitly figured by user preference ontology.

$N_I = \{Apple, Danube, 99, 120, \ldots\}$.

In Herbrand-like interpretation $\mathcal{H}$ we have:

$$Apple^{\mathcal{H}} = \text{Apple}, Danube^{\mathcal{H}} = \text{Danube}, 99^{\mathcal{H}} = 99, 120^{\mathcal{H}} = 120$$
$$cheap_U^{\mathcal{H}}(99) = 0.53, cheap_U^{\mathcal{H}}(120) = 0.42$$
$$price^{\mathcal{H}} = \{(\text{Apple}, 99), (\text{Danube}, 120)\}$$

For simplification we overload our concept $cheap_U$ also for hotels (usually we must create new concepts in this case e. g. $cheapHotel_U$):

$$cheap_U^{\mathcal{H}}(x) = (\exists price.cheap)^{\mathcal{H}}(x)$$

then

$$cheap_U^{\mathcal{H}}(\text{Apple}) = \sup\{cheap_U^{\mathcal{H}}(y) : (\text{Apple}, y) \in price^{\mathcal{H}}\} = 0.53$$
$$cheap_U^{\mathcal{H}}(\text{Danube}) = \sup\{cheap_U^{\mathcal{H}}(y) : (\text{Danube}, y) \in price^{\mathcal{H}}\} = 0.42$$

The supremum affects the case when we have more different prices for one hotel. Then we take the biggest result.

$$(@(cheap_U, close_U))^{\mathcal{H}}(\text{Apple}) = @^{\bullet}(cheap_U^{\mathcal{H}}(\text{Apple}), close_U^{\mathcal{H}}(\text{Apple})) = \frac{(3 \times close_U^{\mathcal{H}}(\text{Apple}) + 2 \times cheap_U^{\mathcal{H}}(\text{Apple}))}{5}$$

Fuzzy description logic f-$\mathcal{EL}$ is the motivation of creating a model for fuzzy RDF. Modelling fuzzy classes by classical RDF we can model a fuzzy instance $cheap_U^{\mathcal{H}}(\text{Apple}) = 0.53$ by RDF triple: $\langle \text{Apple}, cheap_U^{\mathcal{H}}, 0.53\rangle$.

This is an embedding of a fuzzy logic construct into classical RDF, which needs to translate also constructions of DL, namely $\exists price.cheap$ in fuzzy DL turns to composition of roles, where $\exists price.(\exists cheap.\top)$ needs an aggregate max (or top-k) extending DL with a concrete domain (see [1]). It is out of the scope of this paper to describe such DL with a concrete domain and embedding of our fuzzy DL in more detail. What we claim here is an experimental implementation of both our fuzzy DL, special crisp DL with a concrete domain and a related model of RDF with extended syntax and semantics of `owl:someValuesFrom`).

Now, in our model, we can specify user profiles and represent these profiles in Fuzzy RDF triples based on data from domain ontology. Our next task is to find best objects for individual users.

Consider the type of user preferring cheap hotels. There can be many users that prefer cheap hotels. We want to share the same instance of the class *cheap* for them.

However, their notion of "cheapness" can be different. For example one user can say that cheap hotels have price lower than \$30, while for some other user cheap hotel ends at \$50. Fortunately we can easily change the overall evaluation of objects to reflect these individual requirements. In the following theorem $f_1, \ldots, f_n$ are functions that represent local preferences expressed in user preference ontology. Function values $f_1(x_1), \ldots, f_n(x_n)$ are literals from fuzzy RDF expressing the relevance of respective values of an object. We show that for different users with same direction of preference we can use the same functions and adjust the aggregation function only.

Definition 1. *Let D be a subset of real line and f be a bijective (either strictly increasing or strictly decreasing) fuzzy function $f = ax + b$, $a \neq 0$, $f : D \to [0, 1]$ and $g : D \to [0, 1]$. We say, that g preserves the ordering of f over D, if for all $x, y \in D$ we have $f(x) < f(y)$ implies $g(x) \leq g(y)$.*

Theorem 1. *Let $f_1, \ldots, f_n$ be bijective fuzzy functions such that for each $i : f_i = a_i x + b_i$, $a_i \neq 0$, $f_i : D_i \to [0, 1]$. Let $g_1, \ldots, g_n$ be partially linear functions such that they preserve the ordering of $f_1, \ldots, f_n$ over D. Let @ be an n-ary aggregation function. Then there exists n-ary aggregation function @′ such that*

$$(\forall x_1, \ldots, x_n) @(g_1(x_1), \ldots, g_n(x_n)) = @'(f_1(x_1), \ldots f_n(x_n)).$$

Proof. The theorem above says about existence of an aggregation function @′. We will show how to find this function.

We can assume that f_i and g_i are defined on the same unit interval $D = [0, 1]$. Function g_i is linear over certain subintervals of $[0, 1]$. We take one such subinterval and name it K_j. The following holds for f_i and g_i over K_j:

$$(\forall x_i \in K_j) f_i(x_i) = a_i(x_i) + b_i \text{ and } g_i(x_i) = c_i^j(x_i) + d_i^j$$

Now we define new function h_i for each $i \in [1, n]$ as:

$$(\forall i, j) h_i(y) = c_i^j \frac{(y - b_i)}{a_i} + d_i^j.$$

The aggregation function @′ is $@'(y_1, \ldots, y_n) = @(h_1(y_1), \ldots, h_n(y_n))$. If we substitute $f_i(x)$ for every y_i:

$$@'(f_1(x), \ldots, f_n(x)) = @(h_1(f_1(x)), \ldots, h_n(f_n(x_n)))$$

For arbitrary subinterval $K_j \subseteq [0, 1]$ where every g_i is linear we get

$$h_i(f_i(x_i)) = c_i^j \frac{f_i(x_i) - b_i}{a_i} + d_i^j = c_i^j \frac{a_i x_i + b_i - b_i}{a_i} + d_i^j = c_i^j x_i + d_i^j = g_i(x_i)$$

Therefore it holds

$$\begin{aligned} @'(f_1(x_1), \ldots, f_n(x_n)) &= @(h_1(f_1(x_1)), \ldots, h_n(f_n(x_n))) \\ &= @(g_1(x_1), ..., g_n(x_n)) \end{aligned}$$

This theorem allows users to specify their own meaning of "cheapness" exactly by fuzzy function and similarly for other attributes.

2.4 Relevant object search

Now we have orderings of properties from learning of local preferences, rules from learning of global preferences and prepared ordered data stored in fuzzy RDF. Our last task is to find top *k* objects to user. We use the extension of middleware search of Ronald Fagin [5]. The main idea is to browse only necessary data until the system is sure that it has top-*k* objects already. Thus we do not need to calculate with whole data from domain ontology. Retrieving only *k* best objects can save time and is usually sufficient for user. Saving of time grows with the number of objects stored in domain ontology and also with the number of properties we consider.

The model in [5] works with data stored in possibly distributed lists that are ordered from the best to the worst in particular property. Fagin considered two kinds of accesses to lists: sorted and random access. The sorted access gets the next best object from the list after each access, so we can retrieve data ordered from the best to the worst. The random access asks for the value of a particular property it includes for a particular object. We want to minimize the number of accesses to lists and of course the time of searching. The random access has one big disadvantage. Each random access requires searching of the value of specific object. In the case of sorted access the results are prepared immediately (values can be preordered, in fuzzy RDF in our case, to the several orderings before user comes) and, moreover, data can be sent in blocks. Because of this feature in our system we prefer mainly the algorithms, which use only sorted access.

In our application we use 3P-NRA (3 Phased No Random Access) algorithm [7], which is an improvement of NRA [6] algorithm.

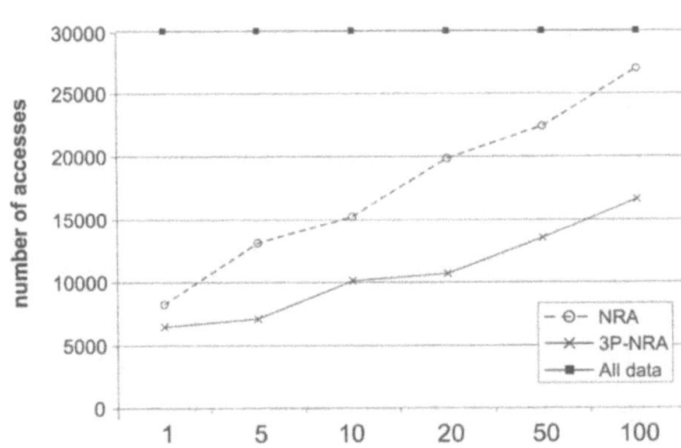

Fig. 3. Number of accesses needed to retrieval of top-*k* objects

As we found in our experiments (see Figure 3), using the techniques of top-*k* search can significantly save number of accesses and accordingly the search time especially in huge sets of objects.

Data for experiments were generated with various distributions of values. We used 2 exponential and 2 logarithmic distributions of properties with 10000 objects and 6 types of aggregation functions. For all experiments we used 3 properties. Using different combination of properties and aggregation functions we compose 25 different inputs for algorithms. In the final results we use the averages of the particular results.

3 Conclusions

In this paper we have described a model of system enabling users to search objects from the same domain and heterogeneous sources. Data are collected from various sources are processed to vector index and to a domain ontology. The system implements both user independent search and personalized search for users with different preferences. Our theoretical model integrates all parts of the system from collected data in classical RDF form to user query answering. We do not have a model for web resource downloading and ontology annotation part.

The model of user dependent search is based on modeling preferences by fuzzy sets and fuzzy logic. We present the semantics of fuzzy description logic f-$\mathcal{EL}$. User dependent search integrates both inductive and deductive approach. By induction we can learn local and global preferences. Results of induction as well as hand-filled orderings can be easily modeled in user ontology and fuzzy RDF. Deductive part of the system uses the preferences to find suitable objects for a user. User can express his preferences precisely by fuzzy functions and aggregation function. Easiest way is to evaluate several objects in a scale and let the system to learn preferences. Repeating this process (evaluating a set of objects and finding new objects) leads to refining the user profile. User, who is content with his profile and with the presented search results, does not have to evaluate objects again. When the domain ontology is actualized, user can just get the best objects according to his profile. Effective identification of suitable user type for a new user is the aim of our future research. It can be done by collecting data from real users. We want to collect some personal information like age, gender, job position, etc. and the explicit specification of preferences from each user and then analyze the data in search for dependency.

Presented model was experimentally implemented and integration was tested using Cocoon and Spring Framework [12] and Corporate Memory of [4]. Both searching methods (user dependent and user independent search) are compound of tools which can be further improved separately. After the phase of gathering data about users, it will be possible to compare the efficiency, speed and complexity of these methods and decide about their practical usage. In future we plan to experiment with infrastructure of [17] and [2], which could possibly replace [12].

Our approach is used also in the Slovak project *NAZOU – Tools for acquisition, organization and maintenance of knowledge in an environment of heterogeneous information resources* [11] to find relevant job offers for the user according to his preferences.

Similar strategy of communication with users (evaluation of sample objects) was used in [10], where the learning part of the system was covered by a neural network.

This approach does not permit to model user preferences. We did not found any other similar approach to the whole process.

Acknowledgements

Partially supported by Czech projects 1ET100300419 and Slovak projects VEGA 1/3129/06 and NAZOU.

References

1. Baader, F., Kuesters, R., Wolter, F.: Extensions to Description Logics. In: Baader, F., Calvanese, D., McGuinness, D. L., Nardi, D., Patel-Schneider, P. F., *Description Logic Handbook*, Cambridge University Press (2003), 219–261
2. Bednárek, D., Obdržálek, D., Yaghob, J., Zavoral, F.: Data Integration Using DataPile Structure. In: *Advances in Databases and Information Systems*, Springer Verlag (2005), ISBN 3-540-42555-1, 178–188
3. Bratko, I., Šuc, D.: Learning qualitative models. *AI Magazine 24* (2003) 107-119
4. Ciglan, M., Babik, M., Laclavík, M., Budinská, I., Hluchý, L.: Corporate memory: A framework for supporting tools for acquisition, organization and maintenance of information and knowledge. In: *ISIM'06*, Czech Republic (2006) 185–192
5. Fagin, R.: Combining fuzzy information from multiple systems, In: *J. Comput. System Sci.* (1999) 58:83–99
6. Fagin, R., Lotem, A., Naor, M.: Optimal Aggregation Algorithms for Middleware. In: *Proc. 20th ACM Symposium on Principles of Database Systems* (2001), 102–113
7. Gurský, P.: Towards better semantics in the multifeature querying. In *Proceedings of Dateso 2006*, ISBN 80-248-1025-5, (2006), 63-73
8. Horváth, T., Vojtáš, P.: Induction of Fuzzy and Annotated Logic Programs. In: *ILP 2006*, LNAI 4455 Springer-Verlag (2007) 260–274
9. Horváth, T., Vojtáš, P.: Ordinal Classification with Monotonicity Constraints. In: *ICDM 2006*, Leipzig, Germany. LNAI 4065 Springer-Verlag (2006), 217–225.
10. Naito, E., Ozawa, J., Hayashi, I., Wakami, N.: A proposal of a fuzzy connective with learning function and query networks for fuzzy retrieval systems. In: *Fuzziness in database management systems*. P. Bosc and J. Kacprzyk (eds.), Physica Verlag (1995), 345–364
11. NAZOU. Tools for acquisition, organization and maintenance of knowledge in an environment of heterogeneous information resources. ⟨http://nazou.fiit.stuba.sk⟩
12. Spring Framework. System for assembling components via configuration files. ⟨http://www.springframework.org⟩.
13. Srinavasan, A.: The Aleph Manual. *Technical Report*, Comp. Lab., Oxford University
14. Vojtáš, P.: Fuzzy logic aggregation for Semantic Web search for the best (top-k) answers. In: *Fuzzy logic and the semantic web*, E. Sanchez (ed.), Elsevier (2006) 341–360
15. Vojtáš, P.: A fuzzy $\mathcal{EL}$ description logic with crisp roles and fuzzy aggregation for web consulting. In: *Proc. IPMU'2006*, B. Bouchon-Meunier et al. (eds.), EDK (2006), 1834–1841
16. Vojtáš, P.: Fuzzy logic programming, Fuzzy Sets and Systems, 124(3) (2001) 361-370
17. Yaghob J., Zavoral, F.: Semantic Web Infrastructure using DataPile. In *WI-IATW '06* (2006), Los Alamitos, California, ISBN 0-7695-2749-3, 630–633

Applying Social Behaviours to Model Trusting

Vincenza Carchiolo, Alessandro Longheu, Michele Malgeri,
Giuseppe Mangioni, Vincenzo Nicosia

Dipartimento di Ingegneria Informatica e delle Telecomunicazioni
Facoltà di Ingegneria - Università degli Studi di Catania
Viale Andrea Doria 6, I95125 - Catania - ITALY

Summary. Social networks study revealed several useful properties, as the small world effect; such networks can be exploited in addressing the question of trusting, which has become a crucial matter inside Internet-based communities. This paper presents such an approach aiming at reproducing the behaviour individuals adopt in their life when they establish and maintain trust relationships, sending queries to collect reputations in order to estimate how much trust in new acquaintances. We consider issues as query network forwarding and lifecycle of trusting relationships, aiming at building an effective and efficient model for trust management.

1 Introduction

Studies on social networks over last years revealed that individuals relationships are characterized by useful properties, as the "small world" effect, according to which a generic person is connected to another through short paths of social relationship. Social networks mechanisms can be positively exploited when addressing the question of trust, a relevant issue in Internet-based communities pushed by the spread of networks and sharing of information over last decades (e.g. as in P2P paradigma).

Several different definition for trust can be found [9], [12], [1], [4]; simply put, trust provides information about with whom we should share information, from whom we should accept information, and what consideration such information should be given.

In this paper we approach the question of trusting asking how a person joins an existing network of trust relationships in real life, and how he builds these relationships over time, in particular reproducing the mechanism of collecting and mediating other's opinions about an unacquainted person to assign him a trust value. Our proposal falls within the context of *reputation-based* systems, allowing an estimation of trust, in opposition to *policy-based* systems, where the hard evidence of owned credentials is used to grant trust[1]. We also consider trust relationships lifecycle, e.g. the evolution from "weak"

Vincenza Carchiolo et al.: *Applying Social Behaviours to Model Trusting*, Studies in Computational Intelligence (SCI) **78**, 105–114 (2008)
www.springerlink.com

(first time impressions) to "strong" (experience based judgement), as well as the removal of links between individuals not having contacts for long periods.

The paper is organized as follows: in section 2 we describe how a person interacts with an existing trust-based (i.e. friendship) community with whom he is initially unacquainted, formalizing our discussions into a trusting algorithm; trust links lifecycle are introduced in section 3, whereas in sections 4 preliminary results of network simulation are shown. Finally, in sections 5 and 6 we present related works and conclusions, respectively.

2 The mechanism of trusting within social networks

2.1 Concepts and Terms

According to the definition given in [9], we say a person (Alice) trusts another one (Bob) if A is persuaded that information coming from B are true, for example, if A needs B to help him in finding a job, and B says that he will help A, A is guaranteed that B says the truth, hence he will actually help A. This definition expresses the same concepts introduced in [3], where trust is intended as the choice A makes when he perceives an ambiguous path and B's actions determine how the result of following the path will be; a similar definition is given in [12], where "trust is a bet about the future contingent actions of others". We quantify how A trusts B with a number $t_B^A \in [-1, 1]$, where -1 means A does not trust B at all, and 1 indicates that A trusts B completely; a value of 0 models the *indifference*, due to either a lack of information about B or when exactly opposite judgements come from the network, hence A is unable to decide whether trust or not in B.

In a social network a trust relationship between A and B is established if A collects some information about B, either with a direct interaction with B or even getting such information from other sources, e.g. his acquaintances, web sites, e-mails, contact board, and so on; indeed, if A does not know anything about the existence B, A also cannot assign any trust value to B. We name these information about B as B's "resume", representing information as study degree, working experiences, personal skills and attitudes, and so on; at this stage, we do not impose any constraint on resume's content or structure We denote the resume about a node X owned by a node Y as:

$$C_X^Y = \{\mathsf{X}, \overline{C_X^Y}\} \tag{1}$$

where X is the identifier for the node X the resume is about, and $\overline{C_X^Y}$ is the set of information cited previously. Note that different nodes actually can know different things about X, as in real world occur, thus we indicate the owner node Y; moreover, resumes evolve over time, as soon as Y acquires information about X, thus updating $\overline{C_X^Y}$. The most updated and complete resume is the one owned by the person it concerns, simply indicated as $\overline{C_X}$.

2.2 A sample scenario

To model trusting mechanisms inside a community, we use an oriented labeled graph $\mathcal{G}$, defined as a pair $(\mathcal{N}, \mathcal{E})$ where $\mathcal{N}$ is a set of nodes and $\mathcal{E}$ is a set of edges (or links) between vertices, i.e. $\mathcal{E} = \{(u,v)|u,v \in \mathcal{N} \wedge u \neq v\}$; a link from A to B is labelled with the pair trust, resume, i.e. (t_B^A, C_B^A)

To show how links are established, let us consider Alice (A), an unacquainted person who wants to establish some contact with an existing community, usually to satisfy some need in her everyday life. For instance A may be a new student that needs to know where classroom are located, hence she asks for some help to other students she meets in a corridor, or she can search for a computer science professor on the University website. In the first case, A directly contacts another individuals, say Bob (B) and Carl (C), thus having links form A to B and from A to C, whereas in the last case A collects information without a direct interaction with individuals, but she is still collecting resumes about some professors (say Danny and Eve, respectively node D and E), so links from A to D and E are present too. Note that arcs are oriented since mutual trust values are not necessarily the same, or one of them could also be not present, for instance A could completely trust in B and C, since she is in a new, unknown context, but B and C might assign a low trust value to A, being an acquaintance they are meeting for the first time; moreover, A will assign trust values to professor D and E based on their resumes, but D and E actually do not know anything about the existence of A, hence neither arc from D to A nor from E to A will be present.

In this scenario, A assigns a first trust value to nodes B,C,D and E. Such values will be refined over time, as soon as other information (resumes) is available, e.g. A can re-evaluate t_D^A and t_E^A as soon as she meets those professors; similarly occurs for trust values assigned to A by nodes B and C.

2.3 Trust evaluation

In addition to direct (personal) interaction between two individuals, in the real world the refinement process of a trust value assigned by a person A to a generic person X, not belonging to the set of A's acquaintances, is often performed by simply asking to X's acquaintances, i.e. to those people who directly know X and can provide a judgement about him. These judgements, together with all resumes about X, are collected and mediated by A, so she can refine her knowledge and assign a proper trust value about X exploiting other people's personal experience with X. Note that in the graph model, the term "acquaintance" correspond to a node that is neighbour to another, i.e. a path of length 1 (an arc) exists between these nodes. In this paper we focus on the trust refinement mechanism exploiting acquaintances opinions, since it better represents social network interactions. The *trustNode* algorithm implements this evaluation of mediated trust performed by a generic node A about a node X; in the following we also comment relevant instructions.

trustNode

Require: $\mathcal{G}$
Require: $\mathsf{A}, \mathsf{X} \in \mathcal{N}$ where A is a node that wishes to trust node X
Require: C_X
Require: τ_{good} that is threshold used to select node for query forwarding
Require: n_r maximum number of request, id_q is a query identifier
1: **if** $\mathsf{X} \notin \mathbf{N}^A$ **then**
2: $\quad addToNet(\mathsf{A},\mathsf{X})$
3: **end if**
4: $\overline{\mathbf{R}} = \{r_i \in \mathbf{R}_X^A | r_i > \tau_{good}\}$
5: $\mathbf{T}_X^A = \emptyset$
6: **for all** $(\mathsf{I} \in \overline{\mathbf{R}}) \wedge (|\, \mathbf{T}_X^A \,| < n_r)$ **do**
7: $\quad t_I^X = forward(I, C_X, id_q)$
8: $\quad \mathbf{T}_X^A = \mathbf{T}_X^A \bigcup \{\mathsf{I}, t_X^I\}$
9: **end for**
10: **if** $|\, \mathbf{T}_X^A \,| < n_r$ **then**
11: $\quad$ **for all** $(\mathsf{I} \in (\mathbf{N}^A - \overline{\mathbf{R}})) \wedge (|\, \mathbf{T}_X^A \,| < n_r)$ **do**
12: $\quad\quad t_X^I = forward(\mathsf{I}, C_X, \mathsf{A}, id_q)$
13: $\quad\quad \mathbf{T}_X^A = \mathbf{T}_X^A \bigcup \{\mathsf{I}, t_X^I\}$
14: $\quad$ **end for**
15: **end if**
16: $t_X^A = trust(\mathbf{T}_X^A, \mathbf{N}^A)$

First (line 1-3), A check whether X does not already belong to the set of her acquaintances (denoted as $\mathbf{N}^A$); in this case the *addToNet* function creates the arc from A to X and randomly assigns an initial trust value; to really make sense, X should not belong to $\mathbf{N}^A$.

Now, A would collect opinions about X to X's acquaintances, but probably she does not know anyone of them, hence A asks *her* personal acquaintances whether they know X, or whether acquaintances of A's acquaintances know X and so on, actually forwarding her request (*query*) through the network by exploiting trusting relationships. Specifically, A initially establish how many opinions (say, n_r) about X she needs to assign X a trust value: the more opinions will ask, the more important for A the evaluation is, the more accurate the trust value to assign will be; the drawback will be the flooding of messages traversing the network. To model real world interactions, A establishes an order according to acquaintances are contacted. In particular, A first considers acquaintances she trusts more, at the same time having a resume similar to X's resume, since A supposes that similarity in resumes increases the possibility that those acquaintances directly know X.

To formalize this discussion, we introduce the *correspondence* function to evaluate resumes similarity and *relevance* function to sort acquaintances based on correspondence and assigned trust level:

- *correspondence* is defined as $corr : C \times C \rightarrow [0;1]$, also noted with $corr(C_I, C_X)$; we call $c_{I,X}$ the result of correspondence between nodes

I and X. At this stage, the implementation of this function simply adopt classical information retrieval techniques aiming at extracting inverse index vector for both resumes. Obviously, $corr(C_I, C_X) = corr(\overline{C_I}, \overline{C_X})$.

- *relevance* function is defined as $r_I^{A,X} = c_{X,I} \cdot t_I^A$ where A is the node that ask an opinion about X to one of her neighbour I; the set of all values is:

$$\mathbf{R}_X^A = \{r_I^{A,X} | \mathsf{I} \in (\mathbf{N}^A - \{\mathsf{X}\})\} \tag{2}$$

Given these definitions, in line 4 of the algorithm A establishes a threshold τ_{good} to find acquaintances she both trusts a lot and also with a resume highly similar to X's. These *qualified* acquaintances are considered first (lines 6-9) when A sends the query (line 7), using the *forward* function to collect their opinions about X, finally storing this trust value together with the neighbour that provided it in the following set (line 8):

$$\mathbf{T}_{\mathsf{X}}^A = \{(\mathsf{I}, t_X^A) | \mathsf{I} \in \overline{\mathbf{N}}_A\} \tag{3}$$

$\overline{\mathbf{N}}_A$ indicates the set of acquaintances (neighbours) who actually answered the query. Note that acquaintances with high relevance are randomly selected when sending queries in order to avoid always contacting the same acquaintances.

A starts sending queries to qualified acquaintances, aiming at getting n_r answers from them, but when their number is not enough or when the number of successfully queries is not enough (line 10), A will continue sending queries to less qualified acquaintances (lines 10-15); note that A might not get all n_r answers, as in real world occurs.

Finally, the collected answers are used to build a mediated opinion about X. When evaluating this average trust level opinions coming from acquaintances should be weighted according to trust level A assigned to them, reflecting the real world behaviour. The formula we adopt is the following:

$$t_X^A = trust(\mathbf{T}_X^A, \overline{\mathbf{N}}_A) = \frac{\sum t_X^i \cdot t_i^A}{\sum t_i^A} \tag{4}$$

This is essentially a weighted average formula where main terms are t_X^i, i.e. acquaintances opinions about X, weights are t_i^A, that is trust values assigned by A to his acquaintances. To understand why t_i^A is also present at the denominator, we impose that acquaintances A assigned low trust values should have less influence over the resulting mean value, in order to reflect the real world behaviour adopted by a person, i.e. he tends to neglect the opinion of an acquaintances he trusts in with a low degree. To do this, at the denominator we place the sum of *contributions*, being each acquaintance's contribution the trust value A assigned him; if A trusts with a low degree someone (i.e. related trust value tends to 0), his contribution as well as also the corresponding term at the numerator will be neglected. The t_X^A is finally associated to the arc from A to X.

A relevant question is that when each of A's acquaintances receives the query, his task is to answer the query with his opinion about X, but the current acquaintance could still get into the same trouble, i.e. he does not know directly X; similarly to A, such node will forward the request to his acquaintances (but A), waiting for a set of trust values about X to be mediated as described previously, hence the resulting mean will be assigned by the node to X and it will be also delivered to A (or generally to the requesting neighbour node). The propagation of the query will occur until one of X's acquaintances is reached. The $forward$ function uses an algorithm (not reported here) very similar to the $trustNode$, except for two mechanisms:

1. Similarly to the first node that generated the query (A), an *intermediate* node needs n_r opinions about X, but differently from A, if n_r is not achieved exploiting his qualified acquaintances, no more answers will be asked to acquaintances having a relevance below the threshold, modeling the absence of a personal interest the intermediate node has in determining an opinion about X. In the worst case, the intermediate node will give back a trust value of 0 (i.e. he will answer he cannot give a significant opinion about X). The n_r and τ_{good} are the same A established; this models the propagation of the *importance* A assigned to the query.
2. The formula adopted by each intermediate node to determine the mediated trust value is the same adopted by A, but the trust value the intermediate node returns is multiplied by a distance factor, i.e. a value in the range]0,1] that models the fact that in the real world we consider more significant an opinion if it comes from a person that directly knows X, decreasing the value of this opinion as soon as increases the number of individuals to be contacted to get to this opinion.

3 Trusting relationships lifecycle

Each time a node A collects information about an unacquaintance node X and consequently assigns X an initial trust value without involving his acquaintances, we name the arc from A to X as *weak* link, since the value of trust has not been verified neither by A's personal experience, nor by collecting acquaintances opinions; as soon as A has a personal experience with X or she gets other opinions and build a refined trust value, the arc will be named as *strong*. A weak link is used to model the first contact a person has with one another, a contact that implies the assignment of a first, temporary and possibly wrong trust level, as in real world actually occurs; for instance, all links A creates with nodes B,C,D and E in section 2 are initially weak.

A strong link can be lowered to a weak one when its associated trust level has been evaluated a very long time ago, hence that link should be not considered affordable anymore, and a new evaluation should be performed. Similarly, if a weak link does not become strong for a long period, it will be removed from the network, modeling the fact that A forgets about X existence if he never contact X during his life. This *aging* mechanism is also useful to avoid a network with an excessive (and probably useless) links.

Finally, note that the mechanism of evaluating trusting by collecting opinions from acquaintances relies both on weak and strong links.

4 Preliminary results

A first set of simulation has been performed using the trusting algorithm described previously, in addition with aging mechanism, in order to test the convergence of the network. Simulation have been performed considering networks with different number of nodes (i.e. 100, 200 and 400), with initially no links; the trusting algorithm is invoked when queries are generated, and it allows the growing of links number, hence the network evolves through a transient state until a stable set of connections is achieved. Input parameters we assigned for trusting algorithm are n_r=3 (in order to avoid the flooding of the network), and distance factor=0.9 (to allow long paths for query forwarding to be considered). Results are quantified by introducing two measures:

1. average trusting requests deepness, denoted as $ATRD$ (where a TR is a query), indicating the deepest path from the requesting node to the node to be trusted, mediated among all TRs; this property is used to express the *efficiency* of the trusting process
2. average number of failed TRs ($AFTR$), where a TR is considered *failed* when no answers are collected, e.g. when no relevant nodes are found for TR forwarding. This property is used to express the *effectiveness* of the trusting process

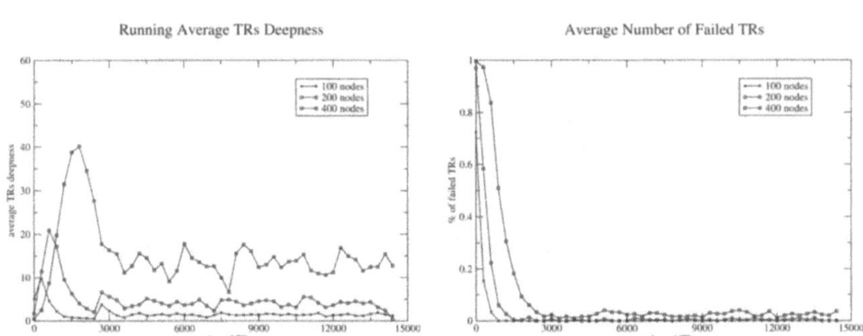

(a) Average TRs deepness

(b) Average number of failed TRs

Fig. 1. Network evolution

In fig. 1(a) and 1(b) running $ATRD$ and $AFTR$ are represented, respectively. These express how many TR are needed for the network to converge, i.e. to provide stable values for both properties. This two experiments clearly point out that the proposed trusting algorithm converge and, in particular, TRs deepness reachs stable values when about 1500–3000 (for networks of 100 to 400 nodes, respectively) TRs are performed. Note that TRs deepness initially increases since when the network is created too few links are present; as soon as links number increases, the deepness decreases since more (possibly shorter) paths are available. The number of failed TRs rapidly decrease to zero when the network converges, meaning that when a trusting request is issued, exactly n_r opinions are received from the source node allowing it trust evaluation according to eq. 4.

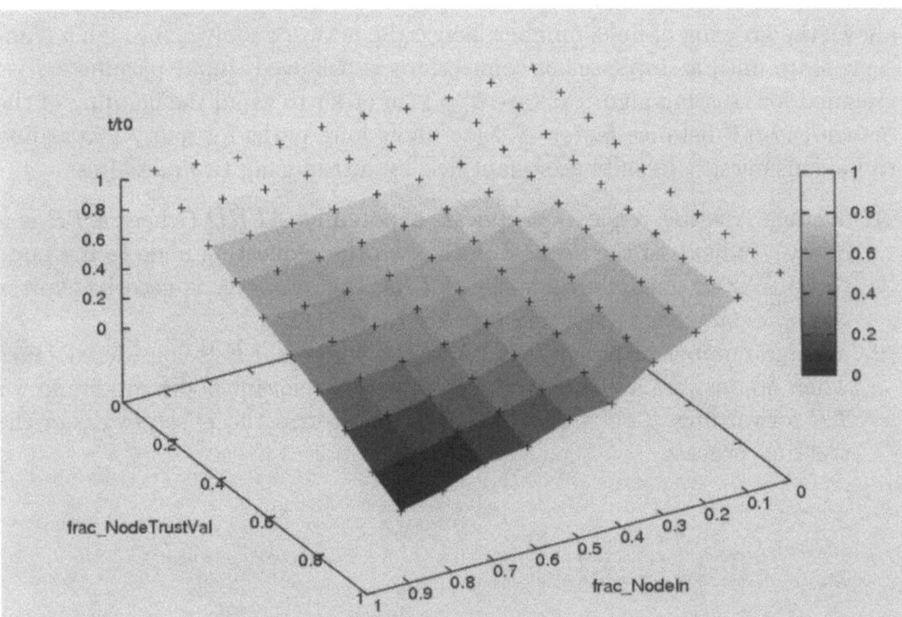

Fig. 2. Resilience to group empowering

In addition to $ATRD$ and $AFTR$ in this paper we shortly discuss the algorithm resilience to group empowering. This measure shows which is the critical cardinality of a cheating group to be able to raise (or equivalently lower) the reputation of a target peer. The experiment is conduct on a network of 100 nodes where 15000 TRs are performed (as previously discussed, in this case the network reaches a stable state). Then, we randomly choose a source node S, a target node T and a TR from S to T is executed; the corresponding trust level $t_0{}_T^S$ is used as a comparison factor during the subsequent TRs. This experiment is repeated lowering from 0% to 100% the trust value ($frac\_NodeTrustVal$ in figure 2) of a given fraction ($frac\_NodeIn$ in figure 2) of the number of nodes having T as neighbour (n ranges from 1 to $|\mathbf{P}^{\mathsf{T}}|$, $\mathbf{P}^{\mathsf{T}} = \{\mathsf{U} \in \mathcal{N} | (\mathsf{U}, \mathsf{T}) \in \mathcal{E}\}$).

For a given pair ($frac\_NodeTrustVal$, $frac\_NodeIn$), we evaluate t_T^S hence the ratio t_T^S/to_T^S. The experiment is iterated changing source and target nodes in order to obtain the average real network behaviour. The ratio gives an idea about the algorithm robustness against group empowering; indeed, as shown in figure 2, a group of cheating nodes can heavly affect the judgement on a target node (i.e. lowering t_T^S/to_T^S) only when $frac\_NodeIn$ is a significant fraction (about 40-50%) over $|\mathbf{P}^{\mathsf{T}}|$, or when $frac\_NodeTrustVal$ grows up to about 40%.

5 Related work

In the last decades, the recommender system is gaining an important role in today's networked worlds because they provide tool to support decision helping in selecting reliable from unreliable. Reliability is often expressed through a trust value with which each agent labels its neighbours; [10, 7] explore this, but they does not investigate in the topic of formation of trust based on real-world social studies. Some recent works have suggested to combine distributed recommendation systems with trust and reputation mechanisms [10, 8].

People surfing the Web has already faced the matter to form opinion and rate them against trust, as far as the well-known reviewers' community *Epinions (http://www.epinions.com)* which collects reviews from the community members, any members can also decide to "trust" or "distrust" each other. All the trust and block relationships interact and form a hierarchy known as the Web of Trust. This Web of Trust (WOT) is combined with rating to determine in what order opinions are shown to readers. However trusting model remains centralized (trust is influenced only by the manager).

Ziegler and Golbeck [5, 13] believe that computational trust model bear several favorable properties from social filtering than they expect notions of trust must reflect user similarity. Therefore a reputation system is an important tool in every network, but assume a central role in emerging P2P networks where many people interacts with many others in a sort of "democratic" fashion. Some author discusses decentralized methods that approximate ranks according to local history and a notion of neighbourhood [2] where trust is calculated trying advantage of small-world properties often emerging in networks that mimic real world. In P2P area EigenTrust [6] propose a distributed implementation of PageRank [11] that needs also a distributed structure to store data and imposes to pre-trust all nodes belonging to the net thus reducing the "de-centralisation".

6 Conclusions and future work

This paper introduced an approach to trust entities hint at human behaviour; we proposed an algorithm based on local behaviour of an human and we believe that several important properties of social network will emerge. The algorithm has been discussed and motivated thruoghout the paper; preliminary promising results about the convergence of the network and its resilience to group empowering have been shown.

Further studies are currently active on testing network behaviour, as well as on investigation about communities that may emerge - of interest, similarity, confidence, etc - and on the importance of *correlation* among resumes to improve the effectiveness and efficiency of the proposed approach.

References

1. Donovan Artz and Yolanda Gil. A survey of trust in computer science and the semantic web. *journal of Web Semantics*, (to appear), 2007.
2. Matteo Dell'Amico. Neighbourhood maps: Decentralised ranking in small-world p2p networks. In *3rd International Workshop on Hot Topics in Peer-to-Peer Systems (Hot-P2P), Rhodes Island, Greece*, April 2006.
3. M. Deutsch. Cooperation and trust, some theoretical notes. In Nebraska University Press, editor, *Nebraska Symposium on MOtivation*, 1962.
4. J. Golbeck. Trust and nuanced profile similarity in online social networks.
5. J. Golbeck and J. Hendler. Reputation network analysis for email filtering, 2004.
6. Sepandar D. Kamvar, Mario T. Schlosser, and Hector Garcia-Molina. The eigentrust algorithm for reputation management in p2p networks, 2003.
7. Michael Kinateder and Siani Pearson. A Privacy-Enhanced Peer-to-Peer Reputation System. In K. Bauknecht, A. M. Tjoa, and G. Quirchmayr, editors, *Proceedings of the 4th International Conference on Electronic Commerce and Web Technologies (EC-Web 2003)*, volume 2738 of *LNCS*, pages 206–215, Prague, Czech Republic, September 2003. Springer-Verlag.
8. Paolo Massa and Bobby Bhattacharjee. Using trust in recommender systems: An experimental analysis. In Christian Damsgaard Jensen, Stefan Poslad, and Theodosis Dimitrakos, editors, *iTrust*, volume 2995 of *Lecture Notes in Computer Science*, pages 221–235. Springer, 2004.
9. Merriam-Webster. http://www.m-w.com.
10. Miquel Montaner, Beatriz Lopez, and Josep Lluis de la Rosa. Opinion-based filtering through trust. In *CIA '02: Proceedings of the 6th International Workshop on Cooperative Information Agents VI*, pages 164–178, London, UK, 2002. Springer-Verlag.
11. Lawrence Page, Sergey Brin, Rajeev Motwani, and Terry Winograd. The pagerank citation ranking: Bringing order to the web. Technical report, Stanford Digital Library Technologies Project, 1998.
12. P. Sztompka. Trust: A sociological theory. 1999.
13. Cai-Nicolas Ziegler and Jennifer Golbeck. Investigating correlations of trust and interest similarity - do birds of a feather really flock together? *Decision Support System*, (to appear), 2005.

Using Ant's Alarm Pheromone to Improve Software Testing Automation *

Ronaldo Menezes[1], Rafael Silva[2], Marcelo Barros[2], and Alexandre M. Silva[2]

[1] Computer Sciences, Florida Tech
Melbourne, Florida, USA
rmenezes@cs.fit.edu

[2] Software Engineering, IVIA Ltda.
Fortaleza, Ceará, Brazil
{rafael.silva,marcelo.barros,alexandre.menezes}@ivia.com.br

Summary. Software testing is the *de-facto* standard for quality control in industry. The complexity of today's applications are becoming so high that our ability to manually test software is diminishing — experts argue that automation is the way forward in the field. Nature-inspired techniques, and in particular the area called swarm intelligence, have got the attention of researchers due to their ability to deal with complexity. In insect societies, and in particular ant colonies, one can find the concept of alarm pheromones used to indicate an important event to the society (e.g. a threat). Alarm pheromones enable the society to have a uniform spread of its individuals, probably as a survival mechanism — the more uniform the spread the better the chances of survival *at the colony level.* This paper describes a model of the aforementioned ant-behavior and shows how it can be integrated as part of a software testing automation methodology thus demonstrating that software testing can also benefit from nature-inspired approaches.

1 Introduction

Software applications are executed following a sequence of well-defined functionalities available in the application. The exact sequence of functionalities that is executed depends on several factors but primarily on the input values, the order these values are presented to the application, and the task the user is trying to accomplish. The majority of software testing tools that perform functionality testing rely on test cases being generated from one's experience with the application. The choice of test cases is driven by the experts' concept of what is important to be tested, leading to bias in the test cases.

In general there are two main approaches to test software. The first, more accepted in the academia, consists of using formal specification to design an application

* This research is funded by the Conselho Nacional de Desenvolvimento Científico e Tecnológico (CNPq), Brazil, under grant number 551761/2005-9.

Ronaldo Menezes et al.: *Using Ant's Alarm Pheromone to Improve Software Testing Automation*, Studies in Computational Intelligence (SCI) **78**, 115–124 (2008)
www.springerlink.com

and then use theorem provers to demonstrate certain properties of the application. The second approach consists of the traditional software engineering models (e.g., waterfall, spiral, prototyping) that have a specific phase for testing generally occurring after the application has been implemented. Despite all the claims of these less formal techniques being effective, the truth is that current approaches are incapable of testing software appropriately.

From a financial point of view, it only makes sense to look at software testing with more attention given that low estimates put the cost of software testing at around 40% of the overall cost of development [14]. Ergo what is needed is a scientific method that allows testing to be carried out effectively by practitioners without requiring the rigorous approach of formal methods.

Nature shows us examples of intelligence in groups of beings (agents) with limited cognitive capacity, such as ants, bees and termites. This type of intelligence that can be observed at the group level but almost inexistent at the individual level is called *swarm intelligence* (SI) [3]. Swarm-intelligent solutions have been shown efficient in dealing with distributed problems where the solution space of the problem is large.

Pheromone-communication is the basis for many algorithm in the Swarm Intelligence field. In nature, they are also used in threat situations. Agents (bees, ants) release an alarm pheromone that can be detected by other individuals in a surrounding area. The result of pheromone release is an *uniform spread of individuals* leading to a high survival rate.

Natus started as a project looking into the usefulness of nature-inspired computing concepts such as genetic algorithms (GA) and swarm intelligence in functional software testing automation. Natus uses GA to generate test plans and the model of alarm pheromone [10] to execute these plans. In this paper, we introduce Natus and show how an alarm pheromone model is used in the methodology.

2 Related Topics and Works

2.1 Software Testing Automation

Software testing is a main area of research for the software engineering community and is also subject of independent study in many universities. The process of software development is far from being trivial. The complexity of the new applications asks for a development process with strict quality control at every phase of development.

In the last years, software testing has been considered one of the most important procedures to assure software quality. A main theme in testing has been the creation of automated tools that are reasonably autonomous to assure that software budgets remain within estimates. An automated testing tool can facilitate considerably the work of a tester and, as the most attractive benefit, it may turn the whole process cheaper by reducing the work that was done manually before.

Research in technologies such as the ones inspired in artificial intelligence and their integration in software testing appears to be the way forward. Artificial intelligence can naturally aid the process of testing by augmenting the ability of a software to discover defects that would not likely be found by experts due to their bias. Experts usually are blind to certain bugs for they attempt to test the unusual functionalities in an application and tend to ignore ordinary functionalities.

It is fair to say that applications will never be free of defects — problems will exist even if all the development phases are taken correctly and cautiously, and the requirement analysis done thoroughly. Software testing aims at eliminating the majority of these problems via a systematic process that identifies components of the software prone to errors.

The identification of components with defects is not trivial given the size of the search space: all the possible execution sequences of the software. This is an unbounded space; there is no systematic process that can explore *all* the search space in a finite amount of time. So, how can we traverse this space in a way that the sample of test plans executed contains an uniform distribution of plans in the search space? This uniformity is necessary because it increases the chance of finding an error in the application. The solution is on the development of automated tools that avoid the bias of experts who tend to prioritize test plans that they *think* to be important but which in practice may not be so. The ALARM model proposed in [10] is an approach to traverse search spaces in a uniform manner and it is an intrinsic part of the Natus methodology described in Section 3.

2.2 Nature-Inspired Approaches in Automated Software Testing

Automated software testing is an area of growing interest but until recently very few researchers had looked at the possibility of using nature-inspired approaches to improve this area. Still today, most of the works focus on the use of genetic algorithms; these works are related to the testing VLSI circuits [2, 4] rather than software applications.

In the area of testing software applications, the research is focused on the automatic generation of testing *data* and not test plans as it is later proposed in this paper. Mantere [9, 1] has done an extensive study on the use of GA in data generation but at the level of mostly toy problems and simple program subroutines. In embedded systems Wegener *et al.* [12] has described how data can be generated for performing structural testing.

At time of conducting this research, we were only aware of one work looking at automatic generation of test plans using GAs. Kasik and George [7] propose to use GA in the generation of test plans but their focus is limited to GUI applications. The idea is to have tests that mimic a novice user manipulating the interface.

Natus is focused on *functionality testing of Web applications* but the idea could be used in other applications. The concentration of Web applications at this stage is due to the fact that we are moving to a Web-based era where most applications run remotely on the Web. Additionally, Web applications use a well defined protocols suitable to the automatic generation of test plans.

2.3 Ant Colony Optimization

The Ant Colony Optimization (ACO) algorithm was proposed by Dorigo *et al.* [5]. Their approach to solving combinatorial optimization problems employs lessons learned from the observation of foraging strategies in ant colonies — in particular ACO makes heavy use of positive and negative feedback as well as indirect communication via the environment (aka *Stigmergy*).

In the real world, ants search for food and return to the nest when they find it, dropping pheromone along the way. These pheromone deposits attract other ants in the colony, biasing their efforts towards the newly located food source (i.e., positive feedback). As several ants search for food, those who find shorter paths to food sources return home faster allowing their trails to be traversed by more ants in a shorter amount of time than the less efficient paths. The shortest path, therefore, gets the strongest concentration of pheromone and the ants in the colony are able to find the shortest route to a new food source. To avoid converging onto paths other than the efficient ones, nature employs evaporation of the pheromone (i.e., negative feedback), thereby gradually removing the incentive for ants to visit inefficient paths (old paths) and improving the chance an ant will find more optimal paths to food sources. Negative feedback also serves the purpose of making the colony behavior adaptable to environment changes (e.g. elimination of old paths to no-longer-interesting food sources). Therefore, through the interaction of very simple agents, a relatively complex behavior has emerged — namely, the ability to find the shortest path to a destination.

2.4 Alarm Pheromone

Pheromones are chemicals used for communications within the same species enabling the member of the species to perform actions such as food location, mating, aggregation, warning other individuals, to name but a few.

An especial kind of pheromone, named *alarm pheromone* was discovered in the early 1960's. There are two primary alarm behaviors: to disperse (flight behavior) or to move towards the source of alarm in an aggressive manner (fight behavior) [6]. It is generally thought that the differentiation between flight and fight behavior depends on the intensity of the alarm pheromone being released.

An example of alarm communication is that of *Acanthomyops* colonies (living exclusively in the USA), they are subterranean ants that are large in size and dense in number and therefore it could be speculated that they would respond to alarm pheromones by the *fight* reaction because of the unlikeliness of dispersal. Workers that are nearby react the fastest and those that are farther away take longer to react. If more pheromone is not released, the signal dies out within minutes [15].

Another example, a *flight* one, can be show by ants of the specie *Lasius alienus*, they are smaller and nests under rocks or wood which allows for fast dispersal when disturbed. The component of their alarm substance is the same as that of *Acanthomyops* but instead of running to the source, they frantically run around in no particular direction. They also react to lower concentrations of alarm pheromones than *Acanthomyops*[15]. The ALARM model [10] deals with the flight behavior only and is used later in this paper to get a uniform execution of test plans.

3 The Natus Testing Methodology

Natus started as a study on the usefulness of nature-inspired metaphors such as genetic algorithms and swarm intelligence in software testing automation. Natus is called a methodology because it goes beyond plan generation and execution; it tells us how these are integrated in the development lifecycle.

Figure 1 shows the entire process that the methodology proposes. It can be divided into five parts: learning, definition of dependency, generation of plans (scripts), execution of the plans, and feedback from execution to improve on the learning. It starts with a process of recording (logging) all possible functionalities the application contains. These logs are inserted in a database as scripts for the recorded functionalities. It follows with users defining dependencies between these operations or letting the system create them. For instance, an operation may depend on another being executed before, and the tool needs to understand this dependency so that plans are generated correctly. Next, we move to the generation of test plans (groups of recorded operations) using a GA approach. The goodness of a plan is measured according to the inconsistency level it can take the application to, so a table of such inconsistencies has to be defined either by the user or randomly by the system (for a further description of all these steps the reader should refer to [11]). After the plans are created, the ALARM model (explained later) is used to prioritize plans according to how many times they have been executed before and how good the plan is. The plans are then executed and the inconsistencies table is updated — if a bug is found, the inconsistency is increased, otherwise its decreased. This allows the further generation and execution of new test plans, thus repeating the cycle.

Note that the use of ALARM is of utmost importance because the process is not limited to just listing the cases and testing from the beginning of the list to the end. Doing so may not give us a diverse list of cases being tested (assuming that not all cases can be tested in the time frame available for testing) and work well if the number of test cases to be executed is small. Basically, a space of test cases generated by the GA can be explored by ALARM agents (ants) by considering that the fewer times the plan has been executed before the more likely it is to be executed now. The fitness of each plan can be used to initialize the environment so that fitter plans are prioritized (using food, as explained later in the paper), thus allowing the amount of executions to be taken into consideration. In a process opposite to ant foraging and inspired in alarm pheromones [8], the ants can make locations in the environment (test plans) less desirable as they select and execute them. This approach makes the ants more inclined to choose tests that have not yet been executed.

Also, one needs to understand that important test plans (according to the fitness value) may be executed and indeed cause problems in the application. This test continues to be open for re-test and should continue to be considered according to its

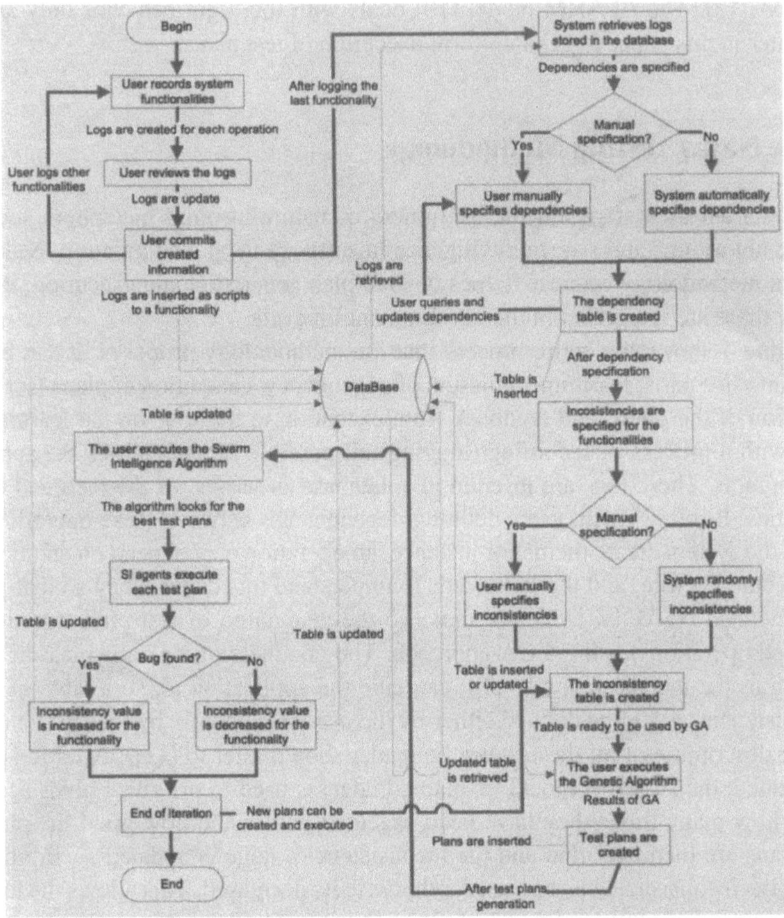

Fig. 1. The flow of the Natus testing methodology that utilizes the GA and SI approaches.

fitness value (or even with an increased fitness value). The use of positive feedback can be used to reinforce the fitness value of such test cases, thus attracting more ants to explore that case again. In case the test does not cause any problem, its attractiveness diminishes. In the ALARM model this is controlled via the mechanism of pheromones and food on the environment.

4 The Alarm Pheromone Model

Given that Vieira *et al.* [11] describes how GA can be used on the generation of test plans, we focus here on the description and evaluation of the second part of the methodology which involves the use of an alarm pheromone model. Ant-Colony Optimization (ACO) has been extensively used as a solution to problems were agents (ants) are required to converge to a path (or few paths). In this model, ants positively reinforce the path they choose with pheromones in order to attract other ants. Better paths tend to be reinforced more therefore attracting more ants. To avoid early convergence, the algorithm uses the concept of negative feedback implemented via evaporation of the pheromone.

If we abstract the search space of test plans as a lattice where ants (test-plan executing agents) walk in search of plans to execute, we require an opposite view of ACO. Our model represent scenarios where ants respond negatively (repulsion) to the existence of pheromones in the locations surrounding its current location. The more pheromone present in a cell the more the ant will be repulsed by it. In terms of the Natus methodology one could say that these cells represent plans that have passed the tests already and should have less priority in future executions. It is also important to point out that we envision several ants dealing with the same search space. So the pheromone left by one influences other agents.

In the proposed model we use the amount of food present in the cells (ϕ). This food acts as an attractor to an ant and is balanced by the repulsiveness of pheromones (τ). The food exist to counter-balance the pheromones. As mentioned above, if a test plan fails we would like for this test plan to be given priority in future executions (even though it has been executed already) to test if the problem has been fixed. The food can be used for this purpose. Food can be used to make agents attracted to the cell (test plan) in spite of the pheromone at that location.

$$P_{ij}(t) = \frac{\phi_{ij}(t)^{\beta}}{\tau_{ij}(t)^{\alpha} \times \sum_{k \in N}[\phi_{ik}(t)]^{\beta}} \tag{1}$$

Equation 1 is the transition rule in the ALARM model and the core of the execution process. P_{ij} represents the chance that a test plan j will be executed after the execution of plan i, where j is a plan in the neighborhood (N) of i (according to the lattice used). The values of α and β control the degree of importance given to the amount of pheromone (number of executions) and food (importance of the plan). The negative feedback via evaporation requires an update rule for all cells. The update rule is exactly as in the ACO model. More details about the ALARM model can be found in [10].

In our simulations in Section 5 an equal distribution of food in the environment is assumed — 20 units of food in each cell of the environment. If the distribution is unequal the coverage will not be perfect in the beginning as it will represent the inequalities of the food distribution (pheromone). As mentioned before this approach can be used to counter-balance the pheromone. In the Natus methodology, extra amounts of food can be deposited in the environment to bias the agents to re-execute a test plan represented by a specific location in the environment. Each time a unit of food is consumed (location where the food is is visited), an amount of alarm pheromone is deposited in the cell.

5 Experimental Results

In order to generate our results, and visualize the behavior of our alarm pheromone model, we have implemented a visual simulator in NetLogo [13]. The simulator contains a 2D lattice where agents are randomly placed. Parameters such as number of pheromone units used by each agent, evaporation rate of the pheromone, amount of food in the environment, quantity of food unit consumed per visit, control variables α and β, etc. can all be controlled in this simulator.

Another important aspect of this simulator is the ability to show results related to the coverage of the environment at real-time: a panel is available to show the standard deviation of the number of visits of the cells of the environment.

5.1 Performance Evaluation of ALARM

We have tested scenarios where ants are placed in the environment behaving in 2 different ways: random walk and ALARM. To show a uniform coverage we are using the standard deviation of how many times each cell is visited by an ant. Lower standard deviations represent a more uniform distribution of visits in the environment. Note that the standard deviation is a global characteristic of the environment. In the environment, each cell represents one test plan previously generated using GA.

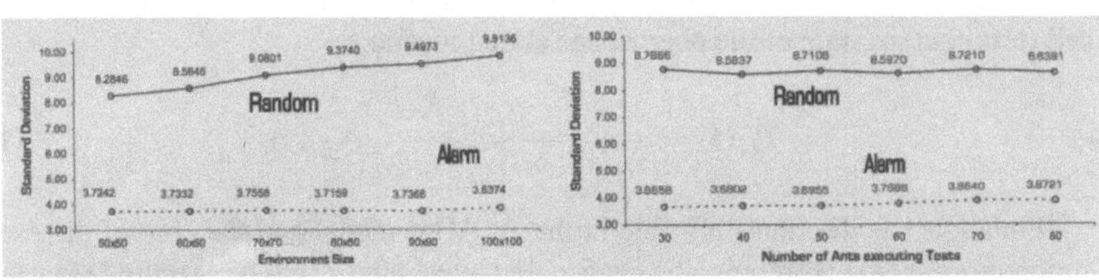

Fig. 2. Comparison between random walk and the execution of test plans in Natus.

In all scenarios tested, the environments are 2D toroidal grids — each cell has 8 neighbors. For Equation 1 we have used $\alpha = 1$ and $\beta = 1$. At each step of the simulation (all ants perform one move) an evaporation of 5% is globally applied to all cells of the grid.

Each cell is initialized with 20 units of food which are consumed one by one as the ants walk to that location. Hence in our simulator we are giving each cell (test plan) the same importance level. The minimum amount of pheromone per cell is one and the same is true for the food otherwise the probability of choosing a cell could get to zero. Every time an ant visits a cell it consumes one unit of food and leaves four units of pheromone. Since the amount of food can represent the importance of a test plan to be executed, we assume in the simulator that errors found during the execution of a test plan are fixed immediately therefore the amount of food is always decreasing — the less food a cell has the more certain we are that the plan has passed and has not caused any problems in the application.

The results shown in Figure 2 are averages of ten runs for each experiment. We have made two types of experiments: scalability in relation to the number of plans (Figure 2(left)) and the number of agents (ants) executing the plans (Figure 2(right)). For the first experiment we released 50 ants in environments of sizes: 50×50, 60×60, 70×70, 80×80, 90×90, and 100×100. At the end of each simulation (all with the same number of execution steps) we use the counter in each cell (number of times an ant passed at the location) and calculate the standard deviation for the entire environment. Another experiment was to see the behavior of the models when the size of the environment is fixed but we vary the number of ants. We used an environment made of 50×50 grid and used 30, 40, 50, 60, 70, and 80 ants respectively. The experiments were compared with random walk and in both cases, ALARM performed significantly better.

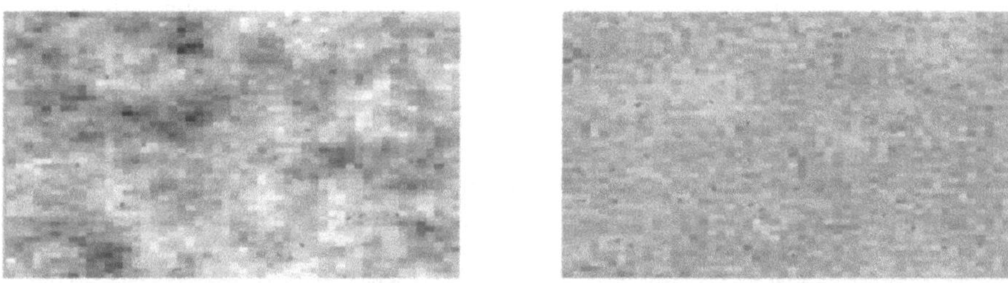

(a) Random Walk (b) ALARM Model

Fig. 3. Snapshots taken at the end of the execution of the simulator of each approach. The shades of grade represent an approximation of the number of times a test case was executed. Darker areas represent more executions.

Figure 3 allows us to visualize the status of the environment at the end of the simulation. One can clearly see that the random walk model does not provide us with a very uniform coverage of the environment — clear patches are formed. On the other hand, ALARM are very uniform with no obvious patch formation.

6 Conclusion

This paper described a methodology for software testing called Natus and concentrated on the evaluation of the test plan executions driven by the ALARM model. We demonstrate the uniformity of our model with experiments that measure the standard deviation of the number of times a cell is visited. We compared that with a well understood metric: random walk.

Natus allows for the elimination of the bias brought by testing experts when choosing test plans for the applications they work with. Our approach works in unison with a test plan generation module based on GA where the result of the execution of the test plans using ALARM can be used by the GA to generate other (more appropriate) test plans. Natus is the first full methodology for software testing automation inspired on its entirety by natural algorithms.

References

1. J. T. Alander and T. Mantere. Automatic software testing by genetic algorithm optimization, a case study, June 02 2005.
2. J. Aylor, J. Cohoon, E. Feldhousen, and B. Johnson. Compacting randomly generated test sets. In *Proceedings of the IEEE International Conference on Computer Design*, pages 153–156. IEEE Press, Sept. 1990.
3. E. Bonabeau, M. Dorigo, and G. Theraulaz. *Swarm Intelligence: From Natural to Artificial Systems*. Santa Fe Institute Studies in the Sciences of Complexity Series. Oxford Press, July 1999.
4. F. Corno, P. Prinetto, M. Rebaudengo, and M. S. Reorda. GATTO: A genetic algorithm for automatic test pattern generation for large synchronous sequential circuits. *IEEE Trans. on CAD of Integrated Circuits and Systems*, 15(8):991–1000, 1996.
5. M. Dorigo, V. Maniezzo, and A. Colorni. The ant system: Optimization by a colony of cooperating agents. *IEEE Transactions on Systems, Man, and Cybernetics Part B: Cybernetics*, 26(1):29–41, 1996.
6. W. O. H. Hughes, P. E. Howse, E. F. Vilela, and D. Goulson. The response of grass-cutting ants to natural and synthetic versions of their alarm pheromone. *Physiological Entomology*, 26(2):165–172, 2001.
7. D. J. Kasik and H. G. George. Toward automatic generation of novice user test scripts. In *CHI '96: Proceedings of the SIGCHI conference on Human factors in computing systems*, pages 244–251, New York, NY, USA, 1996. ACM Press.
8. C. Lloyd. The alarm pheromones of social insects: A review. Technical report, Colorado State University, 2003.
9. T. Mantere. *Automatic Software Testing by Genetic Algorithms*. PhD thesis, University of Vaasa, May 02 2003.
10. R. Menezes, F. Martins, F. E. Vieira, R. Silva, and M. Braga. A model for terrain coverage inspired by ant's alarm pheronome. In *SAC '07: Proceedings of the 2007 ACM Symposium on Applied Computing*, pages 728–732, New York, NY, USA, 2007. ACM Press.
11. F. E. Vieira, R. Silva, F. Martins, R. Menezes, and M. Braga. Using genetic algorithms to generate test plans for funcionality testing. In *Proceedings of the 44th ACM Southeast Conference*, Melbourne, Florida, USA, March 2006. ACM, ACM Press.
12. J. Wegener, K. Buhr, and H. Pohlheim. Automatic test data generation for structural testing of embedded software systems by evolutionary testing. In *Proceedings of the 2002 Genetic and Evolutionary Computation Conference*, pages 1233–1240, New York, 9-13 July 2002.
13. U. Wilensky. Netlogo. Technical report, Center for Connected Learning and Computer-Based Modeling, Northwestern University, 1999. http://ccl.northwestern.edu/netlogo.
14. C. E. Williams. Software testing and uml. In *Proceedings of the 10th International Symposium on Software Reliability Engineering*, Boca Raton, Florida, Nov. 1999. IEEE Press.
15. E. O. Wilson. *The Insect Societies*. Harvard University press, 1971.

Aspects Concerning SVM Method's Scalability

Daniel MORARIU[1], Maria VINŢAN[2], and Lucian VINŢAN[1]

[1] "Lucian Blaga" University of Sibiu, Computer Science Department, Emil Cioran street, no. 4, Sibiu 550025, Romania
[2] "Lucian Blaga" University of Sibiu, Electric and Electronic Engineering Department
{daniel.morariu, maria.vintan, lucian.vintan}@ulbsibiu.ro

Summary. In the last years the quantity of text documents is increasing continually and automatic document classification is an important challenge. In the text document classification the training step is essential in obtaining a good classifier. The quality of learning depends on the dimension of the training data. When working with huge learning data sets, problems regarding the training time that increases exponentially are occurring. In this paper we are presenting a method that allows working with huge data sets into the training step without increasing exponentially the training time and without significantly decreasing the classification accuracy.

Key words: Text Mining, Classification, Clustering, Support Vector Machine

1 Introduction

While more and more textual information is available online, effective retrieval is difficult without good indexing and summarization of documents' content. Documents' categorization is one solution to this problem. The task of documents' categorization is to assign a user defined categorical label to a given document. In recent years a growing number of categorization methods and machine learning techniques have been developed and applied in different contexts. Documents are typically represented as vectors in a features space. Each word in the vocabulary is represented as a separate dimension. The number of word occurrences in a document represents the normalized value of the corresponding component in the document's vector.

In this paper we are investigating how classification accuracy is influenced using only relevant selected input vectors subset belonging to a larger data training set. The answer will show if the classification method is scalable or not. We are using classifiers based on Support Vector Machine (SVM) techniques. They are less vulnerable to degrade when the dimensionality of the

Daniel MORARIU et al.: *Aspects Concerning SVM Method's Scalability*, Studies in Computational Intelligence (SCI) **78**, 125–134 (2008)
www.springerlink.com

feature space is increasing, and have been shown effective in many classification tasks. The SVM is actually based on learning with kernels and support vectors.

We are developing a strategy in seven steps that trains the classifiers on a reduced data set without significant accuracy decrease comparing with training on the larger initial data set but reducing the learning time. In designing this strategy we were inspired by a method presented in [12] which uses a tree structure to hierarchically group similar databases articles, at different abstract levels. Despite this strategy was not recommended by the authors to be used on text documents, we modified it in order to group the similar text documents into a single level.

In this experiment we are using an original implementation of SVM algorithm with some improvements presented by us in [3]. In order to represent the vectors and to select the relevant features we are using a feature selection method based on SVM, already presented in [4], [6]. For training and testing part we are using Reuters' database [10].

Section 2 contains the prerequisites for the work that we are presenting in this paper. In sections 3 we are presenting the proposed scalability method and in section 4 the main results of our experiments. The last section debates and concludes on the most important obtained results and proposes some further work.

2 Experimental Framework

2.1 Support Vector Machine

The Support Vector Machine (SVM) is a classification technique based on statistical learning theory [9], [8] that was applied with great success in many challenging non-linear classification problems processing large data sets.

The SVM algorithm finds a hyperplane that optimally splits the training set. The optimal hyperplane can be distinguished by the maximum margin of separation between all training points and itself. Looking at a two-dimensional space we actually want to find a line that "best" separates points in the positive class from points in the negative class. The hyperplane is characterized by a decision function like:

$$f(x) = sgn(\langle \mathbf{w}, \Phi(x) \rangle + b) \tag{1}$$

where $\mathbf{w}$ is the weight vector, orthogonal to the hyperplane, "b" is a scalar that represents the margin of the hyperplane, "x" is the current tested sample, "$\Phi(x)$" is a function that transforms the input data into a higher dimensional feature space and $\langle \cdot, \cdot \rangle$ represents the dot product. *Sgn* is the sign function. If $\mathbf{w}$ has unit length, then $\langle \mathbf{w}, \Phi(x) \rangle$ is the length of $\Phi(x)$ along the direction of $\mathbf{w}$. Generally $\mathbf{w}$ will be scaled by $\|\mathbf{w}\|$. In the training part the algorithm needs to find the normal vector "**w**" that leads to the largest "b" of the hyperplane.

For extending the SVM algorithm from two-class classification to multi-class classification typically one of two methods is used: "One versus the rest", where each topic is separated from the remaining topics, and "One versus the one", where a separate classifier is trained for each class pair. We selected the first method for two reasons: First, our preliminary experiments show that the first method obtains better performance, which might be explained by the fact that the Reuters' database contains strongly overlapped classes and assigns almost all samples in more than one class. Second, the overall training time is shorter for the first method.

2.2 The Data-set

Our experiments were performed on the Reuters-2000 collection [10], which has 984Mb of newspapers articles in a compressed format. Collection includes a total of 806,791 documents, with news stories published by Reuters Press covering the period from 20.07.1996 through 19.07.1997. The articles have 9822391 paragraphs and contain 11522874 sentences and 310033 distinct root words. Documents are pre-classified according to 3 categories: by the Region (366 regions) the article refers to, by Industry Codes (870 industry codes) and by Topics proposed by Reuters (126 topics, 23 of them contain no articles). Due to the huge dimensionality of the database we will present here results obtained using a data subset. From all documents we selected the documents with the industry code value equal to "System software". We obtained 7083 files that are represented using 19038 features and 68 topics. We represent a document as a vector of words, applying a stop-word filter (from a standard set of 510 stop-words [13]) and extracting the word stem [13]. From these 68 topics we have eliminated those topics that are poorly or excessively represented obtaining 24 different topics and 7053 documents that were split randomly in a training set (4702 samples) and a testing set (2351 samples).

2.3 Kernel Types

The idea of the kernel trick is to compute the norm of the difference between two vectors in a higher dimensional feature space without representing them in this space. We are using in our selected classifiers two types of kernels each of them with different parameters: polynomial and Gaussian kernels [3]. For the polynomial kernel we vary only the degree of the kernel and for the Gaussian kernel we change the parameter C according to the following formulas (x and x' being the input vectors):

Polynomial,

$$k(x, x') = (2 \cdot d + \langle x \cdot x' \rangle)^d \tag{2}$$

where d being the only parameter to be modified and representing the kernel's degree:

Gaussian (radial basis function RBF),

$$k(x, x') = \exp(-\frac{\|x - x'\|^2}{n \cdot C}) \tag{3}$$

where C being the classical parameter and n being the new parameter, introduced by us, representing the number of elements from the input vectors that are greater than 0 [3].

3 Methods for determining the SVM Algorithm's Scalability

The original learning step that uses the Support Vector Machine technique to classify documents is split in the next 7 smaller steps:

1. We normalize each input vector in order to have the sum of elements equal to 1, using the following formula:

$$TF(d, t) = \frac{n(d, t)}{\sum_{\tau=0}^{19038} n(d, \tau)} \tag{4}$$

   where $TF(d, t)$ is the term's frequency, $n(d, t)$ is the number of times that term t occurs in document d, and the denominator represents the sum of terms that occur in the entire document d.
2. After normalization we compute the Euclidian distance between each input vector and each center of the group (*representative vector*) that was created up to this moment. Thus, we compute each distance and we keep the smallest obtained distance (linear clustering). If this distance is smaller than a predefined threshold we will introduce the current sample into the winner group and recompute the representative vector for that group; if not, we will create a new group and the current input vector will be the representative vector for that new group.
3. After grouping, we create a new training dataset with all representative vectors. This set will be used in the classification step. In this step we are not interested in the classification accuracy because the vectors are not the original vectors. Here, we are interested only in selecting relevant vectors (the vectors that have an effective contribution to the classification).
4. We are doing a feature selection step on this reduced set (each of them having 19038 features). For this step we are using SVM_FS method presented in [4]. After computing all weights we select 1309 features because, as we showed in [5], this number of features produces optimal results.
5. The resulted smaller vectors are used in a learning step. For this step we use polynomial kernel with degree equal to 1 and nominal data representation. We use the polynomial kernel because it usually obtains a smaller number of support vectors in comparison with the Gaussian kernel [5]. We use the kernel's degree equal to 1 because in almost all previous tests we obtained better results with this value.

6. After SVM learning step, besides the classification rules (decision functions) that are obtained, we also obtain the elements that have an effective contribution to the classification (named support vectors in the SVM algorithm). We chose all groups that are represented by those selected vectors and we develop a new set only with vectors from these groups. This new set is a reduced version of the original set containing only *relevant input vectors* that are considered to have a significant influence on the decision function.
7. This reduced vectors set will now be used in the feature selection and classification steps as the original input data.

In the second presented step all training data are grouped based on their similarity. To compute the similarity we use two different methods based on the "winner-takes-it-all" idea [1]. First method computes the center of the class (the representative vector) using arithmetic mean and the second method computes the center of the class using LVQ formula (Learning Vector Quantization [2]) in one step. Depending on the method used to compute the representative vector we are using two vector representation types. In the first method, where the representative vector is computed as an arithmetic mean, it contains the sum of all elements that are included in that group, and a value that represents the total number of samples from that group. In the second method the representative vector is computed using equation (5) for each new sample that is included in the group. The formula for computing the representative vector for the LVQ method is:

$$\overrightarrow{w}_i(t+1) := \overrightarrow{w}_i(t) + \alpha(\overrightarrow{x} - \overrightarrow{w}_i(t)) \tag{5}$$

where $\overrightarrow{w}_i$ is the representative vector for the class i, $\overrightarrow{x}$ is the input vector and α is the learning rate. Because we want a one step learning and taking into account the small number of samples, we'll choose a relatively great value for the coefficient α.

3.1 Clustering Data Set using Arithmetic Mean

In our presented results we start with an initial set of 7083 vectors. After the grouping step we reduce this dimension at 4474 representative vectors meaning 63% of the initial set. For this reduction we used a threshold equal to 0.2. On this reduced set a classification step for selecting the relevant vectors was made. After the classification the algorithm returns a number of 874 support vectors. Taking those support vectors we created a dataset containing only 4256 relevant samples meaning approximately 60% from the initial set. Using this new reduced set we make a feature selection step, using SVM_FS method, and we select only 1309 relevant features. The reduced set is split in a training set having 2555 samples and in a testing set having 1701 samples.

3.2 Clustering Data Set using the LVQ Algorithm

In order to compare the two presented methods we tried to obtain approximately the same number of vectors with both methods. With the LVQ algorithm for a value of the threshold equal to 0.15 and a learning rate equal to 0.9 we obtained after the first step 3487 groups (representative vectors). We expect this method to work better when huge data sets will be used. In the next step, we trained a classifier in order to select from these representative vectors only those vectors that are relevant. The algorithm returns 912 support vectors and we selected only those support vectors that have the Lagrange multipliers greater than a certain threshold (in our case 0.25). We considered these support vectors as being the most relevant vectors from all representative vectors. We create a dataset that contains only 4333 samples. This number represents approximately 61% from the initial data set. The obtained set is split randomly into a training set of 2347 samples and in a testing set of 1959 samples.

4 Experimental Results

We are presenting here results only for a reduced vector dimension as number of features (1309 features). We are using this dimension because with this number of features we obtained the best results [6] and because we are interested to quantify the reduced data set's influence on the classification accuracy (scalability of SVM's method). Also in our work we are using three different representations of the input data: binary, nominal and Cornell Smart [3].

In the Fig. 1 we are presenting comparative results obtained for Polynomial kernel and nominal data representation for all four sets - the original set noted as SVM-7053, the set obtained using arithmetic mean to compute the representative vector noted as AM-4256, the set obtained using the LVQ method to compute the representative vector noted as LVQ-4333 and the set randomly obtained noted as RAN-4277. The RAN-4277 set is obtained choosing randomly a specified number of samples from the original set [5].

As it can be observed there is a small difference between the results obtained for AM-4256 and LVQ-4333. The difference in the accuracy obtained between the original set and AM-4256 set is on average equal to 1.30% for all kernel degrees. The same average difference is obtained also between the original set and LVQ-4333. When we work with a small degree of the kernel the difference between the original set and AM-4256 set is smaller than 1% but the difference between the original set and LVQ-4333 is greater (1.60%). When the kernel degree increases the results are better with LVQ-4333 comparatively with AM-4256 but usually the difference can be considered insignificant. At average over all kernels' degree and all data representations the difference between the original set and AM-4256 is 1.65% and the difference between the

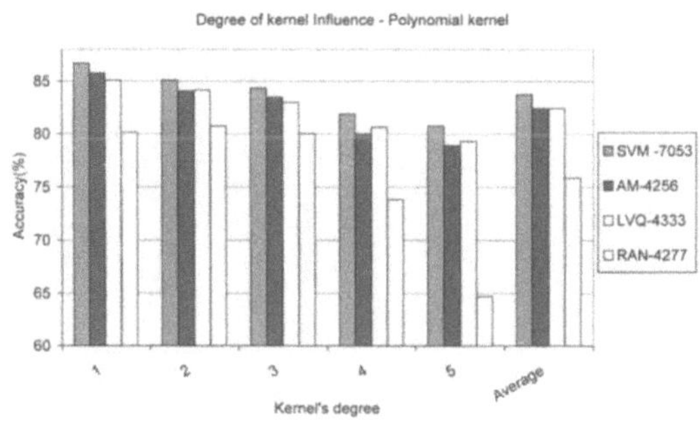

Fig. 1. Comparative results for different set dimensions - Polynomial kernel

original set and LVQ-4333 is 1.64%. We observe that for the same values of kernel's degree, which obtained the best accuracy with the original set, was also obtained the smallest difference between the original set and the reduced set. As we expected, the results obtained using randomly choused set are at average with 8% smaller than the results obtained using the entire set.

In Fig. 2 are presented the results obtained using the Gaussian kernel and Cornell Smart data representation. For this kernel the average accuracy difference between the two sets is greater than for the polynomial kernel case, being at average of 1.89% for AM-4256 and 1.95% for LVQ-4333. The difference between the original set and the randomly reduced set (RAN-4277) is also big for Gaussian kernel (being at average 7%). The smallest difference was obtained with a parameter C equal with 1.8. For this value we obtained the best results in all previous tests (using the original data set). This difference is of 1.5% for AM-4256 and of 1.75% for LVQ-4333. For this type of kernel the method based on LVQ obtains poorly results in almost all cases.

We are reducing the data in the first step at 63% and in the second step at 60% for the first method and respectively to 50% in the first step and 61% in the second step for the LVQ method. With this reduction however the lose in accuracy was about 1% for polynomial kernel and about 1.8% for Gaussian kernel. It is interesting to note that the optimal parameter values (degree or C) are usually the same for the original data set and respectively the reduced one.

Obviously, the time needed for training on a smaller number of samples decreases. For example, for polynomial kernel with degree 1 and Nominal data representation, 1031 seconds are needed to learn using all dataset and 209 seconds using the smallest set. At these 209 seconds we also have to add the time needed to select the support vectors (548 seconds) and the time

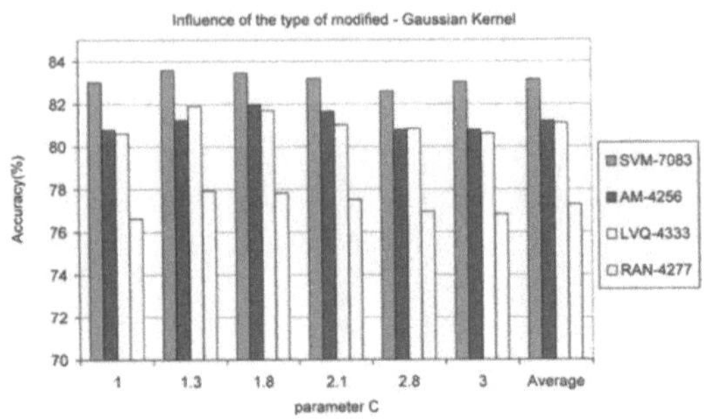

Fig. 2. Comparative results for different set dimensions - Gaussian kernel

needed for grouping data (84 seconds). The last two times occur only once for all the tests with polynomial and Gaussian kernels. The total time for polynomial kernel and degree 1 is 842 seconds. To compute these times, in both cases (with the original set and with the reduced set), we don't take into consideration the time needed for feature selection. Every time the feature selection step starts with 19038 features but in the second case we have a reduced dimension of the set (as number of vectors). Some of these times are also smaller than the first time. These values were obtained using a Pentium IV at 3.2 GHz, 1GB DRAM memory and Win XP. For the second method to obtain the representative vectors (using LVQ) the grouping part takes more time (97 seconds). The time for selecting support vector is 571 seconds. For example the time needed to compute the classification accuracy for polynomial kernel and degree 2 is 232 seconds, so the total time that can be considered is computed as:

$$t_{total\_time} = t_{group\_time} + t_{select\_SV} + t_{classify} \tag{6}$$

where t_{group_time} is the time needed for grouping data, t_{select_SV} is the time needed for classifying data and finding support vectors and $t_{classify}$ is the time needed for classifying the reduced set. This time can also include the time needed for selecting the features. This time is not included in the classifying time using original data set so it will not be included here.

In Table 1 are presented some training times for each data set. We are interested only in training time because after training the testing time depends only of the testing set dimension and the number of support vectors. For only one sample the response is less than one second.

When we'll use the entire Reuters' data set, learning with all dataset will be impossible due to the huge time and memory needed. Our obtained results

Table 1. Training time for each data set and some training characteristics

Data set	Kernel's Characteristics	t_{group_time}	t_{select_SV}	$t_{classify}$	$t_{training\_total}$
SVM-7083	POL, D2.0, BIN	-	-	-	1532.57
SVM-7083	POL, D1.0, NOM	-	-	-	1031.21
SVM-7083	POL, D1.0, CS	-	-	-	1107.64
SVM-7083	RBF, C2.8, BIN	-	-	-	4492.95
SVM-7083	RBF, C3.0, CS	-	-	-	4481.65
AM-4256	POL, D2.0, BIN	84	548.46	263.36	895.82
AM-4256	POL, D1.0, NOM	84	548.46	209.62	842.08
AM-4256	POL, D1.0, CS	84	548.46	215.56	848.02
AM-4256	RBF, C2.8, BIN	84	548.46	511.87	1144.33
AM-4256	RBF, C3.0, CS	84	548.46	513.81	1146.27
LVQ-4333	POL, D2.0, BIN	97	571.43	289.40	957.83
LVQ-4333	POL, D1.0, NOM	97	571.43	232.53	900.96
LVQ-4333	POL, D1.0, CS	97	571.43	250.67	919.10
LVQ-4333	RBF, C2.8, BIN	97	571.43	599.14	1267.57
LVQ-4333	RBF, C3.0, CS	97	571.43	618.04	1286.47

will be useful in choosing the adequate parameters for the learning step. We don't run those tests with all databases because they usually take a lot of time. For example, for the entire Reuters database we only started the first step of feature extraction and after this step we obtained 806791 vectors, each of them having 310033 dimensions (features) and a total of 103 distinct topics.

5 Conclusions and Further Work

In the real life when working with text documents we need to work with huge sets of documents to obtain good classification accuracy. In this paper we are proposing and developing a strategy to work with large text documents sets. This strategy doesn't increase exponentially the training time, actually it decreases and doesn't loose substantially in classification accuracy. A method to reduce the number of vectors from the input set and make two learning steps in order to consider the learning step finished was developed and tested. We noticed that the classification accuracy decreases at average with only 1% for polynomial kernel and about 1.8% for Gaussian kernel when the dataset is reduced at 60% of the entire dataset. If the reduction of the set was randomly made (RAN-4277) at a dimension smaller with 40% than the original set the lost in classification is at average of 8% for the polynomial kernel and of 7% for the Gaussian kernel.

A major issue that occurs in all classification and clustering algorithms is that they are reluctant to fit in the real spaces. For instance they have a problem dealing with new documents for which none of the features are in the previous feature set (the product between the new features set and the previous feature set is an empty set). As a further improvement we will try to develop tests with families of words and use as features only a representative of each family. In this way the number of features will be significantly reduced and thus we can increase the number of files that can be classified further on. In order to achieve this we could use the WordNet database, which contains a part of the families of words for the English language.

Acknowledgements

We would like to thank to SIEMENS AG, CT IC MUNCHEN, Germany, especially to Vice-President Dr. H. C. Hartmut RAFFLER, for his generous and various supports that he has provided in developing this work.

References

1. Hung, C., Wermter, S., Smith, P., Hybrid Neural Document Clustering Using Guided Self-Organization and WordNet, IEEE Computer Society, 2003
2. Kohonen, T., Self-Organizing Maps, Second edition, Springer Publishers, 1997
3. Morariu, D., Vintan, L., A Better Correlation of the SVM Kernel's Parameters, Proceedings of the 5th RoEduNet IEEE International Conference, ISBN (13) 978-973-739-277-0, Sibiu, June, 2006
4. Morariu, D., Vintan, L., Tresp, V., Feature Selection Method for an Improved SVM Classifier, Proceedings of the 3rd International Conference of Intelligent Systems (ICIS'06), ISSN 1503-5313, vol. 14, pp. 83-89, Prague, August, 2006
5. Morariu, D., Relevant Characteristics Extraction, 3rd PhD Report, University "Lucian Blaga" of Sibiu, October, 2006, http://webspace.ulbsibiu.ro/daniel.morariu /html/Docs/Report3.pdf
6. Morariu, D., Vintan, L., Tresp, V., Evaluating some Feature Selection Methods for an Improved SVM Classifier, International Journal of Intelligent Technology, Volume 1, no. 4, ISSN 1305-6417, pages 288-298, December, 2006
7. Chakrabarti, S., Mining the Web. Discovering Knowledge from Hypertext Data, Morgan Kaufmann Publishers, USA, 2003
8. Nello C., Shawe-Taylor, J., An Introduction to Support Vector Machines and other kernel-based learning methods, Cambridge University Press, 2000
9. Scholkopf Bernhard, Smola Alexander, Learning with Kernels, Support Vector Machine, MIT Press, London, 2002
10. Misha Wolf and Charles Wicksteed- Reuters Corpus: http://www.reuters.com/researchandstandards/corpus/ Released in November 2000 accessed in June 2005
11. Vapnik, V., The nature of Statistical learning Theory, Springer New York, 1995
12. Yu, H., Yang, J., Han, J., Classifying Large Data Sets Using SVM with Hierarchical Clusters. In SIGKDD03 Exploration: Newsletter on the Special Interest Group on Knowledge Discovery and Data Mining, ACM Press, Washington, DC, USA, 2003
13. http://www.cs.utexas.edu/users/mooney/ir-courses/ - Information Retrieval Java Application

On Interchange between JBoss Rules and Jess

Oana Nicolae, Adrian Giurca and Gerd Wagner

Department of Internet Technology
Institute of Informatics
Brandenburg Technical University at Cottbus, Germany
{nicolae, giurca, G.Wagner}@tu-cottbus.de

Summary. There is a growing demand for research to provide insights into challenges, and solutions based on business rules related to target PSM (Platform Specific Model in OMG's MDA terms - Implementation Model). As an answer to this needs, this paper argues on the relevance of business rules target platforms for the actual IT and business context, emphazising the important role of business rules inter-operability initiatives, so that rule-system developers can do their work without concern about a vendor-specific format and in particular without concern about the compatibility between the technologies. This paper provides a description of the business rules translation from a particular rule-system such as JBoss Rules to another rule-system representation as Jess, using R2ML as interchange language.

Key words: Business rules, JBoss Rules, Jess, Rete, ReteOO, R2ML, RIF.

1 Business rules and inter-operability initiatives

There is a growing request for business rules technology standardization from both UML modelers and ontology architects communities. For these reasons, business rules aim to express rules in a platform independent syntax.

A number of initiatives on rules inter-operability have been started. They include the RuleML [2], OMG Production Rules Representation [6] , RIF [1], and the REWERSE I1 Rule Markup Language (R2ML[1]) [8]. We mention here the efforts to establish some standards for expressing business rules and their vocabularies in natural language such as OMG's SBVR [7] and Attempto Controlled English (ACE) [4]. SBVR, this human readable format of business rules comes under OMG's Model Driven Architecture (MDA[2]) standards and is defined as Computation-Independent Model (CIM[3]). The second layer in

[1] `http://oxygen.informatik.tu-cottbus.de/rewerse-i1/?q=node/6`

[2] MDA - Model Driver Architecture is a framework for distinguishing different abstraction levels defined by the Object Management Group.

[3] CIM - Computational Independent Model is most frequently used in the context of the Model Driven Architecture (MDA) approach which corresponds the Object

Oana Nicolae et al.: *On Interchange between JBoss Rules and Jess*, Studies in Computational Intelligence (SCI) **78**, 135–144(2008)
`www.springerlink.com`

OMG's MDA is Platform-Independent Model (PIM) [4] where rule interchange formats (i.e. RuleML, RIF, R2ML) try to accomplish their general purpose: a PSM to PSM business rules migration through the PIM level. The third MDA level is Platform-Specific Model (PSM[5]) containing rule specific languages together with their specific engines/platforms like: F-Logic [5], JRules(ILOG[6]), Jess[7] or JBoss Rules[8].

Our rule interchange work address JBoss Rules as source platform and Jess as a target platform, because those languages are actually in business market interest as popular business logic frameworks, used by Java developers to create complex rule-based applications by combining Java platform and business rule technology.

The REWERSE I1 Rules Framework developed R2ML as an interchange language for deploying and sharing rules between different rule systems and tools (e.g. Object Oriented rule languages, Semantic Web rule languages, Artificial Intelligence rule languages). Actually, R2ML (now at version 0.4) is a mature and experienced enough rule interchange language to provide a concrete interchange format for different rule systems and languages (i.e. `http://oxygen.informatik.tu-cottbus.de/rewerse-i1/?q=node/15`).

R2ML has a rich syntax, so it could represent business rules from both JBoss Rules and Jess languages, providing this way the interchange possibility. As an interchange language, R2ML addresses the PIM level. The main idea is to use a model transformation language (MTL), or an application language to transform a PIM model into a PSM as in the Figure 1.

Fig. 1. Interchanging between JBoss Rules and Jess

2 Mapping JBoss Rules to R2ML

In this section we describe the JBoss business rules transformation into R2ML interchange language.

Management Group (OMG) vision of Model Driven Engineering (MDE). The Meta-Object Facility (MOF), is the OMG standard for Model Driven Engineering.

[4] PIM - Platform Independent Model

[5] PSM - Platform Specific Model

[6] `http://www.ilog.com`

[7] *Jess, Sandia Lab.*, `http://herzberg.ca.sandia.gov/jess/`

[8] *JBoss Rules*, `http://labs.jboss.com/jbossrules/docs`

Business rules are built following a business model representation. In many cases, a business model is first represented in a natural language descriptions based on core ontological concepts like classes and variables (CIM). At this stage, we can identify all objects referenced in the rules, and for each object we identify all referenced attributes. For each attribute, we identify all its constraints. Consider the following production rules examples provided in the RIF Public Mailing List [9]:

1. Rule `"Credit Score Adjustments 1"`:*If the* `creditScore` *attribute is between* `580` *and* `679` *, for* `"ACMEPowerBuyerGroup"` *as program group,* `"FIRST_TD;SECOND_TD"` *as link type and* `"Wholesale"` *division, then lower the score attribute with the* `0.3` *value.*
2. Rule `"Occupancy Adjustments 1"`: *For any investor of* `"Wholesale"` *division, grow the occupancy value with* `0.95`.

2.1 Mapping rules vocabularies

Object oriented rules as JBoss Rules and ILOG JRules are build on top of Java vocabularies. JBoss business rules vocabulary consists of Java bean classes. This vocabulary is used by the rules through the `import` declarations, which are specified inside of the rules file (`drl` files or `xml` files). For example, the before JBoss rules use `CreditScore` and `Occupancy` Java beans classes in order to describe the JBoss business rules vocabulary.

An R2ML rule always refers to a vocabulary which can be R2ML own vocabulary or an imported one (RDF(S) and OWL). R2ML vocabulary is a serialization of an UML fragment of class diagrams. Below is an excerpt from `rule "Occupancy Adjustments 1"` vocabulary:

```
<r2ml:RuleBase xmlns:r2ml="http://www.rewerse.net/I1/2006/R2ML"
xmlns:r2mlv="http://www.rewerse.net/I1/2006/R2ML/R2MLV"
xmlns:xsi="http://www.w3.org/2001/XMLSchema-instance"
xsi:schemaLocation="http://oxygen.informatik.tu-cottbus.de/R2ML/0.4/R2ML.xsd">
<r2mlv:Vocabulary>
 <r2mlv:Class r2mlv:ID="Occupancy">
  <r2mlv:Attribute r2mlv:ID="division">
   <r2mlv:range><r2mlv:Datatype r2mlv:ID="xs:string"/></r2mlv:range>
  </r2mlv:Attribute>
 <r2mlv:Attribute r2mlv:ID="occupancy">
  <r2mlv:range><r2mlv:Datatype r2mlv:ID="xs:string"/></r2mlv:range>
 </r2mlv:Attribute>
  <!-- ... -->
 </r2mlv:Class>
 </r2mlv:Vocabulary>
</r2mlv:RuleBase>
```

Any Java qualified class names will be translated into (`xs:QName`) together with the corresponding names declarations. For example, if we assume the JBoss declaration: `org.drools.examples2.Occupancy`, this is translated into the namespace declaration `xmlns:ex="http://drools.org/examples2"` and the qualified name: `ex:Occupancy` for the reference to the class name.

9 `http://lists.w3.org/Archives/Public/public-rif-wg/2007Mar/0083.html`

2.2 Rule mapping

Consider the rule "`Credit Score Adjustments 1`":

```
package examplesRuleSet;
import example1.CreditScore;
rule "Credit Score Adjustments 1"
 date-effective "25-OCT-2001 17:26:14"
 when
  cs : CreditScore(programGroup == "ACMEPowerBuyerGroup",
                   lienType == "FIRST_TD; SECOND_TD",
                   division == "Wholesale", creditScore >= 580 & <= 679 )
 then
  cs(score = cs.score - 0.3);
 end
```

The mapping principles to translate a JBoss rule into an R2ML production rule are:

- A R2ML `r2ml:ruleID` is generated using the JBoss rule ID value. For example `rule "Credit Score Adjustments 1"`, generates the value `"Credit Score Adjustements 1"` for the `r2ml:ruleID` attribute.
- A JBoss rule has a conditions part (the `when` part) and an action part (the `then` part). The condition part of a JBoss rule is mapped into the content of `r2ml:conditions` role element. The RHS part of a JBoss rule containing multiple actions, maps into the content of `r2ml:producedAction` role element.
- Relational operations from JBoss Rules (e.g. `>=`) translates into `r2ml:DatatypePredicateAtom` using SWRL builtins as predicate names (e.g.`"swrlb:greaterThanOrEqual"`).
- In many cases JBoss `column` with field constraints translates into `r2ml:ObjectDescriptionAtom` or into a conjunction of atoms. The attribute `r2ml:classID` corresponds to the column name from JBoss implementation.

  ```
  cs : CreditSCore( .....)
  ```

  translates into:

  ```
  <r2ml:ObjectDescriptionAtom r2ml:classID="CreditScore">
   <r2ml:subject>
     <r2ml:ObjectVariable r2ml:name="cs"
                          r2ml:classID="CreditScore"/>
   </r2ml:subject>
   <!-- ... -->
  </r2ml:ObjectDescriptionAtom>
  ```

- Any JBoss variables (i.e. *declarations* in the JBoss terminology) translates into R2ML variables. The JBoss *fact variable* (e.g. `cs:CreditScore`) is mapped into `r2ml:ObjectVariable` using the value of `r2ml:name` attribute to represent the variable name (i.e. `cs`). The optional `r2ml:classID` attribute specifies the type of the object variable (i.e. `CreditScore`). An `r2ml:ObjectVariable` is a variable that can be only instantiated by objects.
- The rule field constraints i.e `programGroup=="ACMEPowerBuyerGroup"` translates into `r2ml:DataSlot/r2ml:ObjectSlot`, depending on the type

of the involved property. Since **programGroup** is a string value, it translates into **r2ml:DataSlot**, having the name of the attribute as **r2ml:attributeID="programGroup"** and its value translates into **r2ml:TypedLiteral**.

- A JBoss column may contain many field constraints all of them referring to the same context variable. In the example below there is a context variable cs referring the column **CreditScore** together with its field constraints.

```
cs : CreditScore(programGroup == "ACMEPowerBuyerGroup",
                 lienType == "FIRST_TD; SECOND_TD",
                 division == "Wholesale", creditScore>= 580 & <= 679 )
```

- The comma is the implicit connective between constraints and corresponds to the logical **and**. In the above example the field constraints are equalities such as `division == "Wholesale"` or a kind of relational expressions (*literal numeric constraints* in JBoss terminology) such as `creditScore >= 580 & <= 679` which corresponds to the relational `creditScore >= 580 & creditScore <= 679`.
The above **CreditScore** column may be written as a conjunction of the following columns:

```
cs : CreditScore(programGroup == "ACMEPowerBuyerGroup")
cs : CreditScore(lienType == "FIRST_TD; SECOND_TD")
cs : CreditScore(division == "Wholesale")
cs : CreditScore(creditScore >= 580)
cs : CreditScore(creditScore <= 679)
```

Below is depicted the corresponding R2ML serialization of a literal numeric constraint from the rule condition:

```
<!-- cs:CreditScore(programGroup == "ACMEPowerBuyerGroup")-->
 <r2ml:ObjectDescriptionAtom r2ml:classID="CreditScore">
  <r2ml:subject>
    <r2ml:ObjectVariable r2ml:name="cs"/>
  </r2ml:subject>
 <r2ml:DataSlot r2ml:attributeID="CreditScore.programGroup">
    <r2ml:value>
     <r2ml:TypedLiteral r2ml:lexicalValue="ACMEPowerBuyerGroup"
                        r2ml:datatypeID="xs:string"/>
    </r2ml:value>
  </r2ml:DataSlot>
   <!-- ... -->
 </r2ml:ObjectDescriptionAtom>
```

The R2ML markup of equality constraints is an enumeration of `r2ml:DataSlot`(s) inside an `r2ml:ObjectDescriptionAtom`, while the other relational constraints maps into `r2ml:DatatypePredicateAtom`(s) as in example below:

```
<!-- cs:CreditScore(creditScore >= 580) -->
<r2ml:DatatypePredicateAtom r2ml:datatypePredicateID="swrlb:greaterThanOrEqual">
 <r2ml:dataArguments>
  <r2ml:AttributeFunctionTerm r2ml:attributeID="CreditScore.creditScore">
   <r2ml:contextArgument>
    <r2ml:ObjectVariable r2ml:name="cs" r2ml:classID=CreditScore/>
   </r2ml:contextArgument>
  </r2ml:AttributeFunctionTerm>
  <r2ml:TypedLiteral r2ml:lexicalValue="580" r2ml:datatypeID="xs:decimal"/>
 </r2ml:dataArguments>
</r2ml:DatatypePredicateAtom>
```

- R2ML actions are built according with the OMG PRR Specification [6].
  1. InvokeActionExpression - invokes an operation with a list of parameter arguments and changes the state of the context object.
  2. AssignActionExpression - assigns a value to the attribute of the object context.
  3. CreateActionExpression - creates an object from the list of property-value pairs.
  4. DeleteActionExpression - deletes a specific object.

  The JBoss action part of the rule is written in MVEL [10]: `cs(score=cs.score - 0.3)` which corresponds to the following JBoss standard syntax:`cs.setScore(score - 0.3); modify(cs);` and translates into a `r2ml:AssignActionExpression` as below:

```
<r2ml:producedAction>
 <r2ml:AssignActionExpression r2ml:propertyID="score">
  <r2ml:contextArgument>
   <r2ml:ObjectVariable r2ml:name="cs" r2ml:classID="CreditScore"/>
  </r2ml:contextArgument>
  <r2ml:DatatypeFunctionTerm r2ml:datatypeFunctionID="op:numeric-substract">
   <r2ml:dataArguments>
    <r2ml:AttributeFunctionTerm r2ml:attributeID="CreditScore.score">
     <r2ml:contextArgument>
      <r2ml:ObjectVariable r2ml:name="cs" r2ml:classID="CreditScore"/>
     </r2ml:contextArgument>
    </r2ml:AttributeFunctionTerm>
    <r2ml:TypedLiteral r2ml:lexicalValue="0.3"
                       r2ml:datatypeID="xs:decimal"/>
   </r2ml:dataArguments>
  </r2ml:DatatypeFunctionTerm>
 </r2ml:AssignActionExpression>
</r2ml:producedAction>
```

The second rule `"Occupancy Adjustments 1"` translates in a similar manner.

3 Mapping R2ML production rules to Jess language

In this section we describe how R2ML rules are translated into Jess rule language. In Jess, rules are defined using the `defrule` construct. They are in fact written as lists where the head is the special symbol `defrule`.

Jess provides two main categories of rules: forward-chaining rules and backward-chaining rules. Forward-chaining rules are the most common and used rules in Jess, and our translation will obtain Jess forward-rules. Jess is a forward-chaining reasoning engine, backward-chaining rules being simulated in terms of forward chaining [3].

To translate a R2ML rule into a Jess rule we will use Jess `unordered facts`. Any R2ML rule will translate into a Jess rule which use unordered facts, because they are better emulating the internal structure of a R2ML rule (which includes the rule vocabulary).

[10] `http://wiki.mvflex.org/`

3.1 Mapping rules vocabulary

Jess business rules vocabulary consists of **deftemplate**(s) structures.
A **deftemplate** describes a fact, in the same way as a Java class describes an object. In particular, a **deftemplate** is a Jess concept which includes a name, an optional documentation string, an optional "extends" clause, an optional list of declarations, and a list of zero or more member variables (called slot descriptions) with a type qualifier. Each slot description can optionally include a type qualifier or a default value qualifier.

Using vocabulary classes from R2ML[11], corresponding Jess **deftemplates** are generated. The deftemplate[12] structure corresponds to the Occupancy class description from the R2ML vocabulary (i.e. SubSection 2.1) is depicted below i.e.:

```
(deftemplate occupancy
(slot division (type INTEGER))
(slot occupancy (type STRING))
...)
```

They can be placed in the same file as the rules, or in a separate file, which need to be imported into the rules file, using the **import** keyword.

3.2 Rule mapping

Consider the following Jess rule "**Credit Score Adjustements 1**":

```
(defrule CreditScoreAdjustments1
 ?cs <- (CreditScore{programGroup == "ACMEPowerBuyerGroup" &&
         lienType == "FIRST_TD; SECOND_TD" &&
         division == "Wholesale" &&
         (creditScore >= 580 && creditScore <= 679)}(score ?oldScore))
=>
 (modify ?cs (score (- ?oldScore 0.3))) )
```

- The **r2ml:ruleID** attribute value is used to obtain a Jess rule ID. If **r2ml:ruleID=Credit Score Adjustments 1** then, the **defrule CreditScoreAdjustments1** is generated for the corresponding Jess rule.

- An **r2ml:ObjectDescriptionAtom** will translate into a Jess pattern structure, which contains all R2ML's Object/Data Slots translated into Jess constraints i.e.:

```
?cs <- (CreditSCore{programGroup == "ACMEPowerBuyerGroup" &&
        lienType == "FIRST_TD; SECOND_TD" && division == "Wholesale"})
```

- All R2ML variables are mapped into Jess variables: **fact variables** and **variables**. The **r2ml:ObjectVariable** is mapped into the Jess **fact variable** using the value of the **r2ml:name** attribute as the variable name. The **r2ml:classID** attribute specifies the type of the object variable.(e.g. **?cs <- (CreditScore{})**)

---

[11] xmlns:r2mlv="http://www.rewerse.net/I1/2006/R2ML/R2MLV"

[12] herzberg.ca.sandia.gov/docs/71/api/jess/Deftemplate.html

- Each `r2ml:DataSlot` from our R2ML rule representation will translate into appropriate Jess rule constraint. Internally, all Jess values (symbols, numbers, strings etc) are represented by instances of the class `jess.Value`. Its possible values are enumerated by a set of constants in the `jess.RU` class.

```
<r2ml:DataSlot r2ml:attributeID="CreditScore.programGroup">
 <r2ml:value>
  <r2ml:TypedLiteral r2ml:lexicalValue="ACMEPowerBuyerGroup"
                     r2ml:datatypeID="xs:string"/>
 </r2ml:value>
</r2ml:DataSlot>
```

  will translate into:

```
programGroup == "ACMEPowerBuyerGroup"
```

- Our R2ML rule contains an enumeration of `r2ml:DataSlots` inside the `r2ml:ObjectDescriptionAtom`. They will translate into a conjunction of field constraints inside a Jess LHS's pattern, all having the same context: the Jess fact variable `cs`. A Jess conjunction of field constraints is represented using the `&&` operator i.e.:

```
?cs<-(CreditSCore{programGroup == "ACMEPowerBuyerGroup" &&
      lienType == "FIRST_TD; SECOND_TD" && division == "Wholesale"})
```

- All R2ML built-in predicates and operators translate into corresponding Jess constraints. The `r2ml:DatatypePredicateAtoms` having the attributes: r2ml:datatypePredicateID="swrlb:greaterThanOrEqual" and respectively r2ml:datatypePredicateID="swrlb:lessThanOrEqual" translate into the following conjunction of Jess patterns i.e.:

```
?cs <- (CreditScore{creditScore >= 580 && creditScore <= 679})
```

- The possibles actions of a R2ML production rule are defined using OMG's PRR Proposal (the content of `r2ml:producedAction` role element). AssignActionExpression from our R2ML rule example correspondes to a modify() function invocation from Jess language, which updates an unordered fact from working memory. Notice that `?oldScore` is a bound variable to the `score` field value i.e.:

```
(modify ?cs (score (- ?oldScore 0.3)))
```

The second R2ML rule to detail its Jess implementation is `rule "Occupancy Adjustments 1"`:

```
(defrule OccupancyAdjustments1
?o <- (Occupancy{ division== "Wholesale" && occupancy == "Investor" }
                (value ?oldValue))
=>
(modify ?o( value (+ ?oldValue 0.95))) )
```

The conditions part of the R2ML rule `Occupancy Adjustments 1` (the content of `r2ml:conditions` role element) is an `r2ml:ObjectDescriptionAtom` and translates into the LHS pattern with field constraints (the `Occupancy` pattern with appropriate field constraints).

The action part translates into Jess as a function call i.e.:
`(modify ?o( value (+ ?oldValue 0.95)))`

An Excerpt of mapping from JBossRules to Jess		
JBoss Rules	R2ML	Jess

`rule "ruleName"`
`when`
`//conditions`
`then`
`//actions`
`end`
	`<r2ml:ProductionRule`
	`r2ml:ruleID="ruleName">`
	`<r2ml:conditions>`
	`...`
	`</r2ml:conditions>`
	`<r2ml:producedAction>`
	`...`
	`</r2ml:producedAction>`
	`</r2ml:ProductionRule>`
		`(defrule ruleName`
		`patterns`
		`=>`
		`functions calls)`

`import JavaBeans files`	`declare the vocabulary,`
	`appropriate namespaces and`
	`qualified names (xs:QName)`
		`import deftemplate(s)`
		`structures files`

`fact variable`
`(optionally $ symbol)`
`$var:ColumnName()`
	`<r2ml:ObjectVariable`
	`r2ml:name="var"`
	`r2ml:classID="ColumnName"/>`
		`fact variable`
		`?var<-(ColumnName{})`

`field variable`
`(optionally $ symbol)`
`ColumnName($var:prop)`
	`<r2ml:DataVariable`
	`r2ml:name="var"/>`
	`<r2ml:ObjectVariable`
	`r2ml:name="var"/>`
		`field variable (?var)`

`column without` `field constraints`	`r2ml:ObjectClassificationAtom`	`LHS's pattern`
`column with` `field constraints`	`r2ml:ObjectDescriptionAtom`	`LHS's pattern with` `field constraints`
`data constraints` `inside a column`	`r2ml:DataSlot`	`pattern constraints`
`object constraints` `inside a column`	`r2ml:ObjectSlot`	`pattern constraints`
`setter call`	`r2ml:AssignAcionExpression`	`modify()` `function invocation`

4 Limitations of the translation

The problems regarding the translation process refer the PSM systems: JBoss-Rules and Jess.

The translation process from PSM to PSM using intermediate R2ML (PIM) level demands also the mapping of rules vocabulary. The JBossRules to R2ML translator is a Java application that requests access to the JBoss-Rules vocabulary (Java compiled classes) in order to establish the types of objects and primitives. As R2ML supports Production Rules format[6], the JBossRules to R2ML translation of rule conditions part relies naturally.

The problems appear in the translation of the actions part of a JBoss rule, which may contain any Java valid code: variable declaration, non-declarative structures like `if...then` structures or cycling structures(i.e. while, for).

R2ML intends to solve this problem by providing in its future version an `<r2ml:OpaqueExpression>` with the role to encapsulate the code which don't find its semantic equivalent into R2ML `<r2ml:producedAction>` role element.

The purpose of R2ML, as PIM level markup language is to provide PSM to PSM rules translation, by sharing a vocabulary model (actually R2ML vocabulary), which can easily be mapped into Jess vocabulary (deftemplate(s) structures). Therefore, it makes possible the translation from R2ML to Jess language, as the business rules expressed in R2ML format don't require any conceptual changes in order to be implemented in PSM target platforms (e.g. Jess, F-Logic, RuleML, OCL, SWRL, JBossRules, Jena).

The reverse translation, from Jess to JBoss Rules is also possible, as Jess supports a new richer syntax[13] which offers the capability to represent object types using *deftemplate* structures. One limitation of this syntax is that it can only be used with unordered facts.

5 Conclusion and future work

The paper provides a description of rule translation from JBoss Rules, an Object Oriented rule language, into Jess, an Artificial Intelligence rule language using R2ML as interchange format.

Our future work intends to finalize the implementation of all transformation proposed in the paper, which will be followed by the detailed report on the experience.

References

1. H. Boley, M. Kifer (2007). *RIF Core Design*, W3C Working Draft, March 30, 2007 `http://www.w3.org/TR/rif-core/`
2. H. Boley, S. Tabet and G. Wagner (2001) *Design Rationale of RuleML: A Markup Language for Semantic Web Rules*, In Proc. of Int. Semantic Web Working Symposium (SWWS), July 30 - August 1, 2001, Stanford University, California, USA. `http://www.ruleml.org`
3. E. Friedman-Hill (2003). *Jess in Action - Rule-Based Systems in Java*, Manning Publications Co., 2003.
4. N. E. Fuchs, U. Schwertel, R. Schwitter (1997). *Attempto Englisch als (formale) Spezifikationssprache*, In: F. Bry, B. Freitag, D. Seipel (eds.), Proceedings of the Twelfth Workshop on Logic Programming WLP '97, Munich, September 1997.
5. M. Kifer, G. Lausen and J. Wu, Logical Foundations of Object-Oriented and Frame-Based Languages, *Journal of ACM*, May 1995 `ftp://ftp.cs.sunysb.edu/pub/TechReports/kifer/flogic.pdf`
6. OMG (2007). *Production Rule Representation Ver. 1.0* , March 5, 2007 `http://www.omg.org/docs/bmi/07-03-05.pdf`
7. OMG (2006). *Semantics of Business Vocabulary and Business Rules (SBVR)* `http://www.omg.org/docs/dtc/06-03-02.pdf`
8. G. Wagner, A. Giurca and S. Lukichev (2005). *R2ML: A General Approach for Marking up Rules*, Dagstuhl Seminar Proceedings 05371, In F. Bry, F. Fages, M. Marchiori, H. Ohlbach (Eds.) Principles and Practices of Semantic Web Reasoning, ISSN:1862-4405,`http://drops.dagstuhl.de/opus/volltexte/2006/479/pdf/05371.GiurcaAdrian.Paper.479.pdf`

[13] http://herzberg.ca.sandia.gov/docs/71/memory.html

A Cross-Entropy Approach to Solving Dec-POMDPs

Frans A. Oliehoek[1], Julian F.P. Kooij[1], and Nikos Vlassis[2]

[1] Intelligent Systems Lab, University of Amsterdam, Kruislaan 403, 1098 SJ Amsterdam, The Netherlands ({faolieho,jkooij}@science.uva.nl)
[2] Dept. of Production Engineering and Management, Technical University of Crete, Kounoupidiana 73100 Chania, Greece (vlassis@dpem.tuc.gr)

1 Introduction

In this paper we focus on distributed multiagent planning under uncertainty. For single-agent planning under uncertainty, the partially observable Markov decision process (POMDP) is the dominant model (see [Spaan and Vlassis, 2005] and references therein). Recently, several generalizations of the POMDP to multiagent settings have been proposed. Here we focus on the *decentralized POMDP (Dec-POMDP)* model for multiagent planning under uncertainty [Bernstein et al., 2002, Goldman and Zilberstein, 2004]. Solving a Dec-POMDP amounts to finding a set of optimal policies for the agents that maximize the expected shared reward. However, solving a Dec-POMDP has proven to be hard (NEXP-complete): The number of possible deterministic policies for a single agent grows doubly exponentially with the planning horizon, and exponentially with the number of actions and observations available. As a result, the focus has shifted to approximate solution techniques [Nair et al., 2003, Emery-Montemerlo et al., 2005, Oliehoek and Vlassis, 2007].

In this paper we propose a novel approach for approximately solving Dec-POMDPs. We apply the *Cross-Entropy (CE) method* [de Boer et al., 2005], a sampling-based method for solving combinatorial problems, to policy search in Dec-POMDPs. [Mannor, Rubinstein, and Gat, 2003] applied CE for policy search in Markov Decision Processes (MDPs). This work proposes solutions for the problems encountered when going to a multiagent setting with state uncertainty, thereby extending work by [Mannor et al., 2003]. We show experimental results from a toy problem and a standard benchmark, from which encouraging conclusions can be drawn.

Frans A. Oliehoek et al.: *A Cross-Entropy Approach to Solving Dec-POMDPs*, Studies in Computational Intelligence (SCI) **78**, 145–154 (2008)
www.springerlink.com

2 The DEC-POMDP Model

The *decentralized partially observable Markov decision process (Dec-POMDP)* describes a stochastic, partially observable environment for a set of cooperating agents. A Dec-POMDP for m agents can formally be defined as a tuple $\langle \mathcal{S}, \mathcal{A}, T, R, \mathcal{O}, O \rangle$ where:

- $\mathcal{S}$ is a finite set of states.
- The set $\mathcal{A} = \times_i \mathcal{A}_i$ is the set of *joint actions*, where $\mathcal{A}_i$ is the set of actions available to agent i. Every time step one joint action $\mathbf{a} = \langle a_1, ..., a_m \rangle$ is taken.
- T is the transition function, a mapping from states and joint actions to probability distributions over next states: $T : \mathcal{S} \times \mathcal{A} \to \mathcal{P}(\mathcal{S})$.
- R is the reward function, a mapping from states and joint actions to real numbers: $R : \mathcal{S} \times \mathcal{A} \to \mathbb{R}$.
- $\mathcal{O} = \times_i \mathcal{O}_i$ is the set of *joint observations*, with $\mathcal{O}_i$ the set of observations available to agent i. Every time step one joint observation $\mathbf{o} = \langle o_1, ..., o_m \rangle$ is received. We will denote $\mathbf{o}^t$ the joint observation at time t.
- O is the observation function, a mapping from joint actions and successor states to probability distributions over joint observations: $O : \mathcal{A} \times \mathcal{S} \to \mathcal{P}(\mathcal{O})$.

In this paper we assume a finite *planning* horizon of h time steps, and an initial 'belief' $b^0 \in \mathcal{P}(\mathcal{S})$; this is the initial state distribution at $t = 0$. In a Dec-POMDP, an agent only knows its own actions a_i and observations o_i. The *action-observation history for agent i*, $\vec{\theta}_i^t = \left(a_i^0, o_i^1, a_i^1, ..., a_i^{t-1}, o_i^t\right)$, is the sequence of actions taken and observations received by agent i until time step t. The *joint action-observation history* is a tuple with the action-observation history for all agents $\vec{\boldsymbol{\theta}}^t = \langle \vec{\theta}_1^t, ..., \vec{\theta}_m^t \rangle$. The set of all action-observation histories for agent i at time t is denoted $\vec{\Theta}_i$. The *observation history for agent i*, $\vec{o}_i^t = \left(o_i^1, ..., o_i^t\right)$, is the sequence of observations an agent has received. $\vec{\mathbf{o}}^t$ denotes a joint observation history and $\vec{\mathcal{O}}_i$ denotes the set of all observation histories for agent i.

A *pure* or *deterministic policy* π_i for agent i in a Dec-POMDP is a mapping from observation histories to actions, $\pi_i : \vec{\mathcal{O}}_i \to \mathcal{A}_i$. A pure joint policy π is a tuple containing a pure policy for each agent. Solving a Dec-POMDP amounts to finding π^*, the joint policy that yields the highest expected cumulative reward: $\pi^* = \arg\max_\pi E_\pi\left(\sum_{t=0}^{h-1} R(t)\right)$. Bernstein et al. [2002] have shown that optimally solving a Dec-POMDP is NEXP-complete, implying that any optimal algorithm will most likely be doubly exponential in the horizon.

The naive way of going about is to enumerate all joint policies and evaluate their expected cumulative reward, or *value*. The value of a specific (state, joint observation history) pair under a joint policy π is given by:

$$V_\pi(s^t, \vec{\mathbf{o}}^t) = R(s^t, \pi(\vec{\mathbf{o}}^t)) + \sum_{s^{t+1}} P(s^{t+1}|s^t, \pi(\vec{\mathbf{o}}^t)) \sum_{\mathbf{o}^{t+1}} P(\mathbf{o}^{(t+1)}|s^{t+1}, \pi(\vec{\mathbf{o}}^t)) V_\pi(s^{t+1}, \vec{\mathbf{o}}^{t+1}). \quad (1)$$

The total expected reward $V(\pi)$, with respect to the initial state distribution b^0 is then given by $V(\pi) = \sum_s V_\pi(s, \vec{\mathbf{o}}^0) b^0(s)$, where $\vec{\mathbf{o}}^0$ is the initial (empty) joint observation history. For one joint policy this calculation requires evaluation of (1) for each of the $\sum_{t=0}^{h-1} |\mathcal{O}|^t = \frac{|\mathcal{O}|^h - 1}{|\mathcal{O}| - 1}$ observation histories and $|S|$ states, leading to a total cost of $O(|S| \cdot \frac{|\mathcal{O}|^h - 1}{|\mathcal{O}| - 1})$.

3 Cross-Entropy Optimization

De Boer, Kroese, Mannor, and Rubinstein [2005] described the *Cross-Entropy (CE)* method as a general framework to both rare event estimation and combinatorial optimization. We will focus here only on the application to optimization. The cross entropy method can be used for optimization in cases where we want to find a (typically large) vector x from a hypothesis space $\mathcal{X}$, that maximizes some *performance function* $V : \mathcal{X} \to \mathbb{R}$. That is, we are looking for $x^* = \arg\max_{x \in \mathcal{X}} V(x)$. The CE method maintains a probability density function f_ξ over the hypothesis space, parametrized by a vector ξ. The core of the CE method for optimization is an iterative two-phase process:

1. Generate a set of samples $\mathbf{X}$ according to f_ξ.
2. For some $0 \le \rho \le 1$, select the best ρ-fraction of samples $\mathbf{X}_\rho$, and use those to update the parameter vector ξ.

The first step is rather trivial. The second step, however, deserves some more explanation. Let $\gamma^{(j)}$ be defined as the minimum performance within the best ρ-fraction of samples of the j-th iteration. I.e., $\gamma^{(j)} \equiv \min_{\mathbf{x} \in \mathbf{X}_\rho} V(\mathbf{x})$. The CE method requires that this lower bound performance is not allowed to decrease over time: $\gamma^{(j+1)} \ge \gamma^{(j)}$. This implies that $\mathbf{X}_\rho$ can contain less than a ρ-fraction of all samples $\mathbf{X}$. The set $\mathbf{X}_\rho$ is then used to create $\xi^{(j+1)}$, a maximum-likelihood estimate of the parameters. These new parameters can be smoothed using a parameter $0 \le \alpha \le 1$ by interpolating with $\xi^{(j)}$ the parameter vector of the previous iteration: $\xi^{(j+1)} = \alpha\xi^{(j+1)} + (1 - \alpha)\xi^{(j)}$. This reduces the probability that some components of the parameter vector will be 0 or 1 early in the CE process, which could cause the method to get stuck in local optima. Usually, the iterative process is stopped when $\gamma^{(j)}$ has not improved over some predefined number of steps. But other conditions could be used such as a time limit or a fixed number of iterations. When the stop condition is finally met, the best sample $\mathbf{x}$ found in the entire process is returned as an approximation of x^*.

Mannor et al. [2003] showed how the CE method can be applied to MDPs for which the optimal value function is stationary, that is, the expected value of taking a particular action in a particular state does not depend on the time step. The optimal policy for such a MDP is a mapping from states to actions $\pi_{\text{MDP}} : \mathcal{S} \to \mathcal{A}$, which can be represented as an $|\mathcal{S}|$-vector. As above, we want to find the vector that maximizes a performance function, in this case the expected total reward. So we are looking for $\pi^*_{\text{MDP}} = \arg\max_{\pi_{\text{MDP}}} V(\pi_{\text{MDP}})$, where the performance function now is the value of the MDP-policy π_{MDP}. This problem is tackled by maintaining a parameter vector $\xi = \langle \xi_{s_0}, ..., \xi_{s_{|S|}} \rangle$, where each ξ_s is a probability distribution over actions. Using these probabilities it is possible to sample N trajectories: starting from some state actions are randomly selected according to the probabilities as described by ξ until the goal state is reached. Now using the ρ-fraction of best (highest total reward) trajectories $\mathbf{X}_\rho$, the parameter vector can be updated as follows:

$$P(a|s) = \frac{\sum_{\mathbf{x}\in\mathbf{X}_\rho} I(\mathbf{x}, s, a)}{\sum_{\mathbf{x}\in\mathbf{X}_\rho} I(\mathbf{x}, s)}, \tag{2}$$

where $I(\mathbf{x}, s, a)$ is an indicator function that indicates that action a was performed at state s in trajectory $\mathbf{x}$, and $I(\mathbf{x}, s)$ indicates whether s was visited in trajectory $\mathbf{x}$. After updating the parameter vector ξ, a new set $\mathbf{X}$ of trajectories can be sampled, etc. Empirical evaluation shows that this process converges to near-optimal policies in only a few iterations.

4 CE for DEC-POMDPs

In this section we propose an adaptation of the CE method for Dec-POMDP policy optimization. Overall, the approach we describe here follows the algorithm described in the previous section. Unfortunately we cannot apply the above approach directly to Dec-POMDPs. The reason is that, since we consider finite-horizon Dec-POMDPs, there is no stationary value function. Moreover, the policies of the agents are not defined over states but over their *individual* observation histories $\vec{o}_i^{\,t}$. In the Dec-POMDP case, the hypothesis space is the space of joint policies; we need to define a parametrized distribution over this space and an evaluation function for sampled policies. Also we need to show how the parameters can be updated.

4.1 POLICY DISTRIBUTIONS

In the case of Dec-POMDPs, f_ξ denotes a probability distribution over pure joint policies. We will represent this probability as the product of probability distributions over individual pure joint policies: $f_\xi(\pi) = \prod_{i=1}^m f_{\xi_i}(\pi_i)$. Here ξ_i is the vector of parameters for agent i, i.e., $\xi = \langle \xi_1, ..., \xi_m \rangle$. The question is how to represent the probability distributions over individual pure policies. One option is to use a mixed policy [Osborne and Rubinstein, 1994] representation: a distribution over all agent i's pure policies. However, this approach suffers from two drawbacks: the number of pure individual policies π_i might be huge and this representation is hard to parametrize in a meaningful way using some vector ξ_i. That is, it gives no access to the internals of the policies: parameters would specify probabilities for entire pure policies, rather than specifying behavior for particular observation histories. Rather then using a mixed policy representation, we will use a behavioral- [Osborne and Rubinstein, 1994] or stochastic policy [Koller and Pfeffer, 1997] description: a mapping from decision points to probability distributions over actions. We consider two such representations: observation- and action-observation history based.

OBSERVATION HISTORY BASED: The decision points in an MDP are the states. In a Dec-POMDP the decision points for an agent are its observation histories. Therefore, the simplest way to represent a policy distribution is as a probability distribution over actions for each observation history. In particular, for each $\vec{o}_i^t$ we maintain a $\xi_{\vec{o}_i^t}$, that specifies the distribution $\forall_{a_i} \quad \xi_{\vec{o}_i^t}(a_i) \equiv P(a_i|\vec{o}_i^t)$. Consequently, ξ_i is defined as $\xi_i \equiv \langle \xi_{\vec{o}_i^t} \rangle_{\vec{o}_i^t \in \vec{\mathcal{O}}_i}$, and the probability of a policy π_i for agent i as $f_{\xi_i}(\pi_i) = \prod_{\vec{o}_i^t \in \vec{\mathcal{O}}_i} \xi_{\vec{o}_i^t}(\pi_i(\vec{o}_i^t))$. We refer to this policy distribution representation as the OH-based representation.

ACTION-OBSERVATION HISTORY BASED: Defining the parameters as above might be the most straightforward approach, because it is very closely related to the approach for MDPs. Nevertheless, this representation fails to take into account the action history: The choice for action $\pi_i(\vec{o}_i^t)$ has no influence on the choice for the action at the next time step $\pi_i(\vec{o}_i^{t+1})$. To overcome this problem, we propose to make the probability of actions conditional on the entire action-observation history $\vec{\theta}_i^t$. So now the parameter vector for agent i is defined as $\xi_i \equiv \langle \xi_{\vec{\theta}_i^t} \rangle_{\vec{\theta}_i^t \in \vec{\Theta}_i}$ and $\xi_{\vec{\theta}_i^t}$ is a probability distribution over actions: $\forall_{a_i} \quad \xi_{\vec{\theta}_i^t}(a_i) \equiv P(a_i|\vec{\theta}_i^t)$. An action-observation history consists of an action- and an observation history $\vec{\theta}_i^t = (\vec{a}_i^t, \vec{o}_i^t)$. Therefore, in this new representation, which we refer to as AOH-based, the probability of agent i's pure policy π_i becomes $f_{\xi_i}(\pi_i) = \prod_{\vec{o}_i^t \in \vec{\mathcal{O}}_i} \xi_{\vec{\theta}_i^t = (\vec{o}_i^t, \vec{a}_{\pi_i}^t)}(\pi_i(\vec{o}_i^t))$, where $\vec{a}_{\pi_i}^t$ is the action history as specified by π_i for observation history $\vec{o}_i^t$, i.e., $\vec{a}_{\pi_i}^t = \langle \pi_i(\vec{o}_i^0), \pi_i(\vec{o}_i^1), \cdots, \pi_i(\vec{o}_i^{t-1}) \rangle$. Put differently, for each $\vec{o}_i^t$, the parameter $\xi_{\vec{\theta}_i^t}$ used is that of the action-observation history $\vec{\theta}_i^t$ that is *consistent* with $\vec{o}_i^t$ and π_i: it specifies the observations from $\vec{o}_i^t$ and the actions that π_i specifies for those observations. Drawing a sample from this distribution is performed in a root-to-leaf fashion: first an action a_i^0 is sampled for the empty action-observation history $\vec{\theta}_i^0$ according to $\xi_{\vec{\theta}_i^0}$, then, for all possible action-observation histories $\vec{\theta}_i^1 = (\vec{a}_i^1, \vec{o}_i^1)$ at time $t = 1$ *that are consistent with the actions sampled so far*, new actions are sampled according to $\xi_{\vec{\theta}_i^1}$, etc.

4.2 PARAMETER ESTIMATION

The previous section described two ways to represent the probability distribution over policies. This section describes how the set of best policies $\mathbf{X}_\rho$ sampled from the previous distribution $f_{\xi^{(j)}}$, can be used to find new parameters $\xi^{(j+1)}$.

OH-BASED DISTRIBUTION: Let $I(\pi_i, \vec{o}_i^{\,t}, a)$ be an indicator function that indicates whether $\pi_i(\vec{o}_i^{\,t}) = a$. In the OH-based distribution the probability of agent i taking action $a^t \in \mathcal{A}_i$ after having observed $\vec{o}_i^{\,t}$ can be updated using: $\xi_{\vec{o}_i^{\,t}}^{(j+1)}(a^t) = \frac{1}{|\mathbf{X}_\rho|}\sum_{\pi\in\mathbf{X}_\rho} I(\pi_i, \vec{o}_i^{\,t}, a^t)$, where $|\mathbf{X}_\rho|$ normalizes the distribution.

AOH-BASED DISTRIBUTION: Estimating the parameters for the AOH-based distribution is more involved. The indicator function $I(\pi_i, \vec{\theta}_i^{\,t}, a_i^t)$ indicates now whether (i) $\pi_i(\vec{o}_i^{\,t}) = a_i^t$, and (ii) the action-observation history $\vec{\theta}_i^{\,t}$ is consistent with the policy, that is, whether $\vec{\theta}_i^{\,t} = (\vec{o}_i^{\,t}, \vec{a}_{\pi_i}^{\,t})$ and $\vec{a}_{\pi_i}^{\,t}$ is the sequence of actions specified by π_i up to time step t for observation history $\vec{o}_i^{\,t}$. The new distribution parameters can be estimated by

$$\xi_{\vec{\theta}_i^{\,t}}^{(j+1)}(a^t) = \frac{\sum_{\pi\in\mathbf{X}_\rho} I(\pi_i, \vec{\theta}_i^{\,t}, a^t)}{\sum_{a\in\mathcal{A}_i}\sum_{\pi\in\mathbf{X}_\rho} I(\pi_i, \vec{\theta}_i^{\,t}, a)}. \tag{3}$$

However, it may happen that certain action-observation histories $\vec{\theta}_i^{\,t}$ are not consistent with any policy in $\mathbf{X}_\rho$, in which case both the nominator and denominator in (3) are 0. In such a case, one can simply keep the previous parameters, thus take $\xi_{\vec{\theta}_i^{\,t}}^{(j+1)} = \xi_{\vec{\theta}_i^{\,t}}^{(j)}$. We used another approach that defines the next distribution over actions as uniform, indicating that we are indifferent to which action to take since we have no reason to prefer one action above another based on $\mathbf{X}_\rho$.

4.3 APPROXIMATE EVALUATION

Section 2 explained how the value of a pure joint policy can be calculated. Unfortunately, the complexity of this calculation for a single pure joint policy scales exponentially with the planning horizon. Therefore we examine approximate evaluation by simulating a number of episodes, or *traces,* and using the average of outcomes as an estimate for the actual value $V(\pi)$. Although this approximation might introduce errors, notice that the CE method does not discriminate between policies within the set $\mathbf{X}_\rho$ of best samples. Therefore, as long as the relative ordering is preserved, the same policies are used to update the policy distribution, yielding the same results. In fact, only when the ranking of policies is disturbed near the cut-off threshold ρ, will approximate evaluation influence the distribution updating process. There is a second potential source of error, though. When the fraction of best samples $\mathbf{X}_\rho$ is used to update γ, the new γ might in fact be an over-estimation. This could make it very difficult to sample new instances with a higher (approximate) value.

5 Experiments

In this section we present some experimental results. First we describe a single-agent experiment performed in order to determine the influence of using different representations for policy distributions. Next, we show results on the multiagent Dec-Tiger problem, introduced by Nair et al. [2003]. We compare against dynamic programming JESP, a method introduced in the same paper, that works by alternatingly optimizing the policy of a single agent.

To determine the influence of the two policy representations, we devised the test problem illustrated in figure 1. In an effort to isolate the action history's influence as much as possible, the problem includes only one agent. Starting from s_0, the world changes (with equal probability) to either s_1 or s_2 if the agent takes action a_1, and to s_3 or s_4 if it takes action $\bar{a}_1$. When arriving in states s_1 and s_3, the agent (deterministically) receives observation o_1, in s_2 and s_4 this is $\bar{o}_1$. The difficulty in this problem is that the agent has to learn that performing action $\bar{a}_1$ after observation $\bar{o}_1$ is the best thing to do, but only after performing the optimal action a_1 at time step $t = 0$. This is hard because, as long as the distribution over the initial action $\xi_{()}$ has not converged, performing action $\bar{a}_1$ after $\bar{o}_1$ can result in a penalty of -1000. We tested the performance of the CE method for Dec-POMDPs on this problem, sampling four joint policies in each iteration, with $\rho = 0.5$. We used $\alpha = 0.2$, higher values resulted in faster learning, but also in more local optima. When using the CE method for policy search, the goal is to find (i.e., sample) good deterministic policies fast. Instead, here we are interested in the convergence behavior. A good indication of this convergence behavior is given by $V(f_\xi)$, the value of the induced stochastic policy throughout the CE process.

Figure 2(left) shows the mean and standard deviation of 100 runs of the CE method on the toy problem, for both OH- and AOH policy distributions (and each of those runs consists of 60 iterations). Upon convergence of f_ξ[3], the mean $V(f_\xi)$ using OH policy distributions is slightly lower (97.2 vs 103.5), and the standard deviation is greater (21.1 vs 9.12), both of which indicate that it gets stuck in sub-optimal local maxima more often. Also, the learning curve of the AOH based policy distributions grows faster. Even though the AOH representation seems to perform slightly better, the OH representation is still very useful; the latter requires less memory to store and has fewer parameters to update in each iteration, which means that it is faster. Experiments on enlarged versions of this problem showed similar results.

We also tested CE using exact and approximate evaluation on the 2-agent Dec-Tiger problem [Nair et al., 2003], including a variant that involves approximate evaluation without γ. This variant does not require that $\gamma^{(j+1)} \geq \gamma^{(j)}$, but always uses the best ρ-fraction to update the parameters, even if some of the policies have an approximate value worse than γ. Because we found little

[3] Note that the optimal policy always was discovered much earlier than that the distribution f_ξ converged.

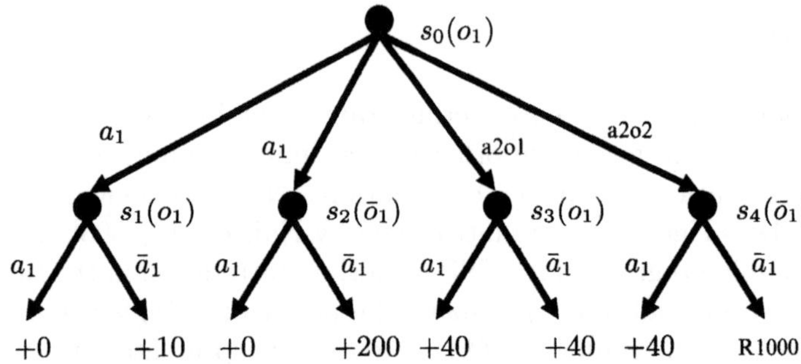

Fig. 1. The 1-agent test problem used to establish the influence of maintaining the action history.

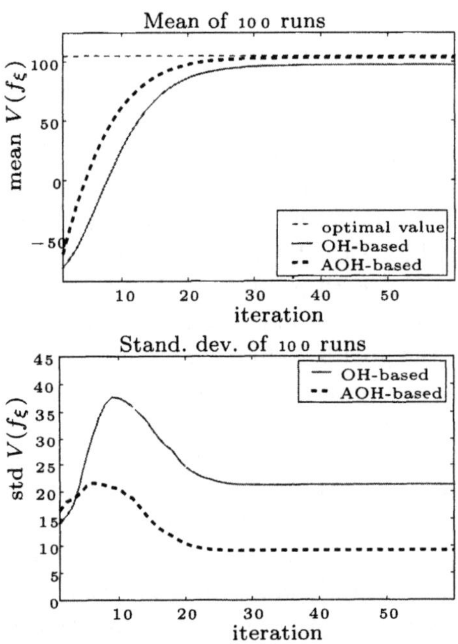

h		CE	CE-A	CE-AW	JESP
2	max	−4.0	−4.0	−4.0	−4.0
	mean	−4.0	−4.0	−4.0	−18.30
	std	0.0	0.0	0.0	7.18
3	max	5.19	5.19	5.19	5.19
	mean	4.48	5.19	5.19	−19.35
	std	3.17	0.0	0.0	13.78
4	max	4.80	4.80	4.80	3.99
	mean	4.33	2.84	4.16	−20.49
	std	0.84	4.05	0.67	13.92
5	max	7.03	6.68	7.03	2.93
	mean	4.50	3.36	3.55	−26.73
	std	1.78	2.13	3.68	14.72
6	max	5.08	5.98	8.32	2.39
	mean	2.45	−2.63	1.66	−24.06
	std	1.28	4.51	3.11	11.86
7	max	-	−5.55	0.81	-
	mean	-	−22.28	−4.19	-
	std	-	14.00	3.12	-
8	max	-	−32.89	−14.22	-
	mean	-	−47.72	−20.79	-
	std	-	12.00	3.66	-

Fig. 2. Results for the toy problem of figure 1 (left), and results on the Dec-Tiger problem (right).

difference between the AOH- and OH-based representation, and the OH-based version is faster, we only used the latter representation in the experiments described here. All the CE variants run for a fixed number of 300 iterations and sampled 50 joint policies in each iteration of which they used a fraction of $\rho = 0.2$ to update the parameters. As before we set $\alpha = 0.2$. The approximate evaluation CE variants used 250 traces to estimate the value of the sampled policies (which was found to be roughly as good as using 1000 traces for $h = 4$).

Figure 2(right) shows the results of CE on the Dec-Tiger problem, using exact evaluation (CE), approximate evaluation (CE-A), approximate evaluation without γ (CE-AW) and dynamic programming JESP. Shown are the maximum, mean and standard deviation of the value of the best pure policy sampled (out of all 300 iterations) over 20 runs. In case of approximate evaluation, the figure reports statistics over the exact values for the best ranked policies (found by an exact evaluation at the end). Empty entries indicate that the method was not able to complete all 20 runs within reasonable time. This immediately illustrates one of the advantages of the approximate evaluation CE methods: they were able to yield results up to horizon 8. For the horizons it did complete, exact evaluation CE performed very well. For horizon 2–5, it achieved the highest maximum, and for $h = 2, 4–6$, it has the highest mean and lowest variance, which indicates it reaches good solutions most frequently. JESP runs into problems for horizons greater than 6, especially memory requirements seem to be the bottleneck here. It is more striking, however, that its mean performance is significantly lower, which implies that it often gets stuck in worse local optima than the CE methods. Finally, the results show that CE-AW outperforms CE-A, implying that overestimating the lower bound γ indeed does negatively influence the search process. Closer analysis of the results confirms this: after an initial period of learning, CE-A has trouble sampling any policies with an (approximate) value higher than γ. This causes the set $\mathbf{X}_\rho$ to be empty, which in turn means there is no updating of the distribution, thus stalling the process.

6 Conclusions

In this paper we have shown how Dec-POMDPs can be approximately solved using an extension of the CE method. We discussed two different representations of probability distribution over joint policies, one based on the full action-observation history and one based only on the observation history, and how their parameters can be updated. Also we introduced an approximate, rather than exact evaluation of the policies sampled in each iteration.

An empirical evaluation showed that the CE approach to solving Dec-POMDPs is competitive with JESP, one of the state-of-the-art methods for approximately solving Dec-POMDPs. Moreover, this evaluation shows that the CE approach using the OH-based policy distributions performed nearly as good as with the AOH-based representation, and that good results can be obtained using approximate evaluation with relatively few traces. A last conclusion that can be drawn is that this approximate evaluation CE performs better when not enforcing a lower bound γ for the set $\mathbf{X}_\rho$ used to update the parameters.

There are a couple of directions for future research. More investigation is required regarding the influence of the action history. There might be problems where the effect of maintaining OH- rather than AOH-based policy distributions are more dramatic. Also the difference between OH- and AOH-based representations can be generalized by making the action distribution conditional on the last k actions taken. Exact evaluation can most likely be accelerated by caching (intermediate) evaluation results of (parts of) joint policies. Finally, and somewhat related, the success of approximate evaluation raises the question whether it is necessary to sample complete joint policies if they are only partially inspected during approximate evaluation. The CE approach could greatly benefit from a construction that samples parts of (joint) policies.

Acknowledgments The research reported here is part of the Interactive Collaborative Information Systems (ICIS) project, supported by the Dutch Ministry of Economic Affairs, grant nr: BSIK03024.

References

D. S. Bernstein, R. Givan, N. Immerman, and S. Zilberstein. The complexity of decentralized control of Markov decision processes. *Math. Oper. Res.*, 27 (4):819–840, 2002.

P.-T. de Boer, D. P. Kroese, S. Mannor, and R. Y. Rubinstein. A tutorial on the cross-entropy method. *Annals of Operations Research*, 134(1):19–67, 2005.

R. Emery-Montemerlo, G. Gordon, J. Schneider, and S. Thrun. Game theoretic control for robot teams. In *Proceedings of the IEEE International Conference on Robotics and Automation*, pages 1175–1181, 2005.

C. V. Goldman and S. Zilberstein. Decentralized control of cooperative systems: Categorization and complexity analysis. *Journal of Artificial Intelligence Research (JAIR)*, 22:143–174, 2004.

D. Koller and A. Pfeffer. Representations and solutions for game-theoretic problems. *Artificial Intelligence*, 94(1-2):167–215, 1997.

S. Mannor, R. Rubinstein, and Y. Gat. The cross entropy method for fast policy search. In *International Conference on Machine Learning*, pages 512–519, 2003.

R. Nair, M. Tambe, M. Yokoo, D. V. Pynadath, and S. Marsella. Taming decentralized POMDPs: Towards efficient policy computation for multiagent settings. In *Proc. Int. Joint Conf. on Artificial Intelligence*, pages 705–711, 2003.

F. A. Oliehoek and N. Vlassis. Q-value functions for decentralized POMDPs. In *Proc. of Int. Joint Conf. on Autonomous Agents and Multi Agent Systems*, Honolulu, Hawai'i, 2007.

M. J. Osborne and A. Rubinstein. *A Course in Game Theory.* The MIT Press, July 1994.

M. T. J. Spaan and N. Vlassis. Perseus: Randomized point-based value iteration for POMDPs. *Journal of Artificial Intelligence Research*, 24:195–220, 2005.

Polyphony, a Knowledge-based Chat System Supporting Collaborative Work

Ciprian Onofreiciuc[1], Alexandru Roşiu[1], Alexandru Gartner[1], and Ştefan Trăuşan-Matu[1,2]

[1] Politehnica University of Bucharest,
313, Splaiul Independentei, Bucharest, Romania
{ciprian.onofreiciuc, alexrosiu, alexandru.gartner}@gmail.com,
trausan@cs.pub.ro

[2] Romanian Academy Research Institute for Artificial Intelligence,
13, Calea 13 Septembrie, Bucharest, Romania
trausan@cs.pub.ro

Summary. This paper presents Polyphony, an advanced chat system, which employs knowledge-based (ontology) based artificial intelligence techniques in order to support collaborative work. There will be pointed out the main differences between Polyphony and other chat systems and the improvements introduced by our solution both at the visualization level and at the knowledge level. The recommender system, one of the knowledge-based modules is described in details.

1 Introduction

The Polyphony project has designed and implemented a knowledge-based chat system with the purpose of facilitating collaborative work and learning. Polyphony is intended to become a support tool for virtual teams solving joint tasks using Computer Mediated Communication. Through the close integration with domain specific ontologies and by analyzing detailed chat logs with natural language techniques, the system's modules generate relevant information (ex. recommendations, summaries, participants competency analysis).

In many cases, communication can prove to be a bottleneck to collaborative work. For example, multiple divergent, opposed threads of discussion in chat conversations many times destroy the collaborative discourse. Therefore, providing support tools for conducting chat conferences, making their discourse well-organized and clear among virtual team members is crucial for efficiently developing their projects.

Maintaining a well-organized conversation is proportionally harder as the number of participants grows [1]. As the discussion develops, it is normal that many points of view appear, each participant being anxious to present his/her own. Unless a conversation policy is respected, this would generally result in a chaotic avalanche

Ciprian Onofreiciuc et al.: *Polyphony, a Knowledge-based Chat System Supporting Collaborative Work*, Studies in Computational Intelligence (SCI) **78**, 155–164 (2008)
www.springerlink.com

of parallel threads of utterances, inevitably leading to misunderstandings. Using traditional chat programs, such a policy cannot be imposed and, therefore, the only chance of having an organized conversation relies on the participants themselves.

Ambiguousness is another real problem that can arise in many situations during a multi-user conference, either as a result of topic unawareness, or caused by multiple interrogative utterances that can have the same answers.

Because it is impossible to define and follow topics within a conversation using most of the current chat media, it is very possible for the participants to be off-topic at some point during the conversation. This can result in triggering parallel, off-topic conversations which not only consume useful time, but also distract the participants attention from the real subject.

Often, confusions can arise when several participants pose different questions and another user types a reply that matches in sense more than one of them. We will illustrate this situation by means of the following example.

Mark: Do you think we can sell the product in it's current form ?

William: Should we improve some aspects of the product ?

Michael: Yes.

Fig. 1. Example of ambiguous conversation

Although the misunderstanding in figure 1 could be clarified by means of a few more utterances, this is the best case scenario and it still leads to important time loss. Another feature that lacks in most of the current chat systems is a structured chat log of each conversation. Chat logs can prove to be very useful for chat participants and thus they should be rich in information. Chat logs are not only the means for finding or regaining information. They can also be the basis for useful and relevant knowledge generation. There are some approaches that are similar with the one presented here. *ConcertChat* [2] is a Java-based implementation of a chat system that supports references to previous chat contributions or text passages and picture excerpts (from within the whiteboard) and multiple references. Also, the content of the chat sessions can be exported to other formats, for example like HTML.

On the other hand, *ConcertChat* has the disadvantage that it doesn't allow an easy design and integration of additional modules that might increase the efficiency of collaborative work. It also does not offer a recommender system or any other tools for generating new and relevant information based on the analysis of the chat log. Another project has implemented a mechanism for human interaction with the Semantic Web (knowledge bases and web services) through the means of Internet Relay Chat (IRC) bots [3]. These bots are easy to program and they provide different services to the users. Also, bots have an intuitive, conversational interface for human users, and fit well with the inputs and outputs of Semantic Web queries. Implementations of bots for interacting with web services and FOAF (Friend Of A Friend [4]) data models are also provided. This approach states that the interaction between humans and web services and knowledge bases should take place in an interactive environment. Because most Semantic Web data and functionality is not accessible through web pages, the authors describe alternative methods for accessing it.

The Polyphony system tries to overcome the problems discussed above, using an ontology and natural language processing techniques. The paper continues with the presentation of the Polyphony system. Section 3 describes the advanced visualization means offered by Polyphony. The following section introduces the knowledge-based modules available in the system. One of these modules, the recommender system, is discussed in detail in section 5.

2 The Polyphony Approach

The Polyphony project aims to solve the problems presented in the introduction by bringing a set of features especially developed for collaborative environments [5]. In such environments, groups of people are working together online in order to find a solution for a certain problem.

Many of the features presented below are based on the ability to reference in an utterance some other one. This option not only allows the system to create innovative and modern views of the discussion but can also be the basis for studying social relationships on the community of Polyphony users.

The solutions envisioned by the Polyphony system aim two levels of improvements. At the knowledge level, the system tries to generate additional information that might be relevant to the participants. The polyphony server will store and transform the data so that it can be readily available for other tools. These can process the existing information in order to produce meaningful results.

The information that could be generated in this manner ranges from outside resources that are related to the context of the conversation to facts that reside inside the discussion or to information deduced regarding the participants themselves.

Important features have been also implemented at the visualization level. The facility to refer in chat messages to previous contributions by means of graphical referencing is more than just a useful feature for a better overview: A current study shows that, especially in small groups, positive learning outcomes can be radically improved by use of such referencing functions [6].

The Polyphony solution is based on a modular and heterogeneous software architecture that can be seen in figure 2. The design of the system allows each module to be written in a different programming language and to run on a different platform. Actually, we believe that the modules should be standalone programs that additionally offer the possibility of remote access, and thus are able to process incoming requests. We would like also to offer the possibility of integrating future and existing third party software modules that deal with analyzing Computer-Mediated Communication, with as little programming effort as possible. Although the whole system could be running on a single site, we were trying to let the option of distributing the system on several stations in a network, up to the system/network administrator. To achieve the goals mentioned above, the communication between the Chat Server and the Chat Modules will take place through the XML-RPC (Remote Procedure Call) protocol. Therefore, all that a module needs to do is to implement some procedures for communicating with the Chat System Server.

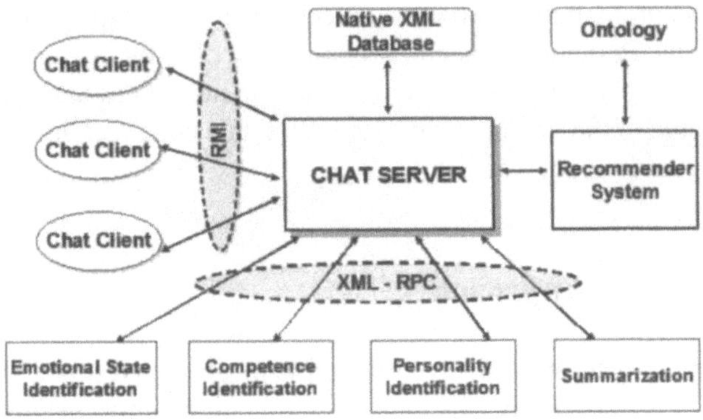

Fig. 2. Architecture of the Polyphony System

On the other hand, since both the Client Application and the Server are written in Java, we have decided to take advantage of the features provided by the Remote Method Invocation (RMI) Communication System. The ability to transfer any kind of serializable Java Object linked with the fact that RMI allows Dynamic (Remote) Class Loading is important for our system since it will allow us to make changes on the server without affecting the client application. In other words, we dont want to force the user to download another version of the client software to often.

We store discussion logs using our own XML format. Therefore, in order to quickly access different chat logs or different parts of a log it seemed like the natural choice to use the *eXist Native XML Database* [7].

3 Advanced visualization means in Polyphony

Polyphony offers the possibility to visualize the chats in different ways, in order to allow each user to have an interaction as close as possible to his/her preferences and profile.

3.1 Classic View with Utterance Linking

This view mode extends a classic view mode, aiming to eliminate ambiguousness. Whenever there is a potentially unclear situation, the users have the possibility to indicate the previous statement that his utterance refers, by clicking on it in the chat corpus. A line will be drawn on the left of the chat, linking the two utterances, as shown in the figure 3

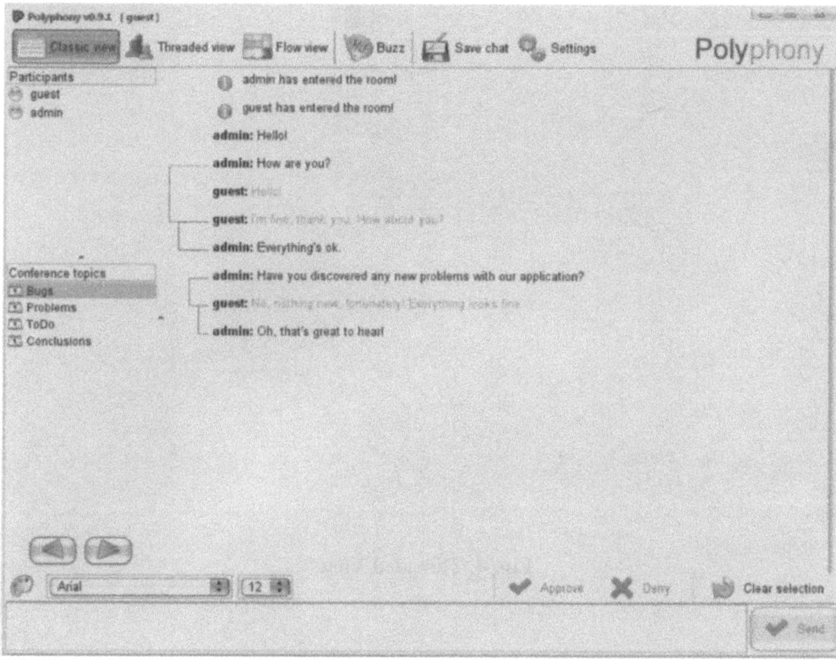

Fig. 3. Classic View with Utterance Linking

This way, the users will no longer have to lose precious time requesting further explanations, as the conversational flow is intuitively rendered right within the chat flow. Polyphony goes a step further and brings two new view modes, in order to provide a better understanding of the information flow within the conversation: the *Threaded View* and the *Flow View*.

3.2 Threaded View

In Threaded View mode [8], using the links indicated by the users, a tree-like presentation mode is used for presenting the chat, a tree in which each utterance is a node. Whenever an utterance refers another, it is being rendered as a child node of the first one. Users can click the utterances from the conversation tree in order to link their own replies to them. As the chat evolves, conversational threads are forming and they are immediately visible to the users, thus finding the right place for their new replies becomes easier.

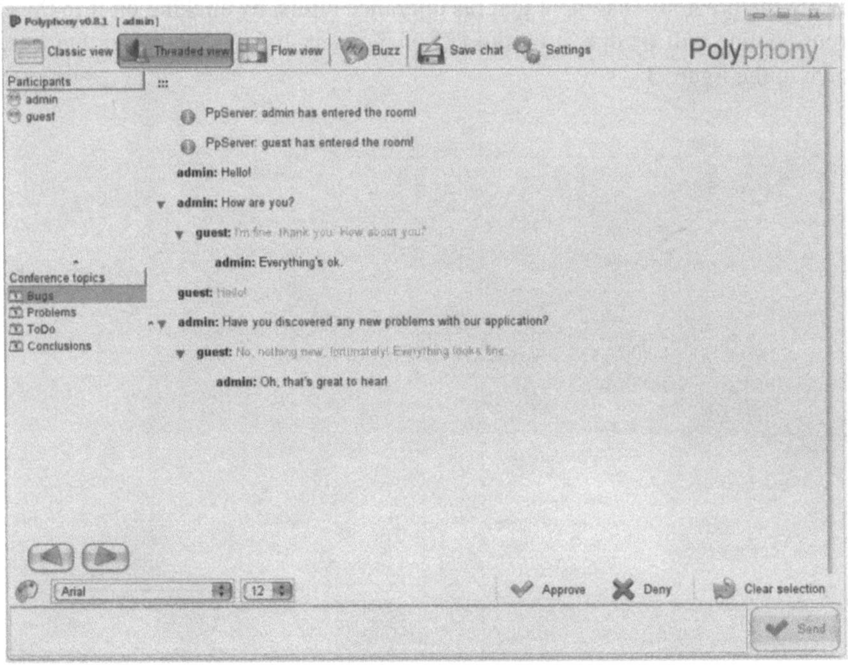

Fig. 4. Threaded View

3.3 Flow View

In the Flow view mode, the chat is rendered in time, using blocks corresponding to each utterance. Block lengths are proportional to the time taken for the statement to be written by a user. Beneath the chat flow, there is a timeline for reference.

4 Knowledge-based Processing of Chat Interactions

A distinctive feature of the Polyphony system is its knowledge-based processing of the chat logs using natural language technology. Several modules available in our system, dealing with the analyses of the chat logs, are listed below. Some of these modules are currently being implemented while others have been already finalized:

- *Recommender System*: tries to generate appropriate documents recommendations that are linked to the subject of the discussion and to the context of the last utterances
- *Summarization*: creates a summary of the discussion
- *Personality Identification*: tries to identify the personality of each user based on the involvement into the discussion
- *Emotional Status Identification*: tries to identify the emotional status by analyzing the words, the emoticons and cue phrases used by a certain user in the dialog

- *Competence Identification*: tries to determine the level of competence of a user by studying the behavior of other users related to him

In the following section we will discuss the recommender system in more detail. This module (like the other mentioned above) has been developed within the Computer Science Department of the Politehnica University Bucharest.

5 The Recommender System

The Rexys recommender module is based on a knowledge base, a parser, an analyzer with a prediction engine and a system that offers the possibility to vote the recommendations provided. There are two big components used as knowledge base for this application. One component is an ontology that brings together concepts in a hierarchical way and the other component is a lexicographic dictionary to obtain senses of the words from the chat.

5.1 The Ontology

The purpose of Rexys is to provide resources about a subject belonging to a domain of science according to a context given. The domain in which the recommender system was tested was mathematics, starting from the MathForum portal and its online library (`http://www.mathforum.org/library`). The ontology has more than 200 concepts taxonomically organized, according to the hierarchy existing on the site. The ontology has been developed using the Protg environment [9].

There are two files for representing the ontology, one for relations between concepts and one for storing the resources of every concept. Concepts are represented by classes. The root concept is Mathematics, the relation between a parent concept (node in the ontology) and a child (another node in the ontology that is a particular aspect on that first concept) is the relation subClassOf, defined in the name-space for RDF.

For example in the next fragment from the file **mathontology.rdfs**, Algebraic Geometry represents a sub-domain of the domain called Geometry:

```
<rdfs:Class rdf:about="&kb;Algebraic_Geometry"
            rdfs:label="Algebraic_Geometry">
  <rdfs:subClassOf rdf:resource="&kb;Geometry"/>
</rdfs:Class>
```

The second file stores the resources for every concept. Each node-concept has at least one child or a resource. For example the concept Algebraic Geometry has as a resource a link from Mathontology:

```
<kb:Algebraic_Geometry
  rdf:about="&kb;mathontology_Instance_10088"
  kb:url=http://mathforum.org/library/topics/algebraic_g/
  rdfs:label="mathontology_Instance_10088"/>.
```

5.2 The Dictionary

For identifying the senses of the words in the chat, the WordNet [10] lexical ontology is used. This kind of dictionary is organized as a graph that has as nodes words with equivalent senses (synset). A word can have multiple senses so it can belong to multiple synsets. Synsets are bound by relations like hyponymy, hypernymy, holonymy and meronymy. The relations from the dictionary are modeling well the relation of sub-classing inside the ontology.

5.3 The Analysis

The words are received from the Polyphony server using the XML-RPC protocol. First there is a filter applied to eliminate those words that appear usually in a chat and don't bring any information in the discussion (stop-words). For each noun or adjective remained a Word object is created. Each Word object is compared against the concepts from the ontology to see the similarities with existing concepts.

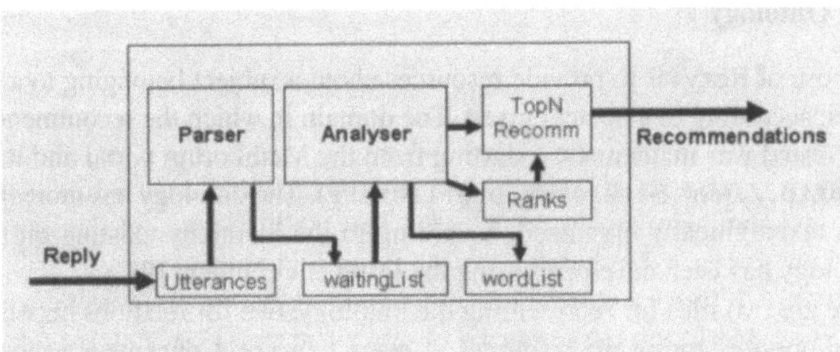

Fig. 5. Recommender System Architecture

For each chat session there is a list of ranks for concepts. The rank of a concept represents how close the concept is to every significant word from the corpus. When analyzing the similarity between a word and a concept, the distance in WordNet between the correspondent synsets is computed. The greater the distance in the graph, the less appropriate is the word to the concept.

The rank has two components: direct rank and indirect rank. Direct rank is increased with the inverse of the distance in WordNet, when this distance is less than a constant, otherwise the indirect rank is increased with the inversed value of the distance. After the rank is adjusted, the rank of all ancestors of concept is adjusted also, with the ratio of the initial value.

There are two kinds of analysis implemented, flat and hierarchical. In the first one, the similarity is computed between every word and every concept in the ontology tree. In the hierarchical one, the similarity is computed on branches, starting with the root concept mathematics, continuing recursively with the most similar child of the node.

5.4 The Collaborative Filtering

The recommendations are given as a top N list, meaning the first N recommendations with the biggest rank obtained. This list is sent to the Polyphony server that forwards the recommendation to the clients. The clients have the possibility to vote a recommendation as relevant or irrelevant. The system starts the voting by according a neutral vote to all the concepts, then the votes of the users are added and the ranks are updated according to this action. If a change occurs in the top N recommendations, a new list of recommendations is sent to the users.

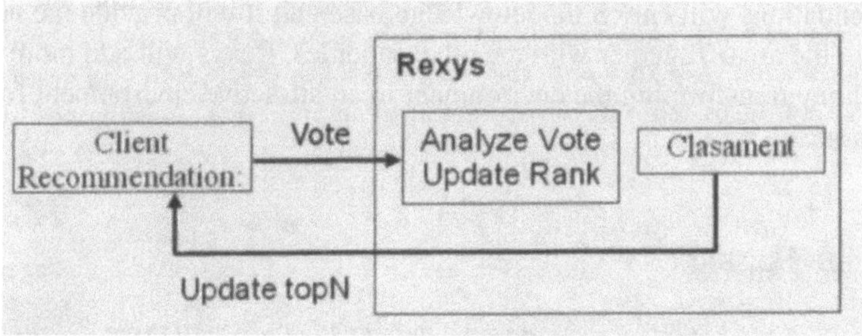

Fig. 6. Collaborative Filtering

6 Conclusions

Virtual, collaborative working environments usually lack necessary information and perspectives that exist in face-to-face meetings. Polyphony tries to improve the user's overview of the dialog, to guide him in solving the problems at hand and to provide him with as much information as possible. It strives also at becoming a knowledge repository that users can access to find the necessary resources.

The Polyphony project uses a number of XML standards and technologies (RDF, XPath, XQuery, XUpdate, XSLT, XML-RPC, etc.), which provide a foundation for a continuous development of modules for chat analysis. Whether we are talking about storing knowledge into an ontology or saving chat logs into an XML native database, XML provides very good means for including meta data beside the actual information. Collaborating software modules work together in order to solve the problems raised by human, collaborative, working environments.

At this date, the recommender system is fully integrated with the Polyphony collaborative environment, offering real-time recommendations for the chat sessions. Extending the recommendations to another domain can be done very easy by changing the files representing the ontology. In the future, the analysis will be improved by tuning the algorithms and the weights given to every rank. Also a phrase analysis will be introduced in the parser so that important words from the phrase to be assigned with a bigger weight.

Another feature that may be added regards the collaborative filtering. In this version there is no binding between the sessions of the chat. So in the future this bound will be added, in a way that if a domain is not relevant to a list of contexts it can be out of context to other domains too. This came as an observation that there are domains like Sports with ancestor nodes: Applications-Connections, Mathematics, that appear very frequent in the list of recommendation due to the fact that being very general, is close to all the domains.

The particularity of this recommender system is that it is designed to work in a specific domain of science. The advantage is that the interaction between users is more accurate and is not a subject of users changing tastes. Letting the users add recommendations will enrich the knowledge base and it will provide the newest resources in the area. Together with the other modules, Rexys will add more usability to Polyphony transforming the environment in an attractive environment for collaborative work.

7 Special Thanks

This paper has been developed within the "NCIT leading to EU IST excellency" FP6 project.

References

1. Vronay, D., Smith, M., Drucker, S.: Streaming Media Interfaces for Chat. Virtual Worlds Group, Microsoft Research
2. Fraunhofer IPSI web site ConcertChat: `http://www.ipsi.fraunhofer.de/concert/index_en.shtml?projects/chat` (Accessed March 2007)
3. Golbeck, J., Mutton, P.: Semantic Web Interaction on Internet Relay Chat. Proceedings of Interaction Design on the Semantic Web, New York, USA (2004).
4. Friend of a Friend web site - `http://www.foaf-project.org/` (Accessed March 2007)
5. Althaus, S.: Computer-mediated Communication in the University Classroom: An Experiment with Online Discussions. The American Political Science Association (1996)
6. Fraunhofer IPSI web site Press Release `http://www.ipsi.fraunhofer.de/ipsi_e/press/press_releases/2006/060321_press_concertchat.html` (Accessed March 2007)
7. Meier, W., eXist: An Open Source Native XML Database. In Web, Web-Services, and Database Systems, Springer LNCS Series (2002)
8. Smith, M., Cadiz, J., Burkhalter, B.: Conversation Trees and Threaded Chats. Microsoft Research.
9. Protege web site open source ontology editor - `http://protege.stanford.edu/` (Accessed March 2007)
10. WordNet web site - a lexical database for English - `http://wordnet.princeton.edu/` (Accessed March 2007)

Decomposition and Processing Strategies of Global XQuery Queries over Distributed Environments

Jong-Hyun Park and Ji-Hoon Kang*

Dept. of Computer Science, ChungNam National University,
Gung-Dong, Yuseong-Gu, Daejeon, 305-764, South Korea
{jonghyunpark, jhkang}@cnu.ac.kr

Summary. The global XML view concept for integrating distributed heterogeneous data sources over the Internet makes it possible to deal with such data as a single XML document. The global XQuery queries based on a global XML view can search heterogeneous data seamlessly, being distributed in the local systems. In distributed relational databases, one of the important techniques for processing global SQL queries was the decomposition. So, it is natural to consider how the decomposition technique for SQL queries can be applied to processing global XQuery queries. Compared to well-structured relational data, however, XML data is not structured since it is like trees. So the XQuery language reflects this characteristic and its syntax is very complex. For example, the structure-based operations such as the FOR clauses are very specific and is not supported by SQL. Therefore, this paper proposes a method for decomposing a global XQuery query into local XQuery queries and also gives query processing strategies based on the decomposition.

1 Introduction

XML has emerged as a leading language for representing and exchanging data on the Internet. In recent years, there have been a number of researches focusing on the integration of XML data and heterogeneous data including relational data [1]. One of the methods for integration is to make non-XML data to be considered as XML data by using XML views [2]. The global XML view integrates the local XML views, and thus users can see and search the distributed heterogeneous data via the global XML view. At this time, one of the standard query languages for searching the integrated heterogeneous data can be XQuery language. The users can consider only the global XML view and can give queries in XQuery and get results, but they need not to know either where the data is or how the data is shaped. Such queries are just the global XQuery queries over the integrated heterogeneous data. XQuery is a standard query language and its expressive power is so rich. As a result, however, the query syntax is very complex and thus it is not easy to process the queries efficiently. There are not so many researches for processing global XQuery queries over distributed environment,

* He is a corresponding author.

Jong-Hyun Park and Ji-Hoon Kang: *Decomposition and Processing Strategies of Global XQuery Queries over Distributed Environments*, Studies in Computational Intelligence (SCI) **78**, 165–175 (2008)
www.springerlink.com

and some of them try to apply the decomposition techniques for processing the global SQL queries into the processing of the global XQuery queries.

The decomposition was one of the popular techniques for processing traditional distributed SQL queries [3]. A global query is decomposed into several subqueries to be sent to the local systems simultaneously. Therefore, it is quite natural to consider how the decomposition technique for SQL queries can be applied to the process of the global XQuery queries over heterogeneous data. Unfortunately, however, the structure of the data handled by the XQuery language is quite different from the one by the SQL. The XQuery language deals with semi-structural data, i.e. tree-structured data, while SQL deals with well-structured data, i.e., the table-shaped data. Therefore, the language features are very different, and moreover the XQuery syntax is so complex. For example, the iteration specified by the FOR clauses in XQuery queries is one of the important features to be considered for processing XQuery queries, but there is no such feature in SQL. Therefore, in order to exploit the decomposition technique for SQL, it is necessary to consider such XQuery-specific features.

Contributions of this paper are threefold. One, we propose two decomposition methods for processing global XQuery queries and describe some considerations for decomposing XQuery query. Two, we propose processing strategies of local queries based on structure of global XQuery queries. Three, we implement a prototype for our strategies and evaluate.

The rest of the paper is organized as follows. In Section 2, we present related works. In Section 3, we introduce some considerations in XQuery query decomposition and show two methods for the decomposition. In Section 4, we introduce processing strategies of global query based on decomposition methods and evaluate these strategies. Finally, we summarize the contributions of this paper in Section 5.

2 Related Work

An increasing number of business users and software applications need to process information that is accessible via multiple diverse information systems, such as data systems, file systems, legacy applications, or Web services. There are some studies on the integration of distributed relational data [4, 5], integration of relational data with semi-structured data [6], and efficient XQuery processing in a single system [7]. How-ever, these approaches are not direct methods for the integration of heterogeneous data with XML data. There are some proposed methods for global query processing in a distributed environment, and query decomposition is one of the popular and basic methods [3, 4, 8]. XQuery is the natural choice for integrating heterogeneous data with XML data because XQuery is standard query language for XML data. However, it involves some problems to apply previous decomposition methods into the global XQuery query because of its structural characteristics [8].

[9] introduces the XQuery-based data integration systems that provide a queryable integrated view of such information. Their system architecture and function are very similar to those of our global XQuery query processor. Moreover, they also describe

some considerations for global XQuery query processing. However, their focus is just the architecture and requirement of the integration system for efficient processing. They do not describe any specific algorithms which are considered the structural features of XQuery. [10] proposes an extension to XQuery language by an additional expression that allows users to express queries over heterogeneous data sources. This paper also introduces the usefulness of decomposition methods for processing global XQuery queries. However, their extended XQuery language is quite complex to users, although they also propose a tool for generating XQuery query. However, XQuery language in our method is purely XQuery language proposed by W3C. introduce the XQuery-based data integration systems that provide a queryable integrated view of such information. Their system architecture and function are very similar to those of our global XQuery query processor. Moreover, they also describe some considerations for global XQuery query processing. However, their focus is just the architecture and requirement of the integration system for efficient processing. They do not describe any specific algorithms which are considered the structural features of XQuery.

[10] proposes an extension to XQuery language by an additional expression that allows users to express queries over heterogeneous data sources. This paper also introduces the usefulness of decomposition methods for processing global XQuery queries. However, their extended XQuery language is quite complex to users, although they also propose a tool for generating XQuery query. However, XQuery language in our method is purely XQuery language proposed by W3C.

[11] proposes some methods for the efficient processing of distributed streaming XML data. Their approaches use the pre-defined interface for interaction between the global XQuery query processor for global query processing and the wrappers for local systems. However, this paper assumes that we can not obtain any information from local systems and just know that all wrappers support XQuery query. This assumption supports the flexibility and extensibility of our system; thus, our system can be applied to any XQuery processor on a distributed environment that focuses on the global XQuery query processing based on the XML view.

3 Decomposition Techniques for Global XQuery Processing

3.1 Considerations for XQuery Decomposition

One of the approaches for decomposing global XQuery queries is to refer methods for decomposing integrated SQL queries. However, the decompositions between the inte-grated SQL queries and global XQuery queries are not same because SQL queries are written on relational data with flat structure and the data source of XQuery queries is semi-structured data. In other words, XQuery queries can contain some special expressions to address the semi-structured data like FOR clauses and these expressions make the decomposition of global XQuery queries hardly [3].

In case of integrated SQL queries, the basic method for decomposing the queries is decomposition method based on clustering which clusters expressions with same

Table 1. An example of decomposition of a Global XQuery query into Local XQuery queries by clustering expressions with the same destination.

<table>
<tr><th>Global XQuery query</th><th>Decomposed Local XQuery Queries</th></tr>
<tr><td rowspan="2">for $item in doc("item.xml")/items/item
for $seller in doc("seller.xml")/sellers/seller
for $item_URI in $item//URI
for $item_feature in $item/feature
for $seller_address in $seller/address
for $item_des in $item_feature/description
for $item_color in $item_feature/color
where $item_color="red" and $item/owner=
$seller/name
return <result>{$item_URI,*$seller_address*,
$item_des}</result></td><td>for $item in doc("item.xml")/items/item
for $item_URI in $item//URI
for $item_feature in $item/feature
for $item_des in $item_feature/description
for $item_color in $item_feature/color
where $item_color="red"
return<SubResult1>{$item_URI,$item_des,
$item/owner}</SubResult1></td></tr>
<tr><td>for $seller in doc("seller.xml")/sellers/seller
for $seller_address in $seller/address
return <SubResult2>{$seller_address,
$seller/name}</SubResult2></td></tr>
</table>

destination in the integrated queries and then to generate the local SQL queries consisting of expressions in cluster. Table 1 shows a sample global XQuery query and two local XQuery queries generated by clustering of expressions. However, this decomposition method is not enough to decompose global XQuery queries because we can not reconstruct the result document of global XQuery queries by the result set of local XQuery queries. For example, even if we can obtain following two results from two local XQuery queries in Table 1, we cannot reconstruct the result of global XQuery query in Table 1 from these local results because of the absence of order information for node combination.

```
"<SubResult1> <URI>URI1</URI><description>A is ~ </description>
      <URI>URI2</URI><description>A is ~ </description>
      <URI>URI3</URI><description>B is ~ </description>
</SubResult1>"
"<SubResult2><address>Daejeon</address><address>Seoul</address></SubResult2>"
```

Decomposed local queries can obtain the sets of nodes for constructing the result of global queries but these result sets do not include any information about the relation-ship for constructing result nodes. Consequently, in order to use decomposition methods for integrated SQL queries written based on table structure for global XQuery queries based on semi-structured data, we need additional information and algorithm for constructing global result from local results. This reason is that the results of local queries are not partial documents keeping the structure of original XML document in local systems and new XML document consisting of nodes in the original XML document. Therefore, we propose the methods to generate local XQuery queries for extracting the minimal part of original document and to construct the result of global queries by using result of the local queries.

3.2 Static Decomposition

Static decomposition method statically decomposes global XQuery queries into local XQuery queries and parallel process the decomposed local queries and then generates the result of global queries in global XQuery query processor by using the results of local queries. This method is generally used to process global XQuery

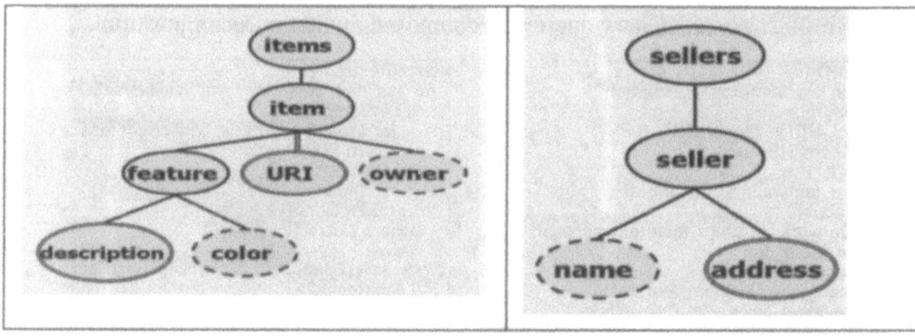

Fig. 1. Information trees for describing the data structure needed to process the global XQuery query in Table 1.

queries which do not contain join operators between different local systems because the method run all join operators in the global XQuery query processor.

In this paper, the goal of our decomposition is to generate local XQuery queries which return minimized parts of nodes stored in local systems needed for processing global XQuery queries and returned results keep the original relationship among the nodes. For this goal, we analyze the structure of nodes needed for processing global XQuery queries and generate local XQuery queries for extracting these nodes. XPath expressions in global queries address nodes which are used to process the global que-ries among nodes stored in local systems. Therefore, we can know the structure information about nodes stored in local systems by using the XPath expressions and can generate local XQuery queries for extracting the nodes. In other words, we do not know the structure information of all nodes stored in local systems but know the structure information of nodes required for processing global XQuery queries. Therefore, the XQuery queries for extracting these nodes are local XQuery queries for processing global XQuery queries.

Figure 1 shows example trees which describe the structure of nodes stored in two local systems for processing the global XQuery query described in Table 1, Nodes circled by a double line express the result nodes which are declared in RETURN clause of the global XQuery query in Table 1. Nodes circled by a dotted line express operand nodes which are used to process the conditions in the global XQuery query. In case of the left tree, the nodes needed for processing the global XQuery query are "description", "URI" and "owner" nodes and nodes needed from the right tree are "name" and "address" nodes. The "description", "URI" and "address" nodes are used to construct the result of the global query and "owner" and "name" nodes are used to process join operation between two local systems.

Table 2 shows two local XQuery queries which extract data needed for processing the global XQuery query in Table 1. In other words, local queries in Table 2 extract data which has the same structure as trees in Figure 1. As previous mentioned, although we are able to obtain enough node set from local systems for constructing the result of global XQuery queries we have to know the relationship and order between nodes for constructing the correct result of global queries. By this reason, we propose

Table 2. Local XQuery queries Decomposed by Static Decomposition.

```
<LocalResult1>{
  for $item in doc("item.xml")/items/item
  let $Wrapper0 :=
    for $item_URI in $item//URI
    return $item_URI
  let $Wrapper1 :=
    for $item_feature in $item/feature
    let $Wrapper2 :=
      for $item_des in $item_feature/description
      return $item_des
    let $Wrapper3 :=
      for $item_color in $item_feature/color
      where $item_color = "red"
      return $item_color
    Where exists($Wrapper2) and exists($Wrapper3)
    return <feature>{$Wrapper2, $Wrapper3}</feature>
  let $Wrapper4 :=
    for $innerVariable0 in $item/owner
    return $innerVariable0
  where exists($Wrapper0) and exists($Wrapper1)
   return
    <item>{$Wrapper0,$Wrapper1,$Wrapper4}</item>
}</LocalResult1>
```

```
<LocalResult2>{
 for $seller in doc("seller.xml")/sellers/seller
   let $Wrapper0 :=
     for $seller_address in $seller/address
     return $seller_address
   let $Wrapper1 :=
     for $innerVariable1 in $seller/name
     return $innerVariable1
   where exists($Wrapper0)
   return
  <seller>{$Wrapper0, $Wrapper1}</seller>
}</LocalResult2>
```

Table 3. Fusion XQuery query.

```
for $item in doc("LocalResult1.xml")/LocalResult1/item
for $seller in doc("LocalResult2.xml")/LocalResult2/seller
for $item_URI in $item//URI
for $item_feature in $item/feature
for $seller_address in $seller/address
for $item_des in $item_feature/description
for $item_color in $item_feature/color
Where $item/owner= $seller/name
return <result>{$item_URI, $seller_address, $item_des}</result>
```

the fusion XQuery query which runs on local results and constructs the correct result of global queries. Table 3 is fusion XQuery query to construct the result document of global query from result documents of local queries in Table 2.

The FOR and RETURN clauses of fusion queries are the same as clauses of global queries. The WHERE clauses of fusion queries just contain only join operations which join between different local systems.

3.3 Dynamic Decomposition

Dynamic Decomposition generates a local query for querying to one of local systems and then collects the result from this local system. Then collected result is used to generate another local query to be sent to next local system. The result returned by the second local system is collected, and so on. Local systems work in sequential. The basic idea for dynamic decomposition is similar to static decomposition. Especially the fist local queries are the same as the queries generated by static decomposition if the destination of these queries equal. However, other local queries except fist local query contain the result of the former local queries.

For example, global XQuery query in Table 1 can be decomposed as two local queries in Table 4 by dynamic decomposition. The left query in Table 4 is the same

Table 4. Local XQuery queries decomposed by Dynamic Decomposition.

<LocalResult1>{ for $seller in doc("seller.xml")/sellers/seller let $Wrapper0 := for $seller_address in $seller/address return $seller_address let $Wrapper1 := for $innerVariable1 in $seller/name return $innerVariable1 where exists($Wrapper0) return <seller>{$Wrapper0, Wrapper1}</seller> }</LocalResult1>	for $item in doc("item.xml")/item for $seller in doc(*<LocalResult1><seller> <address>Daejeon</address><address>Seoul</address> <Seller><LocalResult1>*)/LocalResult1/seller for $item_URI in $item//URI for $item_feature in $item/feature for $seller_address in $seller/address for $item_des in $item_feature/description for $item_color in $item_feature/color where $item_color = "red" and $item/owner= $seller/name return <result>{$item_URI, $seller_address, $item_des}</result>

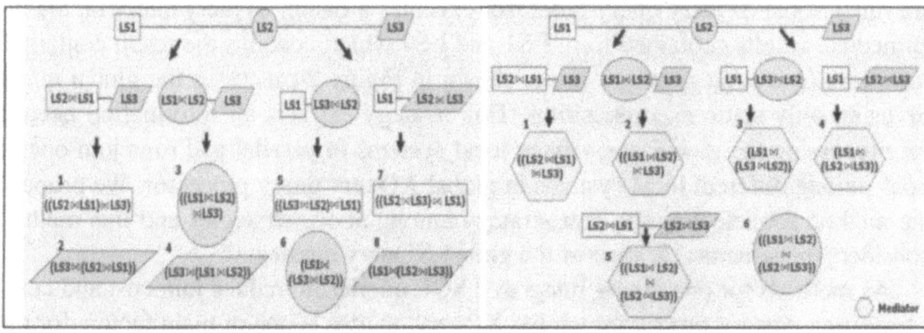

Fig. 2. Processing Orders of Global XQuery query Based on Dynamic Decomposition and Mixed Decomposition with Static Decomposition.

as the right query in Table 2 because the left query is the first local query. The result of left query is used to generate the right query in Table 4. The right query is last query because the number of destination for global query is two. In case of the processing global queries by dynamic decomposition, we can construct the result of global queries by running last local queries in last local systems without fusion queries. Namely the result of the right query in Table 4 is the result of the global query in Table 1.

4 A Processing of Global XQuery Query Based on Structure

4.1 Processing Strategies of Local XQuery Queries

In order to efficiently process global XQuery queries, deciding a processing strategy of decomposed local XQuery queries is important. In other words, which one is more efficient between static and dynamic decompositions for processing of global queries? Among decomposed local queries how to decide the processing order? These questions are of practical importance because wrong processing strategy could produce an enormous number of intermediate results or joins.

Consider the processing a global query which is written on three local systems. Figure 2 shows the fifteen cases of strategies. The eight cases in left figure use only dynamic decomposition and the right figure shows the six cases of processing strategies mixed dynamic and static decomposition for processing the global query. For example, the seventh case of eight cases in left figure first run a local query to querying to LS2 and then executes next local query including the result of previous local query to LS3. Then a local query including intermediate results obtained from two local queries is executed in LS1. The third case of six cases in left figure generates two queries for querying to LS2 and LS3 by dynamic decomposition and generates the local query for LS1 by static decomposition. The processing step between two local queries is that the local query for LS3 is the former and the query for LS2 is the latter. The local query for LS1 is executed in parallel while other two local queries are run. Global XQuery query processor executes a fusion XQuery query on the intermediate results contained from LS1 and LS2 which contains the result contained from LS3.One strategy which is not shown in Figure 2 processes the global query by using only static decomposition. This strategy extracts all information needed for processing the global query from local systems in parallel and runs join operations among different local systems in global XQuery query processor. We propose the method for selecting the best strategy among above strategies and this method considers the structural feature of the global XQuery queries.

As methods for processing integrated SQL queries, to reduce join cost and communication cost for processing global XQuery queries is one of main factors for the efficient processing [12]. Therefore, this paper propose to use dynamic decomposition for reducing the join cost when join operations exist among local systems and to use static decomposition for reducing the communication cost by parallel processing when join operations do not exist. In case of use dynamic decomposition, the result of the former local query is shipped to latter local query and sent to next local system. At this time, reducing communication cost is directly connected with reducing the size of the ship-ping intermediate result which depends on processing order of local queries. To decide processing order of local queries generated by dynamic decomposition, we classify local systems as two cases. One is to store the nodes which are used to process only conditions declared in global queries and another is to store the nodes for conditions and the result nodes for return which are defined in RETURN clause of global queries. Of course, local system can store only result nodes for return. However, the local query for querying to this local system is considered for our classification be-cause this query is not generated by dynamic decomposition and the local system is not joined with other local systems. To decide processing order of local queries for classified local systems, we define following three cases of local queries and propose to execute these queries before other local queries:

1. Return nodes by a local query are only used to process conditions of global XQuery query.
2. A local XQuery which returns nodes needed for processing conditions of global XQuery query and these result nodes are ancestor or self nodes of nodes needed for return of the global query.
3. The number of variables bound in RETURN clause in a local query is less than other local queries.

Table 5. The structure of sample global XQuery queries for evaluation.

Q1	LS1—LS2 LS3	Q3	LS1—LS2 LS3—LS4 LS5
Q2	LS1—LS2 LS3—LS4	Q4	LS1—LS2—LS5 LS3—LS4

4.2 Performance Evaluation

In order to evaluate, we use five local systems which use single hub in LAN. The H/W and S/W of all local systems are the same as follows: the CPU is Intel(R) Pentium(R) M processor 1.6GHz, the memory size is 1.0GB, and the OS is Windows XP. The CPU for global XQuery query processor is Intel(R) Pentium(R) 4 processor 3.2GHz and memory is 2.0GB. XQuery engine for local systems and global query processor is Saxon beta version. Saxon beta version is free and one of popular XQuery engines.

The sample data for evaluation is XMark for "auction" data. To get multiple and related data for local systems, we split the "auction" document into multiple documents each of which is stored in different local systems. The root nodes of sample documents are the second or third level nodes in "auction" document, such as "closed_auctions", "asia", "people" node, and so on. Table 5 shows the join relationship among local systems in sample queries for evaluation. For example, Q2 query joins local system 1 and 2 and system 3 and 4, and then obtains local results from system 2, 3 and 4.

Figure 3 summarizes the processing time of example queries in Table 5. These queries are processed by three kinds of processing strategies, using only static decomposition, only dynamic decomposition and both of them. These queries are decomposed by static and dynamic decomposition and mixed decomposition method which uses two decomposition methods by our processing strategy. The result of the evaluation shows that processing strategy by mixed decomposition outperforms other strategies in generally. However, in some cases, we can see that dynamic or static decomposition is efficient because total processing time is depends on a lot of factors such as processing power of local query processors or global query processors. However, generally the processing time by our strategy is efficient and the evaluation shows that our strategy is one of the efficient strategies for processing global XQuery queries.

5 Conclusion

In this paper, we proposed a processing method of global XQuery query based on decomposition. For this method, we addressed some considerations for XQuery query decomposition and correspondingly proposed a decomposition algorithm. Moreover,

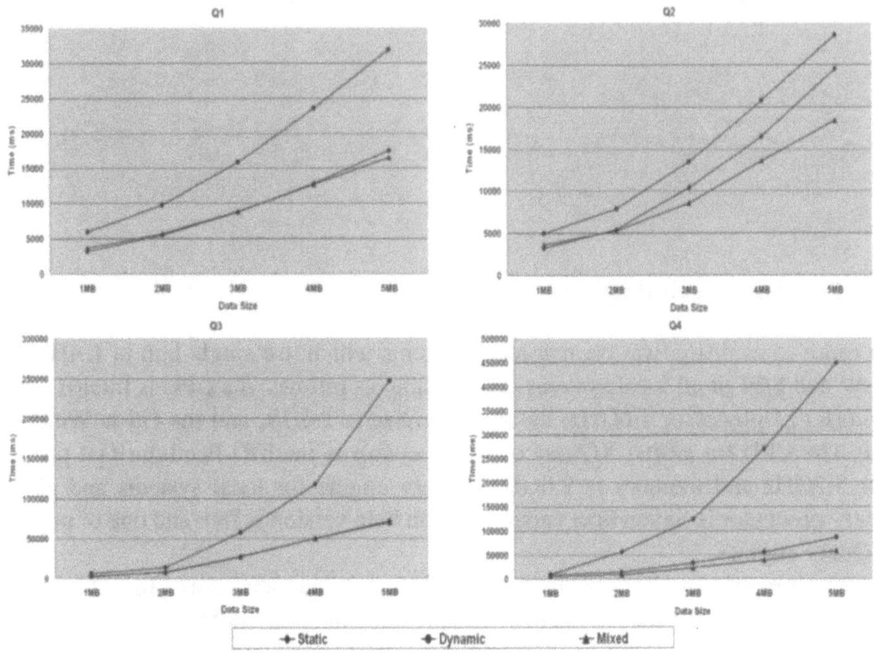

Fig. 3. Comparison of processing times for Q1, Q2, Q3 and Q4.

we have proposed processing strategies of local queries based on structure of global XQuery queries. Three, we have implemented a prototype for our strategies and have evaluated. Our result can obviously be applied to any XQuery processor that focuses on the integration and search of distributed data, since our decomposition methods and processing strategies are a general method and does not refer any information.

Acknowledgement: This research is supported by the ubiquitous Computing and Network (UCN) Project, the Ministry of Information and Communication (MIC) 21st Century Frontier R&D Program in Korea.

References

1. M. Smiljanic, L. Feng, and W. Jonker: Web-Based Distributed XML Query Processing. In: *Proc. Intelligent Search on XML Data 2003*, September (2003).
2. J. Shanmugasundaram, J. Kiernan, E. Shekita, C. Fan, and J. Funderburk: Querying XML Views of Relational Data. In: *Proc. 27th VLDB*, Roma, Italy, September (2001).
3. H. Kozankiewicz, K. Stencel, K. Subieta: Distributed Query Optimization in the Stack-Based Approach. In: *Proc. HPCC 2005*, Sorrento, Italy, September (2005).
4. D. Kossmann: The state of the art in distributed query processing. In: *ACM Computing Surveys*, vol.32(4), pp.422-469, (2000).
5. D. Suciu: Distributed query evaluation on semistructured data. In: *ACM Transaction. Database Systems*, Vol.27, No.1, pp.1-62, (2002).

6. X. Zhang, B. Pielech, and E. A. Rundensteiner: Honey, I shrunk the XQuery: an XML algebra optimization approach. In: *Proc. WIDM 2002*, McLean, Virginia, USA, November (2002).
7. C. Koch, S. Scherzinger, N. Schweikardt, and B. Stegmaier: FluXQuery: An Optimizing XQuery Processor for Streaming XML Data. In: *Proc. VLDB 2004*, Toronto, Canada, September (2004).
8. L. M. MacKinnon, D. H. Marwick, and M. H. Williams: A Model for Query Decomposition and Answer Construction in Heterogeneous Distributed Database Systems. In: *Journal of Intelligent Information Systems*, Vol.11, No.1, July (1998).
9. Y. Papakonstantinou and V. Vassalos: Architecture and Implementation of an XQuery-based Information Integration Platform. In: *IEEE Data Engineering Bulletin*, Vol.25, No.1, (2002).
10. C. Re, J. F. Brinkley, K. P. Hinshaw, and D. Suciu: Distributed XQuery. In: *Proc. IIWeb*, Toronto, May (2004).
11. Z. G. Ives, A. Halevy, and D. S. Weld: An XML Query Engine for Network-Bound Data. In: *VLDB Journal*, Vol.11, No.4, pp.380-402, (2002).
12. X. Lin and M. E. Orlowska: An Efficient Processing of a Chain Join with the Minimum Communication Cost in Distributed Database Systems. In: *Distributed and Parallel Databases*, Vol.3, No.1, pp.69-83, 1995.

Description and Organization of Instructional Resources in an Adaptive Educational System Focused on Learning Styles *

Elvira Popescu[1,2], Costin Badica[1], and Philippe Trigano[2]

[1] Software Engineering Department, University of Craiova, Romania
{popescu_elvira, badica_costin}@software.ucv.ro

[2] Heudiasyc, UMR CNRS 6599, Université de Technologie de Compiègne, France
ptrigano@hds.utc.fr

Summary. In this note we present an intelligent way of organizing learning material in an adaptive educational hypermedia system. We describe the use of instructional metadata which facilitates both the detection of student learning style and the application of various adaptation techniques. The advantage of our approach is that it is independent of a particular learning style model. Furthermore, the author has to supply only the annotated learning content (the static description) while the adaptation logic (the dynamic description) is provided by the system.

1 Introduction

Educational metadata is a special kind of metadata that provides information about learning objects. A learning object represents any reproducible and addressable digital resource that can be reused to support learning [20]. Currently there are several initiatives for standardizing educational metadata, addressing the issues of reusability, interoperability, discoverability, sharing and personalization [4].

IEEE LOM (Learning Object Metadata) [19] is the most prominent standard, being elaborated by the IEEE Learning Technology Standards Committee. IMS Global Learning Consortium also contributed to the drafting of the IEEE LOM and consequently the current version of IMS Learning Resource Metadata specification (IMS LRM v. 1.3 [19]) is based on the IEEE LOM data model. LOM contains nine categories of metadata: General, Lifecycle, Meta-metadata, Technical, Educational, Rights, Relation, Annotation and Classification. The attributes that are relevant from the point of view of instruction and pedagogy are the educational ones, specifically *Learning Resource Type*. Its possible values are: *Exercise, Simulation, Questionnaire, Diagram, Figure, Graph, Index, Slide, Table, Narrative Text, Exam, Experiment, Problem Statement, Self Assessment, Lecture*.

* This research was partially supported by the CNCSIS grant TD 169/2007.

Elvira Popescu et al.: *Description and Organization of Instructional Resources in an Adaptive Educational System Focused on Learning Styles*, Studies in Computational Intelligence (SCI) **78**, 177–186 (2008)
www.springerlink.com

Another widely known standard is *SCORM (Sharable Content Object Reference Model)* [2] which originates from e-learning requirements of the US Armed Forces, being produced by ADLNet (Advanced Distributed Learning Network) initiative. SCORM includes three types of learning content metadata: raw media metadata (that provide information about assets independently of learning content), content metadata (that provide information about learning contents, independently of a particular content aggregation) and course metadata (that provide information about the content aggregation).

Dublin Core metadata standard [12] is a simple yet effective general-purpose metadata scheme, for describing a wide range of networked resources. It was developed within the Dublin Core Metadata Initiative (DCMI). At present, there is a joint DCMI/IEEE LTSC Task Force activity, with the objective of developing a representation of the metadata elements of the IEEE LOM in the Dublin Core Abstract Model.

The main problem with these specifications is that they fail to include the instructional perspective [27]. In case of LOM, the property *Learning Resource Type* attempts to address this issue, but mixes instructional and technical information. Thus some of the values describe the instructional role of the resource (*Exercise, Simulation, Experiment*), while others are concerned with their format (*Diagram, Figure, Graph, Slide, Table*). Moreover, some important instructional types are missing, such as *Definition, Example* or *Theorem*. In order to overcome this issue, Ullrich introduced an instructional ontology, which is domain independent and pedagogically sound [27]. One of the most important advantages of this ontology is its pedagogical flexibility, being independent of a particular instructional theory. Moreover, as we will show in section 3, the ontology can also be enhanced to serve adaptivity purposes, from the point of view of various learning styles.

The rest of the paper is structured as follows: the next section gives a short overview of adaptive educational systems that focus on the learning style of the students and sketches our approach. Section 3 describes the suggested organization of the learning resources and introduces the educational metadata used. Sections 4 and 5 illustrate the use of these metadata for learner modelling and adaptation provisioning respectively. Finally, some related works are presented in section 6 and conclusions are drawn in section 7.

2 Learning Style-based Adaptive Educational Systems

One of the most important goals of today's research in e-learning refers to the provision of an adaptive educational experience, that is individualized to the particular needs of the learner, from the point of view of knowledge level, goals or motivation. The purpose of this adaptation is to maximize the subjective learner satisfaction, the learning speed (efficiency) and the assessment results (effectiveness) [3].

Learning style-based adaptive educational systems (LSAES) are a special case of adaptive educational systems (AES), which focus on students' learning preferences as the adaptation criterion. According to [21], learning styles represent a combination of cognitive, affective and other psychological characteristics that serve as relatively stable indicators of the way a learner perceives, interacts with and responds to the learning environment. At present there is a large number of learning style models proposed in the literature (over 70 according to [10]), which differ in the learning theories they are based on, the number and the description of the dimensions they include. There are also a few educational systems that deal with them [25]. Some examples include: INSPIRE [23] (based on Honey and Mumford learning style model [18]), EDUCE [22] (based on Gardner's theory of multiple intelligences [15]), CS383 [8], Heritage Alive Learning System [9] and ILASH [5] (all based on Felder-Silverman learning style model [14]).

The main problem of the above systems is that they only take into account a single learning style model. Moreover, most of them use an explicit learner modelling method, asking the student to fill in a dedicated psychological questionnaire. The resulted membership to a particular learning style is stored once and for all in the student model kept by the system and is subsequently used for adaptation. A few systems also adopt an implicit learner modelling method, trying to identify the student learning preferences dynamically, by monitoring and analyzing student behavior in the system. The approach that we propose in [24] is included in the latter category; furthermore it is not tied to a particular learning style model, but it integrates the most relevant characteristics from several models proposed in the literature, such as:

- perception modality: visual, auditory, kinesthetic
- processing information (abstract concepts and generalizations vs. concrete, practical examples; serial vs. holistic; active experimentation vs. reflective observation; careful vs. not careful with details)
- reasoning (deductive vs. inductive)
- organizing information (synthesis vs. analysis)
- motivation (intrinsic vs. extrinsic; deep vs. surface vs. strategic vs. resistant approach)
- pacing (concentrate on one task at a time vs. alternate tasks and subjects)
- social aspects (individual work vs. team work; introversion vs. extraversion; competitive vs. collaborative).

Our first objective is to dynamically model the learner: identify the learning preferences by analyzing the behavioral indicators and then, based on them, infer the belonging to a particular learning style dimension. The second objective is to consequently adapt the navigation and the educational resources to match the student learning preferences (see figure 1 for a schematic description of the process). In order to achieve these two objectives, we need an intelligent way of organizing the learning material as well as a set of instructional metadata to support both the learner modelling and the adaptation.

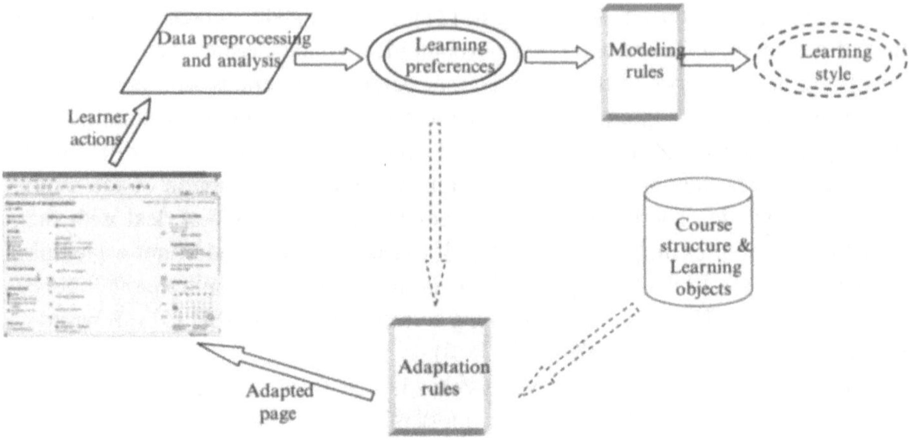

Fig. 1. Schematic representation of our LSAES

3 Organizing the Educational Material in an LSAES

According to [20], learning objects represent any digital resources that can be reused to support learning. In our case, the most complex learning object (with the coarsest granularity) is the course, while the finest granularity learning object is the elementary educational resource. We have conceptualized the learning material using the hierarchical organization illustrated in figure 2: each course consists of several chapters, and each chapter can contain several sections and subsections. The lowest level subsection contains the actual educational resources. Each such elementary learning object corresponds to a physical file and has a metadata file associated to it.

Based on our teaching experience, this is the natural and most common way a teacher is usually organizing his or her teaching materials. Additionally, this hierarchical approach presents several advantages, facilitating:

- good reuse of the educational resources
- detailed learner tracking (since we know all the information about the learning resource that is accessed by the learner at a particular moment) - see section 4
- fine granularity of adaptation actions - see section 5.

As far as the educational metadata is concerned, one possible approach (which is used in [16]) would be to associate to each learning object the learning style that it is most suitable for. One of the disadvantages is that this approach is tied to a particular learning style. Moreover, the teacher must create different learning objects for each learning style dimension and label them as such. This implies an increase in the workload of the teacher, and also the necessity that she/he possesses knowledge in the learning style theory. Furthermore, this approach does not support dynamic learner modelling, since accessing a learning object does not offer sufficient information regarding the student (a learning object can be associated with several learning styles).

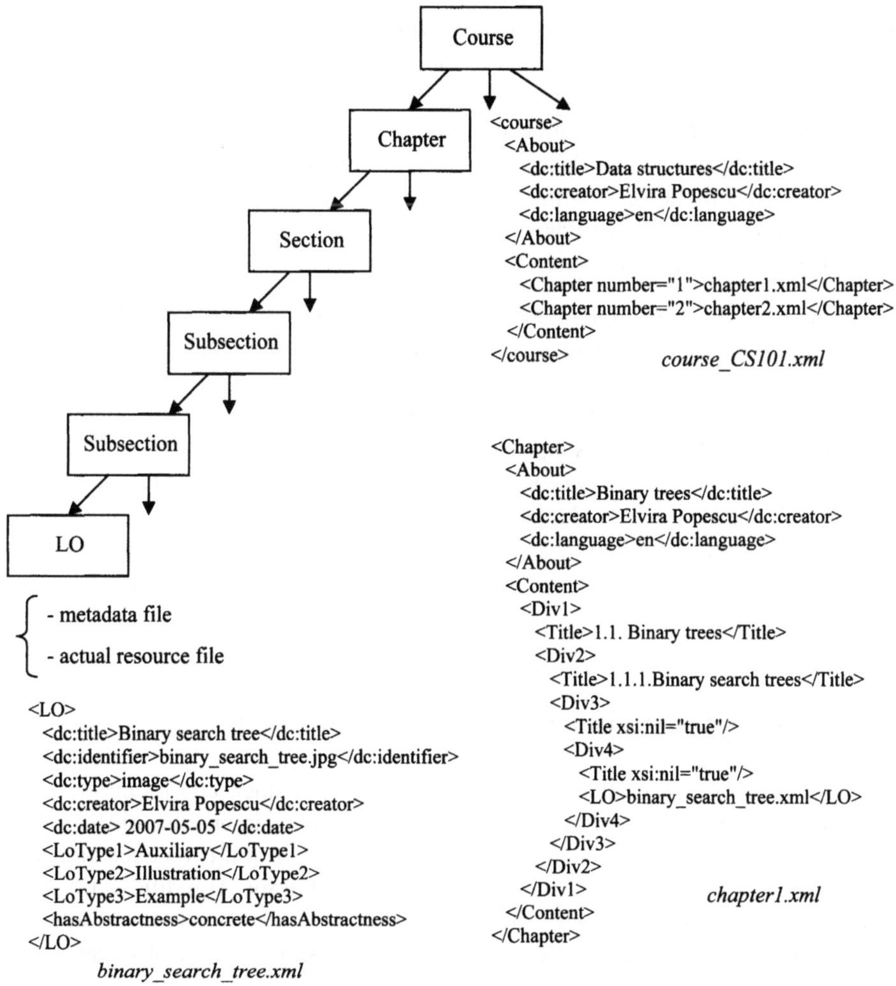

Fig. 2. Suggested organization of the learning content in an LSAES

Instead, we propose a set of metadata that describe the learning object from the point of view of instructional role, media type, level of abstractness and formality, type of competence etc. These metadata were created by enhancing core parts of Dublin Core [12] and Ullrich's instructional ontology [27] with some specific extensions to cover the requirements of an LSAES. Thus some of the descriptors of a learning object are:

- title (the name given to the resource) → dc:title
- identifier (a reference to the actual resource, such as its URL) → dc:identifier
- type (the nature of the content of the resource, such as text, image, animation, sound, video) → dc:type

- format (the physical or digital manifestation of the resource, such as the media-type or dimensions of the resource) → dc:format
- instructional role, either i) fundamental: definition, fact, law (law of nature, theorem) and process (policy, procedure) or ii) auxiliary: evidence (demonstration, proof), explanation (introduction, conclusion, remark, synthesis, objectives, additional information), illustration (example, counter example, case study) and interactivity (exercise, exploration, invitation, real-world problem) → LoType1, LoType2, LoType3, LoType4.

Obviously, these descriptors are independent of any learning style. However, by analyzing the interaction between the student and the learning objects described by these metadata (time spent on each learning object, order of access, frequency of accesses), the system can infer a particular learning preference of the student. Furthermore, the teacher has to supply only annotated learning content (the static description) while the adaptation logic (the dynamic description) is provided by the system. This means that the adaptation rules are independent of the learning content and that they can be supplied by specialists in educational psychology. The next two sections illustrate the use of these metadata for modelling the learner and providing adaptation respectively.

4 Educational Metadata and Learner Modelling

The first step towards dynamic learner modelling is to track and monitor student interactions with the system. Student observable behavior in an educational hypermedia system includes: i) navigational indicators (number of hits on educational resources, navigation pattern); ii) temporal indicators (time spent on different types of educational resources proposed); iii) performance indicators (total learner attempts on exercises, assessment tests). Based on the interpretation of these observables, the system can infer different learning preferences. Knowing the media type, the instructional role as well as other characteristics of the learning object the student interacts with is essential for an accurate identification of the learning preferences. Figure 3 illustrates the possible use of some of the learning object metadata.

5 Educational Metadata and Adaptation Logic

Modelling the learner is not a goal in itself. The value of a student model lies in its usability for providing a learning experience which is most beneficial for the student. Specifically, this could mean several things: in some cases, the most suitable attitude is to offer the student the educational resources that match their learning preferences, in terms of media type, order of resources, communication and collaboration facilities, level of navigation guidance etc. In other situations, students could benefit more from being faced with a mismatched learning environment, which provides the necessary challenge to boost learning [22]. Moreover, when learners are firstly offered

Learning preference	Behavioral indicators	Corresponding metadata tag
Visual preference	High amount of time spent on contents with graphics, images, video	<dc:type>image</dc:type> <dc:type>video</dc:type>
Verbal preference	High amount of time spent on reading text	<dc:type>text</dc:type> <dc:type>audio</dc:type>
Abstract concepts and generalizations	Access of abstract content first (concepts, definitions) High amount of time spent on abstract content	<LoType2>Definition</LOType2> <LoType2>Law</LOType2> <hasAbstractness>abstract</hasAbstractness>
Concrete, practical examples	Access of concrete content first (examples) High amount of time spent on concrete content	<LoType2>Illustration</LoType2> <LoType2>Fact</LOType2> <hasAbstractness>concrete</hasAbstractness>
Active experimentation	Access of practical content (simulations, exercises, problems…) before theory	<LoType2>Interactivity</LoType2>
Reflective observation	Access of theoretical content before practical content	<LoType1>Fundamental</LoType1>
Synthetic	High performance on exercises requiring synthesis competency	<hasCompetency>synthesis</hasCompetency>
Analytic	High performance on exercises requiring analysis competency	<hasCompetency>analysis</hasCompetency>

Fig. 3. Correspondence between learning preferences and educational metadata

the educational content that doesn't match their learning preferences, they will usually not limit themselves to that particular resource, being inclined to access more of the available resources on the subject.

The application of one or the other of the above methods depends on the intended pedagogical objective and the characteristics of the target students (knowledge level, motivation, goals). The advantage of our approach is that it allows complete independence between the learner model and the pedagogical model: various adaptation actions can be associated with each learner preference. Furthermore, we could combine several pedagogical goals, using some of the identified learning preferences to improve the efficiency of the learning process (matching), others to provide the needed challenge and variety or to develop weaker skills (mismatching) and others to increase student's self-awareness about her/his strengths and weaknesses in the learning process (open model approach). Figure 4 illustrates a possible use of the detected learning preferences for a particular student. The adaptation techniques suggested are classified according to the levels of adaptation identified in [7] and [3].

As we can see, the adaptation techniques can be decomposed into elementary adaptation actions (sorting/inserting/removing learning objects) based on various criteria, all of which are included in the metadata:

- media type → dc:type
- instructional role → LoType1, LoType2, LoType3, LoType4
- level of abstractness → hasAbstractness
- type of competency required (in case of exercises) → hasCompetency

Learning preference	Matched learning experience	Adaptation techniques
Visual	The course should include plenty of multimedia objects based on video and images; the content will be presented as much as possible using graphics and schemas.	Content level adaptation (specific media type filtering)
Concrete, practical examples	The course should be focused more on facts, practical aspects and examples. Each new concept will be first illustrated by an example and only then the theoretical aspects will be covered.	Content level adaptation (content hiding, specific item filtering) Presentation level adaptation (sorting fragments, dimming fragments)
Holistic	The course will include outlines and summaries for each course item, which will be presented at the beginning and end of each chapter and will be permanently accessible through a menu. The links to related or complex topics will be integrated in the content, to help situate the learnt subject and contribute to create the big picture. The exercises will be placed at the end of the chapter, not after each course item, in order to give the users the opportunity to holistically understand the subject first	Navigation level adaptation (link annotation, link generation) Content level adaptation (additional explanations) Presentation level adaptation (inserting fragments, sorting fragments)

Fig. 4. Ways of providing adaptivity for different learning preferences

6 Related Works

Currently there are several works that address aspects related to ontologies and metadata for personalized e-learning, such as: [1], [6], [13], [16], [17], [26]. A few of them, that we will briefly discuss here, also take into consideration learning styles.

In case of [16] the ontology is tied to a particular learning style model, namely Felder-Silverman (FSLSM) [14]. There is a special class, *LearningStyle*, which represents the FSLSM dimension associated to a particular learning object (active-reflective, visual-verbal, sensing-intuitive, sequential-global). Thus all learning objects have to be indexed according to FSLSM in order to allow for delivering of adapted content.

Paper [6] proposes a learning style taxonomy, based on Curry's onion model [11]. In the LAG adaptation model, each learning style can be associated with a specific instructional strategy, which can be broken down into adaptation language constructs, which in their turn can be represented by elementary adaptation techniques. It is the role of the author to specify not only the annotated learning content (the static description) but also the adaptation logic (the dynamic description).

Finally, paper [26] introduces the concept of Open Learning Objects, which represent distributed multimedia objects in SVG format. They incorporate inner metadata in XML format which is structured on several levels (content, adaptation, animation...). Each Open Learning Object is tied to a particular learning style dimension; however any learning style model can be employed, by configuring the adaptation markup.

7 Conclusions

In this note we sketched an intelligent way of organizing the learning resources in an LSAES. Based on Dublin Core metadata [12] and Ullrich's instructional ontology [27], we introduced a set of educational metadata that are independent of any learning style. We then showed how these metadata can be employed for modelling the learner and applying various adaptation techniques. Currently, an adaptive educational system based on the proposed approach is under development, which will provide a validating framework for our conceptualization.

References

1. Al-Khalifa, H. S., Davis, H. C.: The Evolution of Metadata from Standards to Semantics in E-Learning Applications. *Proceedings of Hypertext'06*, Odense, Denmark (2006).
2. ADL SCORM. `http://www.adlnet.gov/scorm/index.aspx`.
3. ALFANET - D8.2 Public final report. `http://rtd.softwareag.es/alfanet/` (2005).
4. Anido, L., Fernandez, M., Caeiro, M., Santos, J., Rodriguez, J., Llamas, M.: Educational metadata and brokerage for learning resources. *Computers and Education*, Elsevier, 38 (2002) 351-374.
5. Bajraktarevic, N., Hall, W., Fullick, P.: Incorporating learning styles in hypermedia environment: Empirical evaluation. *Procs. Workshop on Adaptive Hypermedia and Adaptive Web-Based Systems* (2003) 41–52.
6. Brown, E., Cristea, A., Stewart, C., Brailsford, T.: Patterns in Authoring of Adaptive Educational Hypermedia: A Taxonomy of Learning Styles. *Educational Technology and Society*, 8, (2005) 77–90.
7. Brusilovsky, P.: Adaptive navigation support: from adaptive hypermedia to the adaptive Web and beyond. *PsychNology Journal*, Vol. 2, No. 1 (2004) 7–23.
8. Carver, C. A., Howard, R. A., Lane, W. D.: Enhancing student learning through hypermedia courseware and incorporation of student learning styles. *IEEE Transactions on Education*, 42 (1999) 33–38.
9. Cha, H. J., Kim, Y. S., Park, S. H., Yoon, T. B., Jung, Y. M., Lee, J. H.: Learning styles diagnosis based on user interface behaviors for the customization of learning interfaces in an intelligent tutoring system. *Proc. 8th Intl. Conf. on Intelligent Tutoring Systems*, LNCS, Vol. 4053, Springer (2006).
10. Coffield, F., Moseley, D., Hall, E., Ecclestone, K.: *Learning styles and pedagogy in post-16 learning. A systematic and critical review*. Learning and Skills Research Centre, UK (2004).
11. Curry, L.: *Integrating Concepts of Cognitive or Learning Style: A Review with Attention to Psychometric Standards*. Ottawa, ON: Canadian College of Health Service Executives (1987).
12. Dublin Core Metadata Initiative `http://dublincore.org/`.
13. Dolog, P., Henze, N., Nejdl, W., Sintek, M.: Personalization in distributed eLearning environments, *Proc. Intl. Conf. WWW2004* (2004).
14. Felder, R. M., Silverman, L. K.: Learning and teaching styles in engineering education. *Engineering Education*, Vol. 78, No. 7 (1988).
15. Gardner, H.: *Multiple Intelligences: The Theory in Practice*. Basic Books, New York (1993).

16. Gascuena, J., Fernndez-Caballero, A., Gonzalez, P.: Domain Ontology for Personalized E-Learning in Educational Systems. *Proc. 6th IEEE Intl. Conf. on Advanced Learning Technologies - ICALT 2006*, IEEE Computer Society Press (2006).
17. Geser G. (Ed.): *OLCOS Roadmap 2012 - Open Educational Practices and Resources*, Austria (2007).
18. Honey, P., Mumford, A.: *The learning styles helper's guide*. Maidenhead: Peter Honey Publications Ltd. (2000).
19. IEEE LOM. `http://ltsc.ieee.org/wg1/`
20. IMS MD. `http://www.imsglobal.org/metadata/index.html`.
21. Keefe, J.W.: Learning style: an overview. *NASSP's Student Learning Styles: Diagnosing and Prescribing Programs* (1979) 1–17.
22. Kelly, D., Tangney, B.: Adapting to intelligence profile in an adaptive educational system. *Interacting with Computers*, Elsevier, 18 (2006) 385–409.
23. Papanikolaou, K.A., Grigoriadou, M., Kornilakis, H., Magoulas, G.D.: Personalizing the interaction in a Web-based educational hypermedia system: the case of INSPIRE. *User-Modeling and User-Adapted Interaction*, 13 (2003) 213–267.
24. Popescu, E., Trigano, P., Badica, C.: Towards a Unified Learning Style Model in Adaptive Educational Systems. *Proc. 7th IEEE Intl. Conf. on Advanced Learning Technologies - ICALT 2007*, IEEE Computer Society Press (2007) (in press).
25. Popescu, E., Trigano, P., Badica, C.: Adaptive Educational Hypermedia Systems: A Focus on Learning Styles. *Proc. 4th IEEE Intl. Conf. on Computer as a Tool - EUROCON 2007* (2007) (in press).
26. Shi, H., Rodriguez, O., Chen, S., Shang, Y.: Open Learning Objects as an Intelligent Way of Organizing Educational Material. *International Journal on E-Learning*, 3(2) (2004) 51–63.
27. Ullrich, C.: The Learning-Resource-Type is Dead, Long Live the Learning-Resource-Type!. *Learning Objects and Learning Designs*, 1 (2005) 7–15.

JTangPS: Content-Based Publish/Subscribe over Structured P2P Networks

Dongcai Shi, Jianwei Yin, Jianfeng Qian, Yiyuan Li, Jinxiang Dong

Department of Computer Science and Technology, Zhejiang University, Hangzhou, China
{shidcai, jfqian, morkran}@163.com, {zjuyjw,djx}@zju.edu.cn

Summary. Publish/subscribe systems have been adopted by many distributed applications to deliver data from publishers to subscribers in an anonymous fashion. However, existing content-based publish/subscribe systems suffer from limitations in scalability. This paper introduces a novel architecture for implementing a content-based publish/subscribe system, called JTangPS, on top of structured P2P networks. JTangPS achieves scalability by mapping subscriptions and events to rendezvous nodes in terms of the combination of domain schema identifier, number of attributes and the first attribute name in the subscription or event. JTangPS enhances routing efficiency by exploiting the embedded multicast tree on P2P networks and by aggregating event delivery messages. Experimental results using a variety of synthesis datasets demonstrate the scalability of system with large numbers of users.

1 Introduction

The publish/subscribe system is well suited to loosely-coupled large-scale distributed applications. In such a system, publishers can be completely unaware of subscribers and simply publish events into the system. Subscribers can register their interests with the system in the form of subscription. The publish/-subscribe system manages subscriptions and delivers events to the matched subscribers. Many distributed applications such as online stock quotes, internet games, and sensor networks can benefit from the presence of publish/subscribe systems.

Traditional distributed content-based publish/subscribe systems, such as Siena [1] and Gryphon [2], typically are built on a small number of trusted brokers that are inter-connected in an overlay network. However, such systems do not provide adequate support for dynamically joining and leaving of the brokers in the overlay network. Specially, when a single broker in the overlay network crashes, it may result in the partition of overlay network and a large number of state transfer operations during recovery. Hence, traditional distributed publish/subscribe systems lack the adaptive ability to node failures or changing of topology.

Dongcai Shi et al.: *JTangPS: Content-Based Publish/Subscribe over Structured P2P Networks*, Studies in Computational Intelligence (SCI) **78**, 187–196 (2008)
www.springerlink.com

Recently, structured peer-to-peer (P2P) networks (e.g. Pastry [3]) based on distributed hash table (DHT) gain more and more visibility due to their high scalability, fault-tolerance and self-organization. Such characteristics make structured P2P networks attractive for building distributed publish/subscribe systems. They generally employ P2P routing substrate to dynamically construct the event dissemination tree, or map subscriptions and events to the rendezvous nodes (RNs) which match the events to the subscriptions and deliver the events to relevant subscribers. Scribe [4] and Bayeux [5] are two typical publish/subscribe systems built on top of Pastry and Tapestry respectively. However, both of them are topic-based publish/subscribe systems, thus offering limited expressiveness to users.

To address these limitations mentioned above, we present a new approach to building a distributed content-based publish/subscribe system on top of structured P2P networks. Our proposed system called JTangPS can serve as a general architecture that not only host an application with one or more domain schemas, but also hold many publish/subscribe services simultaneously. JTangPS maps subscriptions and events to rendezvous nodes in terms of a combination of domain schema identifier, number of attributes and the first attribute name in the subscription or event. Events do not have to be published to all RNs, only to the RNs at which the number of attributes in the stored subscriptions is equal to or less than that in the event to be published and the access point attribute of the event is the same as that in the subscription. This reduces the quantity of event publication messages. In addition, JTangPS exploits the implicit multicast tree on DHT routing substrate and aggregates event notification messages to deliver events, thereby imposing little overhead on substrate. Moreover, JTangPS is resilient to node failures and provides reliable event delivery by leveraging the fault-tolerance and self-organizing nature of DHT.

The rest of the paper is structured as follows. Section 2 provides a survey of related work. Section 3 presents the basic design of our proposed approach. Section 4 presents experimental detail and results. We conclude this paper and suggest directions for future work in Section 5.

2 Related work

Much research has been done to build content-based publish/subscribe systems over structured P2P networks, which exploit the underlying DHT infrastructure to different degrees. Here we just present the most related work due to space limitation.

Tam et al. [6] propose P2P-ToPSS, a publish/subscribe system built from Scribe. It provides support for mapping a subscription or event to multiple topics. A potential limitation of the system is that it has some restrictions on the expression of content. There are also problems of partially matching and duplicated notification of events.

Meghdoot [7] based on CAN imposes no restrictions over subscriptions and allows them to be specified in terms of range predicates over all the attributes in a schema. It addresses the load balancing issue resulting from skewed real world datasets by zone splitting and zone replication. The major drawback of Meghdoot is that the overlays dimension is twice the number of event attributes.

Baldoni et al. [8] present three different methods for mapping subscriptions and events to overlay keys, namely Attribute-Split, Key Space-Split and Selective-Attribute. These approaches either map the subscription to a large number of keys or publish the event to all RNs, which impedes system scalability.

Terpstra et al. [9] present a publish/subscribe system featuring many event multicast trees over Chord. Since every node in the overlay network has an event delivery tree rooted at itself, subscriptions are propagated up simultaneously on all multicast trees. Hence, it is not an easy task to install and update the subscriptions when they are added or removed frequently.

3 JTangPS Design

In this section, we first discuss the publish/subscribe model that represents the rules and constraints which content-based publish/subscribe systems follow when providing content. Then, we present the design of JTangPS over DHT-based networks in detail.

3.1 Publish/Subscribe Model in JTangPS

We consider a variant of schema based on the model proposed by Fabret et al. [10] for content-based publish/subscribe systems. Differently from the schema in [10], ours can be described as a sequence $\varsigma =< A_1, A_2, ..., A_n >$, where each A_i represents an attribute. An attribute can be defined as a tuple $(Name, Type, Value_{max}, Value_{min})$ and each has a name, type and value domain. A subscription is a conjunction of predicates where a predicate specifies one continuous range or constant value over one or more attributes. An example subscription is $s =< A_1 = v_1, v_2 < A_2 < v_3 >$. An event can be represented as $e =< A_1 = c_1, A_1 = c_1, ..., A_n = c_n >$ which is a sequence of equalities over attributes. Note that a subscription or an event is also a sequence in which the precedence of attributes is in accordance with the corresponding schema. An event e matches a subscription s if each predicate of s is satisfied by the value of the corresponding attribute specified by the event e.

Every application domain may have a different schema with a unique identifier. The system is able to simultaneously handle multiple domain schemas.

3.2 Subscription Installation

In distributed content-based publish/subscribe systems, there are two major approaches to solving the subscription propagation problem: 1) source-based forwarding approach, where the subscription is flooded from one node to all the others, and then the event is routed to all entries on a destination list generated through matching against all subscriptions; and 2) rendezvous-based forwarding approach, where subscriptions and events are first sent to some of RNs, and unwanted events are filtered out at these RNs before delivery. These approaches have different tradeoffs in routing efficiency and load balancing.

We propose an approach of mapping a subscription to the RN in terms of the combination of domain identifier (or domain name), number of attributes and the first attribute name in the subscription to carefully balance the tradeoffs mentioned above. In the proposed approach, a subscription is represented by a triple (*did, sid, f*), where *did* is the unique identifier of domain, *sid* is a subscriber's node ID and *f* is a filter that specifies the subscriber's interest in specific events. A subscription is installed at the RN whose node ID is equal to or most immediately precedes the key $k = hash(did : num_a : name_a)$, where *hash()* is a consistent hash function(e.g. SHA-1), *did* is the domain identifier, num_a is the number of attributes over which constraints are defined, and $name_a$ is the name of the first attribute in the subscription. Note that the first attribute in the subscription is also called the access point which is used in the publication and matching of events. In addition, we assume the means of subscription management at RNs and the event matching algorithm which are similar to those in [10].

When a subscriber wishes to unsubscribe from the system, it simply sends an unsubscribing message to the corresponding RN to remove its subscriptions. When a new node joins the system, the subscriptions mapped to its partition in the key space have to be moved to it from other nodes. While a node leaves, the subscriptions that it stores should be relocated to its neighbors in the key space. As DHT-based P2P networks generally maintain the correct state by exchanging messages, it is convenient for us to move subscriptions between neighbors when a node joins or leaves. On the other hand, the subscriptions are replicated on a small number of neighbors to avoid losing them if there is a RN failure.

3.3 Event Publication

In the proposed approach, the event is not sent to all RNs but only to some of them. The basic idea is to direct the event to the RNs at which the number of attributes in each stored subscription is equal to or less than that of the event, and the access point of event is the same as that in the subscription. Note that the access point in the subscription is the first attribute while in the event it may be any one.

When a node wishes to publish an event, the event is first sent to the RN following the same scheme as described in section 3.2 On receiving the event, the RN first matches it with the subscriptions to find matched ones. Then it continues to publish the event according to the following steps:

$$if \quad NAS(RN) > 1 \quad then \quad k = hash(did : (NAS(RN-1)) : AP) \tag{1}$$

$$if \quad NAS(RN) + Index(AP) == NAE(Event)\&\&NAS(RN) > 1$$
$$then \quad k = hash(did : (NAS(RN-1)) : Next(AP)) \tag{2}$$

The functor *NAS* in the first step stands for "Number of Attributes in the stored Subscription". The RN figures out whether the number of attributes in the stored subscription is greater than one or not. If true, it forwards the event to the RN whose node ID is numerically equal to or most immediately precedes the key $k = hash(did : (NAS(RN-1)) : AP)$, where AP denotes the access point name of the event.

In step (2), *Index(AP)* indicates the index of access point in the event, and the functor *NAE* stands for "Number of Attributes in the Event". The RN will go on with event publication if the condition is satisfied that not only the summation of *NAS(RN)* and *Index(AP)* is equal to *NAE(Event)* but also *NAS(RN)* is greater than one. The event is directed to the RN whose node ID is equal to or most immediately precedes the key $k = hash(did : (NAS(RN-1)) : Next(AP)$, where *Next(AP)* refers to the name of the attribute next to the current access point of the event.

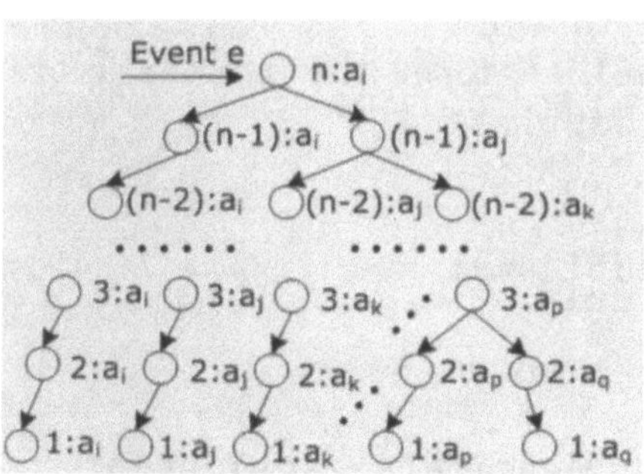

Fig. 1. Event publication

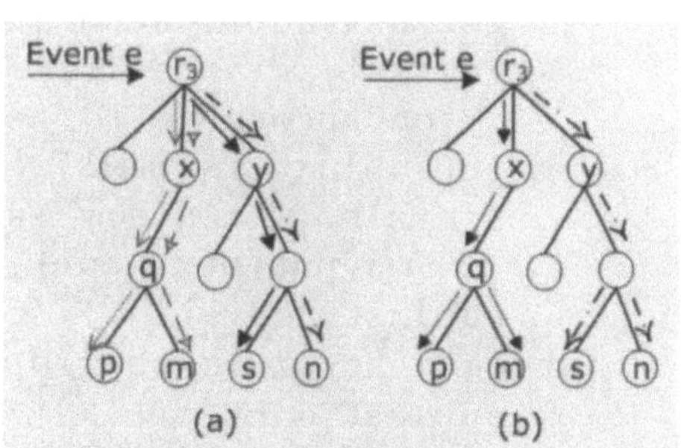

Fig. 2. Event delivery

We illustrate the publication of event $e =< a_i, a_j, a_k, ..., a_m, a_p, a_q$ with n attributes in Fig. 1. The publisher first publishes e to the RN whose node ID is numerically equal to or most immediately precedes the key $k = hash(did : n : a_i)$. After the RN, labeled as $n : a_i$, receives the event, it follows the steps above to forward the event to others, labeled as $(n-1) : a_i$ and $(n-1) : a_j$. Each RN processes the received event in the same way until the event is published to all the corresponding nodes. Since the event publication tree is dynamically constructed according to the event and subscriptions at RNs, it is not necessary for nodes to provide extra support for the maintenance of publication tree, therefore imposing little overhead on the system.

It is worth pointing out that an event may be published to the almost entire RNs at which the number of attributes in each stored subscription is equal to or less than that in the event. In the worst case that the event has all the attributes in the corresponding schema, it has to be sent to all RNs in its domain. As a matter of fact, since most real world datasets tend to be highly skewed rather than uniformly distributed, the proposed approach can efficiently reduce the number of exchange messages for event publication in the system.

3.4 Event Delivery

We use the routing schema on structured P2P substrate to implement event delivery. However, if we do not leverage the primitive routing schema on the substrate, it may result in a serious message redundancy on links. For example, as shown in Fig. 2 (a), when *r3* node delivers event e to all the matched subscribers such as *q*, *p* and *s*, link *r3-x* and *r3-y* have two copies of event e respectively, which inflicts unnecessary overhead on them.

```
Algorithm deliverEvent(e, mslist) {
  for each subscriber ID sid in the mslist
        if(sid == local node ID)
              deliver event e to local subscriber;
              continue;
        get nexthopid according to sid;
        get subscribers list subList form the router map
           rm according to nexthopid;
        if(subList is null)
              allocate a new subList;
              rm.put(nexthopid, subList);
        subList.add(sid);
  for each key k in the router map rm
        subList = rm.get(k);
        route(k, new NotificationMessage(e, subList));
}
```

In order to alleviate the link press, all event delivery messages to the subscribers that share common ancestor nodes on the embedded multicast tree are aggregated into one single notification message along the path from the root node to their lowest common ancestor node, thereby minimizing the flow of messages. As outlined in the algorithm *deliverEvent()*, for each subscriber ID *sid* in the matched subscribers list *mslist*, we compare *sid* with the node ID of *r3*. If sid is equal to *r3*'s ID, it delivers the event e to its subscriber, otherwise it chooses a neighbor whose node ID *nexthopid* is equal to or most immediately precedes sid and add sid into the list to which *nexthopid* is mapped. And then, it forwards all the neighbors in the router map the event notification message which carries a corresponding subscribers list *subList* and the event e. On receiving the message, the neighbor node extracts the subscribers list and the event, and enforces the same algorithm *deliverEvent()* to dispose of the remaining subscribers. Fig. 2 (b) shows the aggregate event delivery in JTangPS.

Note that the aggregate event delivery algorithm has two main features: 1) to utilize and leverage the self-organization and scalable routing capabilities provided by overlay networks so that it is not necessary to dynamically construct the event multicast tree according to the reverse path of subscriptions and makes event routing resilient to node joining, departure and failures; 2) to assume the aggregation of messages to deliver events so that no redundant messages are transmitted on overlay networks.

4 Experimental Evaluation

We evaluate the performance of system along the following dimensions: (1) Total Notification Cost (TNC): a basic metric for the publish/subscribe system, calculated by summing the costs of all peer-to-peer event notification message traffic; (2) Average Event Delay (AED): the average time taken by the system to deliver an event to all of its matched subscribers. The absolute value of average event delay is meaningless due to underlying network latency, but its relative value gives a useful characterization.

The metrics above captures an important aspect of scalability and routing efficiency by revealing how performance is impacted by increases to the load presented to the system. Before we discuss the results, we describe the setup of the experiments.

4.1 Experimental Setup

We developed a prototype of JTangPS on top of FreePastry [11], an open source implementation of Pastry. We exploited the Pastry infrastructure for node management and message routing. The key space is 2^{160}.

Experiments are conducted by generating subscriptions and events over a twelve attributes event space. All the attributes are integers, ranging from 0 to 1000. Each subscription or event specifies 1 to 12 predicates. We use either an uniform or a zipf distribution ($\alpha = 0.8$) to select attributes, while the operator and value fields are chosen uniformly randomly.

If not specified differently, the parameters used for experiments are as follows: the number of nodes is 5000; the number of subscriptions is 10,000; the number of subscriptions per subscriber is 20. Event publishers are randomly selected from the nodes in the simulation overlay network. Events are injected with an average interval of 10ms and 10,000 events in all are published.

4.2 Experimental Results

In the plots below we use the aliases shown in Table 1.

Table 1. Aliases for subscription distribution, event distribution and event delivery mode

Aliases	Subscription Distribution	Event Distribution	Event Delivery Mode
UA	uniform	uniform	aggregate
UNA	uniform	uniform	non-aggregate
ZA	zipf	zipf	aggregate
ZNA	zipf	zipf	non-aggregate

Fig. 3 shows the total notification cost under four conditions. The horizontal axis gives the number of events or of subscriptions in a logarithmic scale ranging from 1 to 10,000. The main vertical axis (left) gives a logarithmic measure of cost while the secondary axis (right) gives a linear measure of cost. The first observation from two main coordinates of Fig. 3 is that all the total notification cost under four conditions grows linearly with the number of subscribers and of events; the zipf distribution generates more notification messages than the uniform distribution. Another observation from two secondary coordinates is the aggregate event delivery mode largely reduces notification messages. For the subscriptions of 10,000 under zipf distribution, the number of messages generating by aggregation mode is five hundred million less than non-aggregation mode. Hence, the scalability of system can benefit from the aggregate event delivery mode. As shown in Fig. 4, at a fixed event

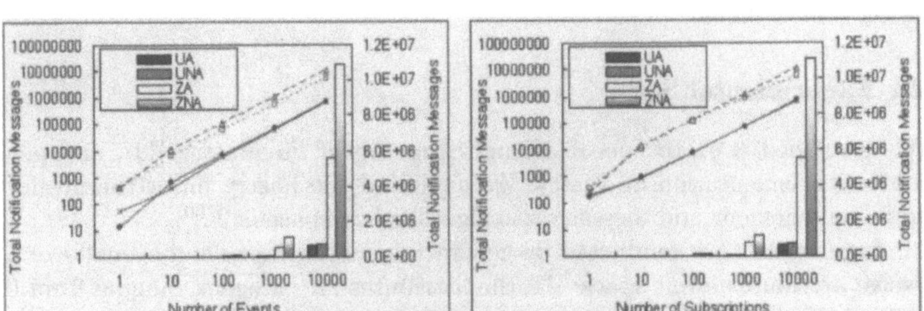

Fig. 3. Total notification cost

publication rate, increases to the subscriptions have a slight influence on the average event delay, whereas increases to the events have little impact on it. It is clear that more subscriptions registered, more time taken to match events to subscriptions. In contrast, event increments do not affect the matching time per event. The results point out that the efficiency of matching algorithm is very important for the scalability of system.

We discuss the delay penalty of aggregation here. As shown in Fig. 4 (a), under zipf distribution, when the subscriptions are less than 20,000, the average event delay with aggregation is greater than non-aggregation. With the increments of subscriptions, the positive efficacy due to aggregation is more and more obvious. Interestingly, as plotted in Fig. 4 (b), no matter how the number of events varies, as long as the number of subscriptions is fixed, the average event delay is almost constant at the certain value which approximates to that under the same number of subscriptions in Fig. 4 (a).

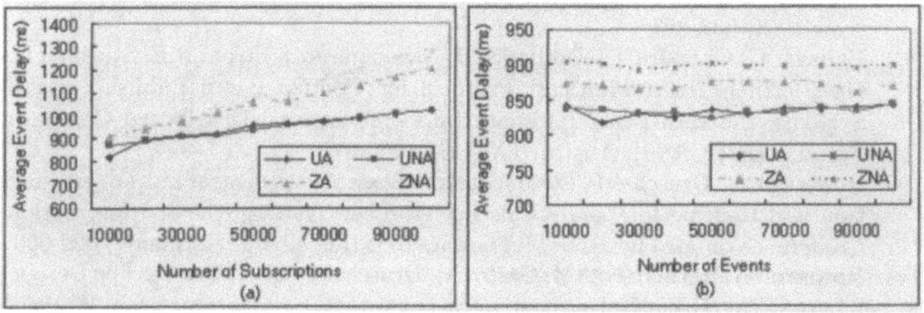

Fig. 4. Average event delay

5 Conclusion and Outlook

We presented JTangPS, a large-scale and fully decentralized content-based publish/subscribe system built on top of structured P2P network. JTangPS has several unique features: (1) JTangPS maps subscriptions and events to rendezvous nodes in terms of the combination of domain schema identifier, number of attributes and the first attribute name in the subscription or event. Events do not have to be published to all RNs, only to the RNs at which the number of attributes in the stored subscriptions is equal to or less than that in the event to be published and the access point attribute name of event is the same as that in the subscription. This decreases the number of exchange messages for event publication. (2) JTangPS exploits the implicit multicast tree on DHT routing substrate and aggregates event delivery messages to disseminate events, thereby imposing little extra overhead on substrate. Moreover, by leveraging the fault-tolerance and self-organizing nature of DHT, JTangPS is resilient to node failures and provides reliable event delivery. (3) JTangPS can serve as a general architecture that not only hosts the application with one or more domain schemas, but can also hold many publish/subscribe services simultaneously. In addition, JTangPS can be implemented on any DHT-based P2P networks.

In this paper, we assumed content-based data model to support syntactically heterogeneous data. Currently, we work on a semantic data model to enhance the expressiveness of JTangPS. Besides this, we are investigating semantic matching algorithms to improve matching efficiency. As a future research direction, we intend to study semantic event routing algorithms.

References

1. Carzaniga A, Rosenblum D.S., Wolf A.L (2001) Design and Evaluation of a Wide-Area Event Notification Service. ACM Transactions on Computer Systems 19(3) 332-383
2. Banavar G, Chandra T, Mukherjee B, Nagarajarao J, Strom R.E, Sturman D.C (1999) An Efficient Multicast Protocol for Content-Based Publish-Subscribe Systems. the 19th IEEE International Conference on Distributed Computing Systems (ICDCS'99), Austin, TX, USA :262-272
3. Rowstron A, Druschel P (2001) Pastry: Scalable, Decentralized Object Location and Routing for Large-Scale Peer-to-Peer Systems. the 3rd International Conference on Middleware (Middleware'01), Heidelberg, Germany : 329-350
4. Rowstron A, Kermarrec AM, Castro M, Druschel P (2001) Scribe: The Design of a Large-Scale Event Notification Infrastructure. the 3rd International Workshop on Networked Group Communication (NGC2001), LNCS, London, UK 2233: 30-43
5. Zhuang S.Q, Zhao B.Y, Joseph A.D, Randy H, Katz J.D.K (2001) Bayeux: An Architecture for Scalable and Fault-tolerant Wide-area Data Dissemination. the 11th International Workshop on Network and OS Support for Digital Audio and Video (NOSSDAV'01)
6. Tam D, Azimi R, Jacobsen HA (2003) Building Content-Based Publish/Subscribe Systems with Distributed Hash Tables. the 1st International Workshop on Databases, Information Systems and Peer-to-Peer Computing, Berlin, Germany :138-152
7. Gupta A, Sahin O.D, Agrawal D, Abbadi A.E (2004) Meghdoot: Content-Based Publish/Subscribe over P2P Networks. the 5th ACM/IFIP/USENIX International Middleware Conference (Middleware 2004), Toroto :254-273
8. Baldoni R, Marchetti C, Virgillito A, Vitenberg R (2005) Content-Based Publish-Subscribe over Structured Overlay Networks. the 25th IEEE International Conference on Distributed Computing Systems (ICSCS'05)
9. Terpstra W.W, Behnel S, Fiege L, Zeidler A, Buchmann A.P (2003) A Peer-to-Peer Approach to Content-Based Publish/Subscribe. the 2nd International Workshop on Distrib-uted Event-Based Systems(DEBS'03). ACM Press, New York, USA :1-8
10. Fabret F, Jacobsen H.A, Llirbat F, Pereira J, Ross K.A, Shasha D (2001) Filtering Algo-rithms and Implementation for Very Fast Publish/Subscribe. the 20th International Conference on Management of Data (SIGMOD01), Santa Barbara, CA, USA :115-126
11. FreePastry: http://freepastry.org/FreePastry, (2006)

Using the color set back-projection algorithm in retrieval and evaluation of endoscopic images

Liana Stanescu,[1] Dan Burdescu,[1] Anca Ion[1] and Eugen Georgescu[2]

[1] University of Craiova Faculty of Automation, Computers and Electronics Craiova, Romania stanescu@software.ucv.ro
[2] University of Medicine and Pharmacy Craiova efgmed@yahoo.fr

Summary. This paper presents an original method of implementation of the color set back-projection algorithm, algorithm that allows the automated detection of the color regions from a color medical image. For the implementation of the color set back-projection algorithm, the image is transformed from RGB color space to HSV color space and quantized at 166 colors. At the end of this process, the color set of the image is obtained and used in color region detection. The software was tested on a gastroenterologic imagistic database that included 202 patients with peptic ulcer disease. One endoscopic image was taken at inclusion and another two at visit 1 and 2 for each patient, in order to evaluate the cicatrisation process under treatment. It was compared the percentage of correct diagnosis attended by using the classic observational method with the results obtained with the computer-based one in terms of reliability and reproducibility of this method. There were evaluated the advantages of using a computer-based retrieval system in terms of reducing the time spent on this operation. The inter-observer disagreement between human observer and the computer software was significantly low and the speed of the computerized method was higher, proving benefits in terms of time and cost efficiency.

1 Introduction

Medicine is one of the domains where a large quantity of gray-scale or color images, produced by medical apparatus for establishing a more correct diagnosis, is accumulated.

In the medical departments, hundreds or even thousands of images are collected, and a big part of them are color images, like the ones produced by endoscope - images that are taken into account in this paper.

Among color images from the digestive domain collected by an endoscope, a big part of the diseases is indicated by the abnormal aspect of the color and texture of tissue. There were effectuated a series of studies, there were created applications that permit the medical image processing and the complex query taking into consideration many criteria and characteristics [2], [3], [4], [5], [6].

Liana Stanescu et al.: *Using the color set back-projection algorithm in retrieval and evaluation of endoscopic images*, Studies in Computational Intelligence (SCI) **78**, 197–206 (2008)
www.springerlink.com

The application of an automated algorithm for detecting the color regions in medical images has two important utilizations:

1. in content-based region query on medical images collections, the specialist chooses one or more of the detected regions for querying the database, the purpose being the retrieval of images similar by color and texture; this can be useful for clarifying some uncertain diagnosis. This problem was studied in [1], [9] interesting results being presented.
2. during the evolution in time of the disease of patients that follow a certain treatment. This second direction is the object of study of this paper and is detailed as follows. The database is populated by color images captured and collected by an endoscope from a number of patients with ulcer disease.

The color set back-projection algorithm is applied on each image for detecting the single color regions, and the region(s) that represents the sick zone were marked as relevant. The color, the minimum bounding rectangle and the number of pixels that roughly indicates the dimension of the sick region, were stored in the database for each region. By applying a certain drugs treatment on some patients, at some time intervals strictly specified by the physician, the images were again collected by the endoscope from the same patients and the same algorithm is performed for detecting the color regions. The relevant region(s) that represents the sick tissue is marked.

The comparison between the new and old regions detected as relevant for the same patient, taking into consideration the number of pixels, can help the physician to establish in what percentage the sick region is reduced because of the administrated drugs. This approach may lead to a more rapid estimation and correct enough of the percentage in which the medication has a good effect in the ulcer diagnosis.

This process may be repeated several times, in conditions established by the physician. Based on the information stored in the database for each patient from the studied lot of sick persons, the statistics about the treatment efficiency can be generated. This may come in the help of the patients, specialists, and the drugs producers that intend to test a medical product.

2 The Color Set Back-Projection Algorithm

For detecting color regions it was chosen the color set back-projection algorithm, introduced initially by Swain and Ballard and then developed in the research projects at Columbia University, in the content-based visual retrieval domain [7]. This technique provides the automated extraction of regions and the representation of their color content.

The extraction system for color regions has four steps [7], [8]:

1. the image transformation, quantization and filtering (the transformation from the RGB color space to HSV color space and the quantization of the HSV color space at 166 colors)
2. back-projection of binary color sets
3. the labeling of regions
4. the extraction of the region features

The algorithm reduces the insignificant color information and makes evident the significant color regions, followed by the generation, in automatic way, of the regions of a single color, of the two colors, of three colors. To conclude with, the second step of the color set back-projection algorithm is the following [7]:

- Detection of single color regions
  1. Having the image histogram, H[m], all the values m'=m for which $H[m] \geq p_0$ are detected.
  2. For each m' the color set c having the property c[k]=1 for k=m and c[k]=0 in other cases is found. On the image R[m,n] the back-projection algorithm for each color set c is applied and the color regions are found. For each region n the local histogram Ln[m] is stored.
  3. The residue histogram Hr[m]=H[m]-$\sum_n L_n[m]$is computed.
- Detection of two colors regions
  1. The values l'=l and m'=m, $H[l] \geq p_0$, $H[m] \geq p_0$ and $Hr[l] \geq p_1$, $Hr[m] \geq p_1$ are found.
  2. For each set l', m' the color set c having the property c[k]=1 for k=l' or k=m' and c[k]=0 in other cases is found. On the image R[m,n] the back-projection algorithm for each set c is applied and the color regions are found. For each region the local histogram Ln[m] is recorded.
  3. The residue histogram Hr[m]=H[m]-$\sum_n L_n[m]$ is updated.
- Detection of the three colors regions. For each detected regions the color set that generated it, the area and the localization are stored. All the information is necessary further on in studying the evolution of the patients. The region localization is given by the minimal bounding rectangle (MBR). The region area is represented by the number of color pixels, and can be smaller than the minimum bounding rectangle.
  For the implementation of the color set back-projection algorithm, the image is read in a .bmp format. Each pixel from the initial image is transformed in HSV format and quantized. At the end of this process, the color set of the image is obtained [8].

On the matrix that memorizes only the quantized colors from 0 to 165 it is applied a 5x5 median filter, which has the role of eliminating the isolated points [7]. Having the HSV quantized matrix it is possible to begin the process of regions extraction presented above. It may be observed that this process is in fact a depth - first traversal, described in pseudo-code in the following way:

```
procedure FindRegions (Image I, colorset C) is: InitStack(S)
 for *each node P in the I do
     if *color of P is in C then
         PUSH(P)
         Visited = Visited union {P}
         while not Empty(S) do
               CrtPoint <- POP()
               Visited = Visited union {CrtPoint}
               for *each unvisited neighbor S of
               CrtPoint do
                              if *color of S is in C then
                                Visited = Visited union {S}
                                PUSH(S)
                          end if
               end for
          end while
          * Output detected region
     end if
 end for
```

The total running time of a call of the procedure FindRegions (Image I, colorset C) is $O(m2 * n2)$, where m is the width and n is the height of image [10]. In figure 1 there are two images for which there were detected the color regions using the presented algorithm.

The relevant color regions for the diagnosis detected by the method are marked as relevant. For each relevant region the following information is presented: the area (the number of pixels), the color and the dimensions of the minimal bounding rectangle (MBR).

3 Experiments and Results

In order to evaluate the efficiency of the color set back-projection algorithm, it was tested on a gastroenterologic imagistic database at Filantropia University Hospital of Craiova.

A double blind, double dummy, placebo controlled study was performed in order to evaluate the efficacy and safety of a new proton pump inhibitor in the treatment of peptic ulcer.

Two hundred and two patients were included and allocated in two groups: a placebo group (102) and a treatment group (100). One endoscopic image was taken at inclusion for each patient, in order to evaluate the cicatrisation process under treatment.

During the 6 weeks of follow-up, another endoscopic image of the ulcer was taken after the first 3 weeks and another one at the end of the treatment (6 weeks). Both images were stored in the database. The endpoint of the study

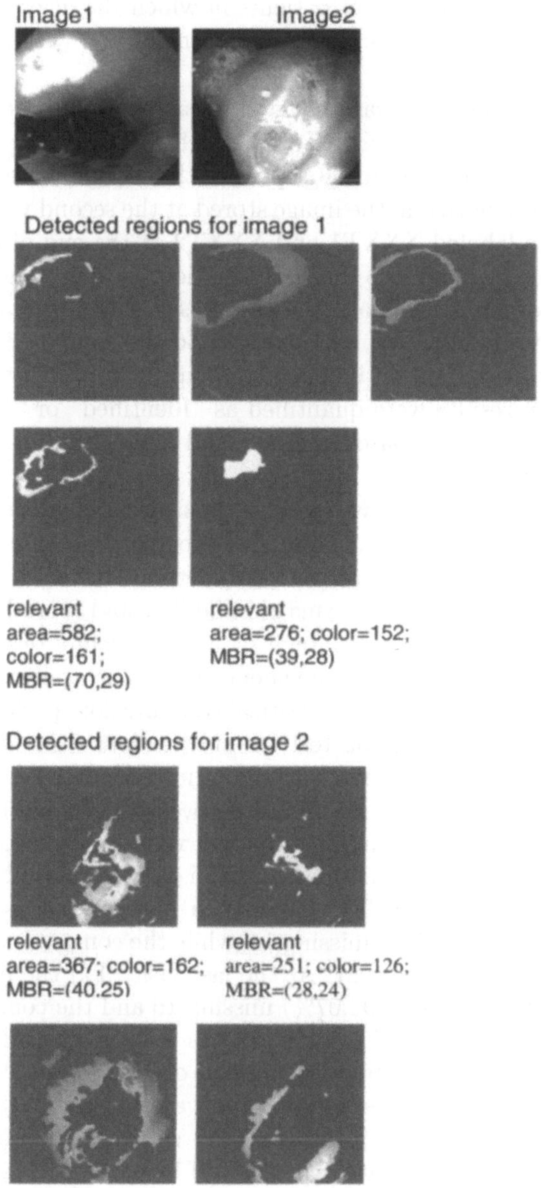

Fig. 1. Two images with detected color regions using the color set back-projection algorithm

was to estimate the percentage of patients in which the ulcer size diminished with more than 50% after three weeks of treatment and with more than 80% at the end of the treatment.

A randomization number was allocated to each patient. Each image stored at randomization was labeled XXXT0, in which XXX was the randomization number and T0 designed the initial aspect of the ulcer, before treatment.

Using the same algorithm, the image stored at the second visit (three weeks of treatment) was labeled XXXT1 and XXXT2 at the end of the treatment. Only the endoscopist made the initial evaluation, i.e. the diagnosis of peptic ulcer at T0. In order to evaluate the colour set back-projection algorithm, at the following visits both the endoscopist and the computer performed the search in the database, and an independent expert validated the results.

The validation results were quantified as "Identified" or "Non-Identified" both for the observational-based method and the computer-based one and included in statistics.

The duration of the retrieval process was also taken into account and included in the statistics. It is compared the percentage of correct diagnosis attended by using the classic observational method with the results obtained with the computer-based one in terms of reliability and reproducibility of this method. It is evaluated the advantage of using a computer-based retrieval system in terms of reducing the time spent on such an operation.

For this purpose it was used a statistical computation, performing the non-parametric two-sample Wilcoxon test for independent data and the analysis of variance. In all, both the observer and the computer software correctly identified the healing staging at T1 (three weeks visit) in 170 patients at T1, i.e. 84.15% while the attended statistical value was 168 patients, and in 158 patients at T2 (end-of-treatment visit) i.e. 78.21% while the attended statistical value was 155. At T1, the human observer correctly staged the ulcer in 188 patients (93.07%), missing 14, while the computer software staged correctly 181 (89.60%) missing 21. At the last visit (T2) the human observer staged correctly 186 patients (92.07%) missing 16 and the computer provided a good staging in 169 (83.66%) missing 33 as seen in graphic from figure 2. There were slight differences between human observer and computer system in order to appreciate healing staging in the way that the endoscopist missed the correct staging in certain cases and the software in others, but in all the inter-observer agreement was 84.15% at T1 and 78.21% at T2 as seen in table 1.

Despite this, no significant statistical difference between human observer (HO) and computer system (CS) was noticed by using the Chi-square test and the Wilcoxon nonparametric test, arguing for a good liability of the method (table 2).

The speed of the retrieval process was also tested, comparing the time spent by the observer and the computer to find each patient's record in the database.

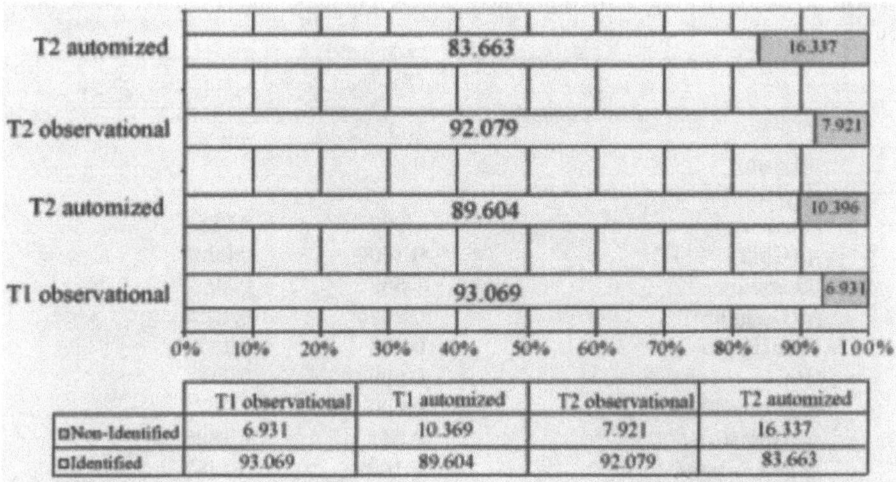

Fig. 2. Comparison between automatized and observational detection of gastroduodenal ulcer

Table 1. Evaluation of inter-observer agreement in human- and computer-based query in the database

Patients: 202		Correctly staged		Missed by both computer and observer		Missed by computer but not the observer		Missed by observer but not the computer	
		Nr	%	Nr	%	Nr	%	Nr	%
	HO	188	93.0	3	1.48	11	5.44	8	3.96
*T*1	CS	181	89.6	3	1.48	15	7.42	18	8.91
	Both	170	84.1	3	1.48	18	8.91	11	5.44
	HO	186	92.0	5	2.47	11	5.44	6	2.97
*T*2	CS	169	83.6	5	2.47	23	11.38	28	13.86
	Both	158	78.2	5	2.47	28	13.86	11	5.44

This process was electronically measured and stored in the computer for statistics. The result is that the software has a significantly higher speed than human observer with no significant decrease of the retrieval quality (figure 3).

In average, at T1 visit, the observer spent 37.52 seconds for each patient to find and evaluate the record, while the computer needed only 22.14. At visit T2 the human used 37.80 seconds to retrieve and evaluate the record, and the computer only 18.88 seconds (table 3). These differences had statistical significance as showed the t and Z test with a 95% confidence interval (table 4).

Table 2. Statistic analysis of differences between human- and computer-based query in the database

	Observer-based	Computer-based
N	202	202
Missing	0	0
DOL	1	1
Chi2	1.966	2.828
p(Chi2)	0.1609	0.0927
G-square	1.589	2.396
p(G-square)	0.2075	0.1217
Contigence	0.098	0.117
Phi	0.099	0.118
corrected Chi2	0.899	1.767
corrected p	0.341	0.1838
Fisher's exact test	0.1661	0.1482
Mann-Whitney U	19695.00	18685.00
U'	21109.00	22119.00
Wilcoxon T	40198.00	39188.00
Significance level for one-sided test	11.11%	0.52%
Significance level for two-sided test	22.22%	1.03%

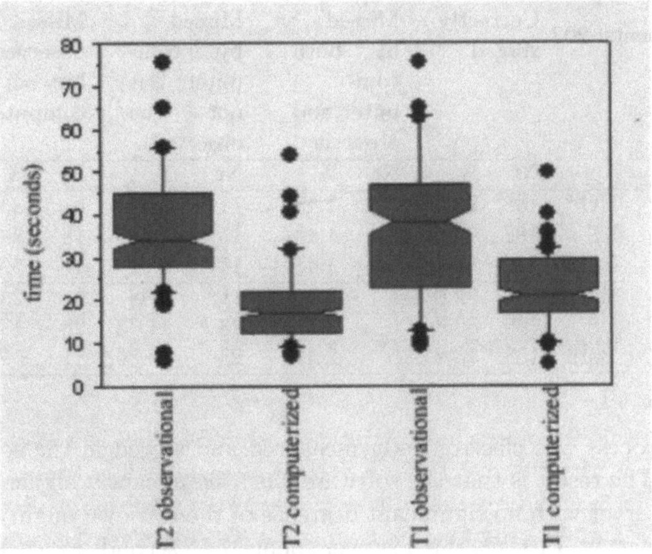

Fig. 3. Comparison between the retrieval speed

Table 3. Descriptive statistics of time spent by human observer and computer software to retrieve and evaluate a patient's record

Retrieval Time	Human observer t1	Software t1	Human observer t1	Software t1
n	202	202	202	202
Mean	37.520	22.149	37.807	18.881
Std. Deviation	16.962	8.878	14.499	9.914
Std. Error	1.193	0.625	1.020	0.698
Min.	9.000	5.000	6.000	7.000
Max	76.000	50.000	76.000	54.000
Missing	0	0	0	0

Table 4. Evaluation of statistical significance of differences observed in terms of retrieval speed between human observer and computer software

Human observer vs. Software	Mean diff.	DOL	t	p	95% inf.	95% sup
T1	15.371	201	11.937	<0.0001	12.832	17.910
T2	18.926	201	15.094	<0.0001	16.453	21.398

4 Conclusion

Endoscopic measurement of lesions is of great importance in the design and performance of clinical trials, as, in this case in peptic ulcer disease.

Although endoscopes are constructed with wide-angle lenses that significantly distort the image by creating a relative compression of points in its periphery, the presented computerized method showed a remarkable capacity to retrieve similar images in a quite large database giving an acceptable reproducibility for independent evaluations. Some authors [11] have already developed a computer program to correct the distortion of the wide-angle lens, which permits reducing errors to 1.8% +/- 2.2% ($p < 0.05$).

On the other hand, evaluations of ulcer healing by measuring video-endoscopy images as a dynamic process could not be realized up to now. The documentation of the dynamic healing process shattered either on the patient's compliance or on the inconstancy of the image cut due to wobbling [12].

In this work, instead of presenting a continuous film, instant takes of ulcer healing were processed. The inter-observer disagreement between human observer and the computer software was significantly low and the speed of the computerized method was higher, proving benefits in terms of time and cost efficiency.

In the future, the computerized methods will be developed in order to increase the capability of the system to recognize more complex lesions than ulcers and to diminish the percentage of errors.

References

1. Stanescu L, Burdescu D, Mocanu M (2004) Detecting Color Regions and Content-based Region Query in databases with Medical Images. Periodica Politechica, Transactions on Automatic Control and Computer Science, Vol 49(63) Timisoara
2. Mller H et al. (2004) Comparing feature sets for content-based medical information retrieval. SPIE Medical Imaging, San Diego, CA, USA
3. Mller H. et al. (2004) A review of content-based image retrieval systems in medicine-clinical benefits and future directions. International Journal of Medical Informatics 73: 1–23
4. Lehmann TM et al.(2004) IRMA - Content-based Image Retrieval in Medical Applications. In Proceedings of the 14th World Congress on Medical Informatics. IOS Press, Amsterdam, 842-848
5. Lehmann TM, et al. (2004) The IRMA Reference Database and Its Use for Content-Based Image Retrieval in Medical Applications.
6. Shyu C et al. (1999) ASSERT, A physician-in-the-loop content-based image retrieval system for HRCT image databases. Computer Vision and Image Understanding 75: 111–132
7. Smith JR (1997) Integrated Spatial and Feature Image Systems: Retrieval, Compression and Analysis. Ph.D. thesis Graduate School of Arts and Sciences, Columbia University
8. Stanescu L, Cojocareanu C (2000) The Color Set Back-Projection Method for Color Image Retrieval, Sintes 10, Craiova
9. Stanescu L, Burdescu D (2003) IMTEST-Software System For the Content-based Visual Retrieval Study, 14th International Conferece On Control Systems And Computer Science, Bucuresti
10. Burdescu, DD (1998) Analiza complexitatii algoritmilor, Ed. Albastra Cluj-Napoca
11. Vakil N, Smith W, Bourgeois K (1994) Endoscopic measurement of lesion size: improved accuracy with image processing. Gastrointest Endosc 40(6):779–80
12. Jaspersen D, Keerl R, Weber R, Huppmann A, Hammar CH, Draf W (1996) Dynamic imaging of gastric ulcer healing using the most modern Morph-Software, Bildgebung 63(2):101-4

Granularity of Medical Software Agents in ICU - Trade-off Performance versus Flexibility

Kristof Steurbaut[1], Sofie Van Hoecke[1], Kirsten Colpaert[2], Chris Danneels[2], Johan Decruyenaere[2], Filip De Turck[1]

[1] Ghent University, Department of Information Technology, Gaston Crommenlaan 8 Bus 201, 9050 Gent, Belgium. Filip.DeTurck@intec.UGent.be
[2] Ghent University Hospital, Department of Intensive Care, De Pintelaan 185, 9000 Gent, Belgium. Johan.Decruyenaere@UGent.be

Intelligent computing opens new opportunities to assist physicians with automatic medical decision support. Today, physicians still experience manual time-consuming data analysis of medical records while evaluating the patient's outcome. Especially in the Intensive Care Unit (ICU), large amounts of data per patient per day are generated, making manual data analysis a very complex task. Early detection of changes in patients' conditions is a major challenge in health care. To achieve this, agent technology and a service-oriented architecture have been deployed for medical decision support. In this paper, the choice of the level of service granularity is studied in detail. Fine-grained granularity of services seems an attractive idea to establish extreme reuse and flexibility but we need to be aware of possible performance implications. The modular design of an antibiotic switch therapy agent is presented and trade-offs between performance and flexibility are thoroughly evaluated.

1 Introduction

Highly critical hospital departments, such as the Intensive Care Unit (ICU), can benefit from intelligent computing to assist physicians with adequate decision support and immediate interpretable data reports of the patient's conditions. Early diagnosis and early detection of changing patient's conditions have become extremely important to improve the quality of care. Large health centers were already using information systems to capture and store data from patient monitors, laboratory results, ventilators and pumps in databases. Although this data collection contains valuable information to suggest a diagnosis or to alert physicians, current clinical information systems still have limitations to process the large amount of data and to offer alerts and adequate, complete reporting based on patient's information. As a result, the

Kristof Steurbaut et al.: *Granularity of Medical Software Agents in ICU - Trade-off Performance versus Flexibility*, Studies in Computational Intelligence (SCI) **78**, 207–216 (2008)
www.springerlink.com

physician experiences labor-intensive and time-consuming analysis of the patient's data. Furthermore the increased complexity of medical protocols and clinical decision logic shows a strong need to design flexible software applications.

Agent technology offers the abstraction of medical knowledge units. Using an agent-based approach each software element provides some functionality or decision logic and can act autonomously upon data. Furthermore, the combination of agent technology with data mining techniques leads to the discovery of unknown information or patterns from the databases [1]. Agents improve the quality of care as they are helpful to physicians and nurses in diagnosis, treatment and evaluation of data [2]. Typical examples of medical agents, that were already implemented in the ICU department of Ghent University Hospital, include agents monitoring the daily consumption of antibiotics, an agent detecting the kidney dysfunction and an agent calculating the SOFA score as a predictor for the patient's outcome. With an increased acceptance and deployment of the software agents, it is noticed that specific parts of one agent should be designed as reusable modules for the other agents. Reusing existing parts of an agent not only offers more flexibility to the developer but accelerates also the design process because the existing parts are already tested and evaluated.

In a multi-agents system several distributed agents are interacting and are working together to establish a well-defined goal. Complex problems can be dealt with by a group of autonomous agents. A modular platform using web service technology has been constructed consisting of basic agent data modules, calculation units and decision logic in order to build complex agents. This platform, the Intensive Care Agent Platform (ICAP) [3], simplifies the management of the several medical agents by providing platform functionalities such as data retrieval, delivery of patient reports or alarm messages and resource management. Moreover, using a modular approach provides a flexible development for future agents by enabling composition and reuse of existing functionality. Because high computational load can be imposed by the agents, these agents will typically run in a distributed environment [4]. Service-orientation is widely accepted as an important technology for enabling distributed computing. It provides an easy distribution of services and allows easy integration with both internal and external services using web service standards.

Software design has evolved from monolithic applications to highly modular distributed services. Functional decomposition is a powerful technique for reducing the complexity of software into smaller partitions. However, exposing each building block as an independent service can impact the overall performance. The decision of the level of granularity plays an important role. Advantages of fine-grained units are the flexibility and reuse of these units in other applications without redevelopment. On the other hand, coarse-grained units provide ready-to-use functionality without numerous intermediate interactions. To estimate the impact of the decompositions, the trade-off between flexibility and performance for agents has to be investigated. In this paper an example agent suggesting a switch from intravenous (IV) to oral (PO) antibiotics is used as a case study to examine extended granularity of medical agents. Each basic agent building block is wrapped in a service and the imposed overhead on the overall application is measured.

This paper is structured as follows. In section 2, we describe the concepts of agent technology, service-orientation, the ICAP platform and granularity. The case study of the IV-PO switch agent is presented in section 3, followed by a performance evaluation in section 4. Finally, in section 5 the main conclusions of the study are discussed.

2 Agents in Intensive Care

2.1 Multi-agents and service-orientation

A software agent is characterized by its autonomous, flexible, proactive and intelligent behavior. Intelligent agents are software elements capable of making rational decisions [5]. The medical software agent takes its own decisions on behalf of the medical expert, and monitors changes in the retrieved patient data. As such, the agent contains some clinical decision logic and is seen as an intelligent software element providing the physician with medical suggestive reports. In [1] it is argued that data mining working in association with agent-based applications offers a strong autonomous and adaptable framework. In a multi-agent system (MAS) loosely coupled agents communicate and interact to fulfill a more advanced task. To enable communication in a distributed environment service technology is used.

Service-orientation (SO) expresses an architectural style that defines the use of loosely coupled services, which serve as building blocks for distributed software applications. It encourages a style where connectivity or interaction among software pieces is central and where the software design is broken down in interacting services. These services can be seen as facades that encapsulate the inner implementation and expose functionality to the outside through their service description.

Web service technology is one way to implement a service-oriented architecture, providing platform interoperability, simplicity and platform neutral behavior. The web service (WS) has an interface description (Web Service Description Language or WSDL) and other systems interact using SOAP-messages, typically using eXtensible Markup Language (XML)-serialization in conjunction with standard Internet protocols. In the next sections we give a description of a medical software agent, the antibiotic switch agent, in which web services are implemented in the Microsoft .NET environment, using Visual Studio 2005.

2.2 ICAP

The Intensive Care Agent Platform (ICAP) [3], is designed based on the principles of service-orientation, wherein all components are implemented as services. The platform prototype has been implemented using web service technology and is currently being evaluated by the ICU department of Ghent University Hospital. This agent execution platform covers the patient subscriptions and management of running agents. It handles the data and notification of results from the agents and is responsible for the management of available computational resources[3]. ICAP has the following platform functionalities: database management, trigger handling, event handling, agent management, subscription management and communication management. For a detailed description we refer to [3],[4]. As show in figure 1, some agents are activated by incoming triggers from the laboratory. These triggers are handled and forwarded to the agents. Data can also be requested for the agent through usage of a database manager, which handles the communication with the external patient database. The external communication, to the PDAs and monitors of the physicians is handled by the platform. A physician can subscribe to receive agent results from several patients. Therefore the physician subscribes for an agent/patient combination and chooses how the alerts should be delivered, by mail or delivered as an alert on the PDA. All processed data results are stored in the platform and accessible through a web interface to the physicians. Agents can be distributed across different workstations. All agents are managed by the ICAP platform. When the designed agent has a web service interface, it is possible to integrate and interact with other services offered by the framework. Examples of already deployed agents are a kidney failure recommendations agent and agents for prescribing antibiotics. In this paper the focus is on the design of a specific agent.

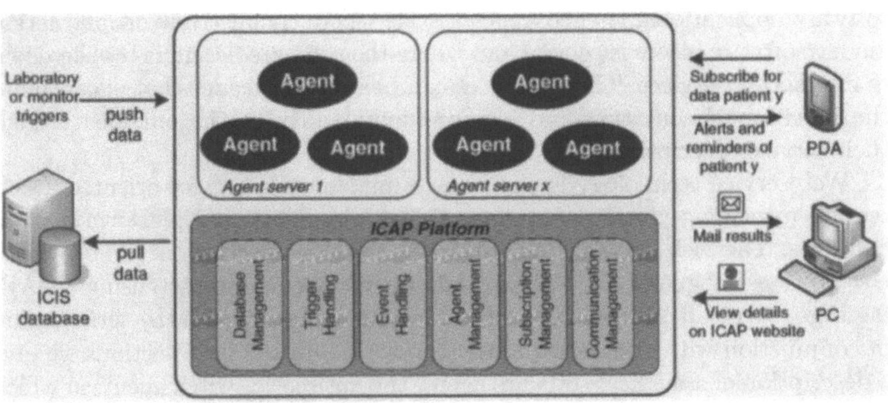

Fig. 1. Concept of the ICAP platform

2.3 Granularity of services

The term granularity refers to the level of modularity of a software application. Granularity is associated with a higher degree of flexibility and reuse of existing components. However, measurement of these exact benefits is difficult. Fine-grained building blocks offer greater flexibility because they can be reused easily in the design of new applications. Moreover, this enables the rapid design of new complex composed distributed software applications. On the other hand, a coarse-grained building block can be used to group similar operations.

In a service-oriented architecture the functional building block is the service. Earlier design of modular software applications included object-orientation and component based design. Functional decomposition is a basic technique to reduce the complexity of a problem by dividing the problem in smaller, more manageable subtasks. By exposing a functional building block as an individual basic service, functionality can be reused across multiple distributed applications. A basic service provides one functional task such as processing data, encapsulating calculation logic, sending an alert. Advanced applications can be composed out of these simple basic services. However determining the independent basic services and the level of granularity can be hard decisions to make.

Every functional operation can be transformed in a web service interface, but this imposes that with each invocation of the operation XML serialization takes place and that numerous SOAP-messages are sent across the network. Fine granularity may lead to a slow performing application. When the services are all on a local machine the overhead might be small, but when distributed across different machines and locations a performance overhead is measured due to the high number of web service calls. Modularization of software reduces the risk of building black box agent prototypes, in which the whole data and business logic can't be reused by other agents. Moreover, it can reduce time and cost in the design of similar prototypes. From a business perspective it is only useful to create fine-grained services when this offers an extra value to the users. When a sequence of similar operations is constantly invoked, it is better to create more coarse-grained services. This avoids a lot of network communication and each service takes a large amount of the work, to execute independently. In a study about design strategies to improve performance of Geographic Information System web services, it was argued that coarse-grained services should be preferred to fine-grained services, even if there is a a tendency in favor of fine-grained for flexibility [6]. The given argument was to avoid thousands of web service requests when building a geographic map because the communication overhead would hinder the performance.

3 Case study: IV-PO switch Agent

3.1 Description

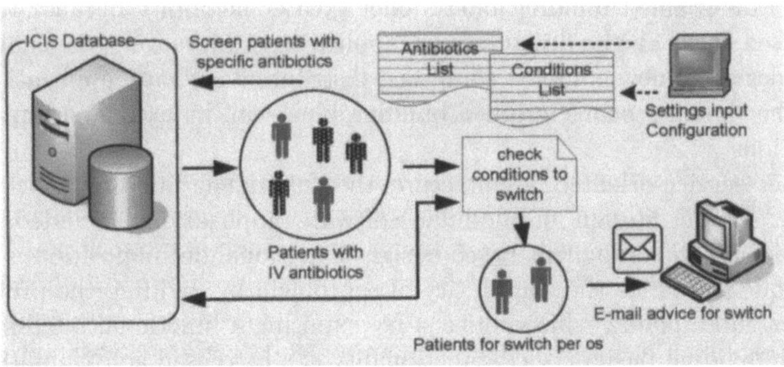

Fig. 2. Case study: IV-PO switch agent

The purpose of this agent is to identify patients at ICU who are candidate for an early switch from intravenous (IV) antibiotic therapy to oral (per os, PO) antibiotic therapy. Medical studies already proved similar therapeutic outcome of patients who made an early switch [7]. In addition, the switching therapy has many benefits. First, an early switch to oral antibiotics can reduce the length of stay at ICU, and can result also in a decrease of the number of nosocomial intravenous-line infections. Besides these patient benefits, there is an important economic cost savings argument as well, in favor of oral antibiotics therapy [8].

A software agent, identifying candidate patients, can help physicians making the switch decision in an early stage, by giving electronic alerts. Therefore switching criteria were derived from the knowledge of medical experts, and implemented in the software agent. Based on the criteria, which include the food indications, septic detection and the patient's medications, the agent will automatically alert the physician of a possible switch based on the patient's medical data. Without this piece of software, the physician would face a time-consuming and routine data analysis process for each patient. Figure 2 shows the general setup of the IV-PO switch agent. Patient data is extracted daily from the Intensive Care Information System (ICIS) database and analyzed in the agent. The switch suggestions are sent as an electronic report to the physician.

3.2 Implementation Details

The IV-PO switch agent is activated by a timer to capture the patients with IV antibiotics from the database. As shown in figure 2, the requested data,

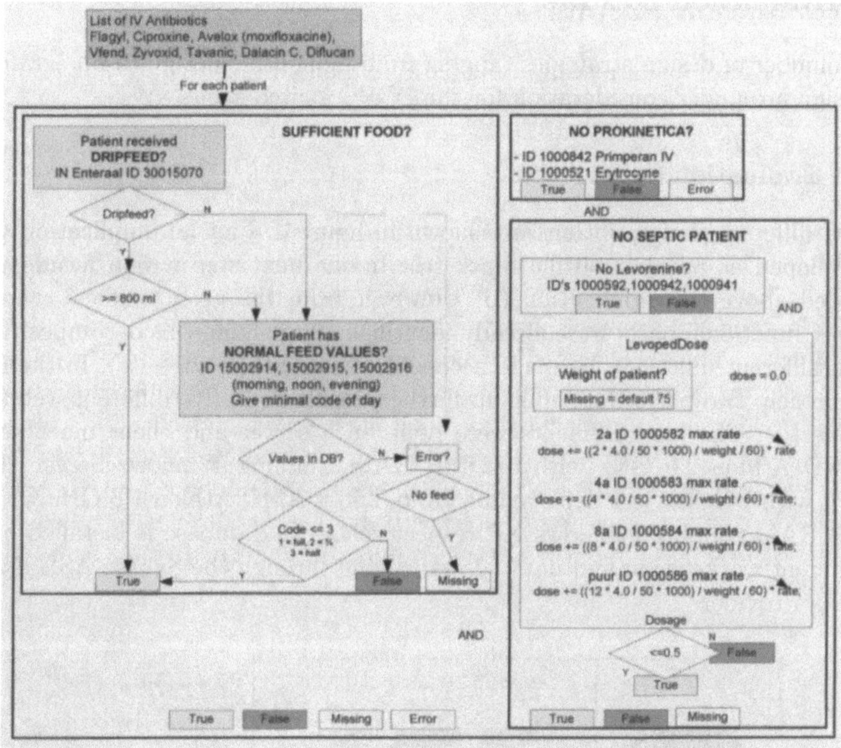

Fig. 3. Internal workflow of switch agent

available from patient's monitors, lab results and diagnosis, are retrieved using a pull-based approach. Figure 3 shows the detailed medical decision support algorithm as obtained from medical experts. Based on the functionality we can decompose the agent in three building blocks. The first functional part determines if a patient is able to receive oral medication, based on detected values of drip-feeds or normal oral feed. The second part checks the absence of certain prokinetics medications and a third part determines whether it is a septic patient. These three units constitute the decision to switch from intravenous to oral antibiotics therapy. Although these decompositions seem logic, the feed part and septic part can be split up further in two independent building blocks. This decision depends on the desired level of granularity. When choosing for the finest level of granularity in the design, building blocks for drip feed, normal feed, prokinetics, levorenine medication, weight detection are used.

4 Evaluation Results

A number of design strategies ranging from monolithic design to fine-grained design are under consideration for the IV-PO switch agent.

4.1 Evaluation Setup

The different design options are shown in figure 4. First an application was developed as one monolithic block (a). In the next step a web facade was added above this application (b). However from the given medical schema three functional parts were already identified. By making this decomposition, the different units can be made accessible using web services (c). In the last approach, two parts were split up further, resulting in five different web services (d). The measurements were done on a server and client machine, a AMD Athlon 2.0 GHz with 512 MB RAM, running Windows Server 2003 SP1 with web services deployed on IIS 6, and a AMD Athlon 1.8 GHz with 1 GB RAM running Windows XP respectively. The database is installed on a different workstation and data access is done through the Sybase ASE .NET Data Provider.

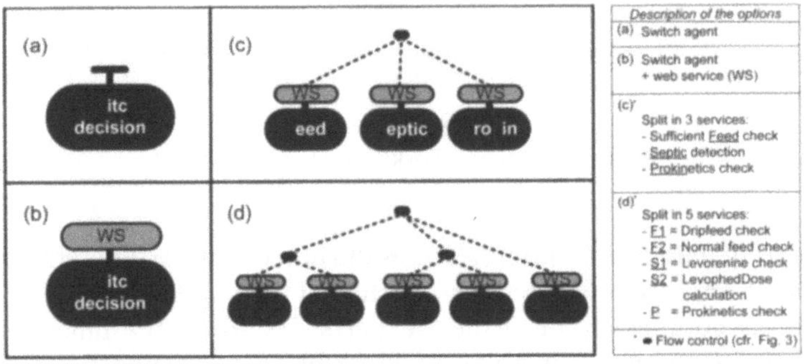

Fig. 4. Options with different granularity level

4.2 Results Description

This section details the performance results of the antibiotic switch agent. The execution times were measured for different levels of service granularity. In figure 5 the different granularity options are compared. A higher execution time is noticed when the level of service granularity increases. The agent code of the monolithic switch agent was split up in different units, each provided with a web service interface. Each option gives the same end result, only the internal construction of services is different. The total execution times are shown in figure 5.

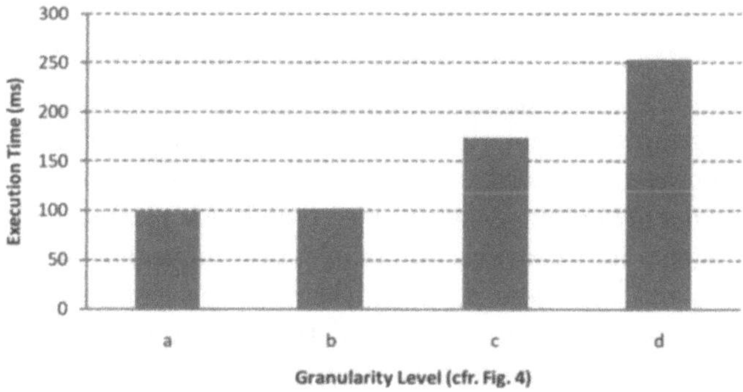

Fig. 5. Performance results of decomposition in services

For each option the execution time of the different invoked services is measured. In figure 6 two options are compared. Option c contains three web services, while the other option d has five web services. Both the feed part and the septic part are split up in two different services. In a future agent these services can be reused as functional building blocks.

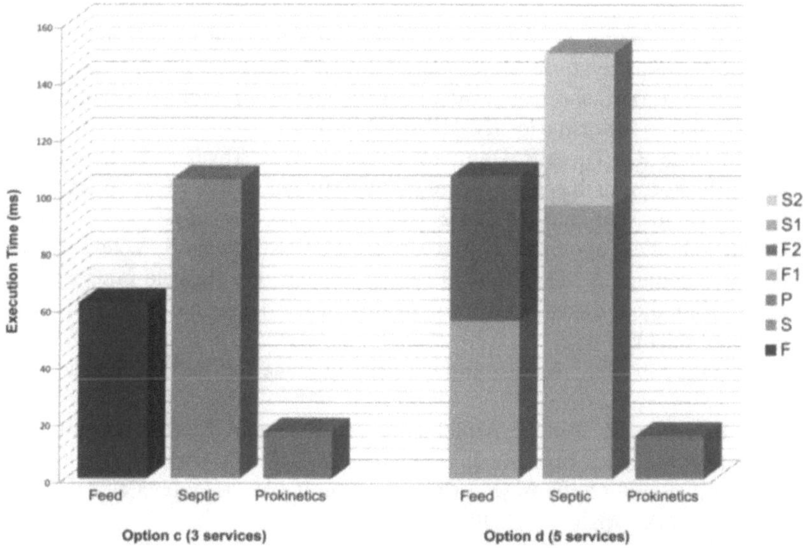

Fig. 6. Timing results of the option c and d

5 Discussion and conclusion

The aim of this paper was to isolate autonomous reusable building blocks and to examine the trade-off between flexibility and performance. The choice between fine-grained granularity and coarse-grained granularity is difficult. Due to the complexity of medical software agents it is necessary to design the agents consisting of flexible reusable units. The ICAP platform offers the management of agents and provides communication management between database, platform and end-users. As such, the developer can focus on the design of the agent functionality. It is expected that future medical agents will be developed with already implemented, tested and validated agent features. As motivated, in a granular approach the designer can enforce flexibility and high reusability, but performance characteristics should be considered carefully when numerous web services are used. In this paper, we compared the results of an antibiotic switch agent in the ICU and showed the influence of the agent granularity on the execution time for a specific real-live use case in ICU.

References

1. Symeonidis AL et al. (2007) Data mining for agent reasoning: A synergy for training intelligent agents. Engineering Applications of Artificial Intelligence, Elsevier
2. Foster D, McGregor C, El-Masri S (2005) A survey of agent-based intelligent decision support systems to support clinical management and research. First International Workshop on Multi-agent systems for medicine, computational biology, and bioinformatics (MAS*BIOMED'05).
3. Van Hoecke S, De Turck F, Danneels C, De Proft K, Taveirne K, Decruyenaere J (2006) Platform for Intelligent Agent subscription in ICU. 11th World Congress on Internet in Medicine (MEDNET'06), Toronto, Canada.
4. De Turck F, Decruyenaere J, Thysebaert P, Van Hoecke S, Volckaert B, Danneels C, Colpaert K, De Moor G (2007) Design of a flexible platform for execution of medical decision support agents in the Intensive Care Unit. Comput Biol Med 37(1):97-112
5. Wooldridge M (1998) Agents and software engineering. AI*IA Notizie XI(3):31-37
6. Tu S, Flanagin M, Wu Y, Abdelguerfi M, Normand E, Mahadevan V (2004) Design strategies to improve performance of GIS web services. Proceedings of the International Conference on Information Technology (ITCC'04) VolumeII, Las Vegas, NV, April 11-13,2004
7. Fischer MA, Solomon DH, Teich JM, Avorn J (2003) Conversion from intravenous to oral medications: assessment of a computerized intervention for hospitalized patients. Arch Intern Med 163(21):2585-9
8. Cunha BA (1997) Intravenous-to-oral antibiotic switch therapy: A cost-effective approach. Postgrad Med 101(4):111-124

APP: Agent Planning Package

Saša Tošić, Miloš Radovanović, and Mirjana Ivanović

University of Novi Sad
Faculty of Science, Department of Mathematics and Informatics
Trg D. Obradovića 4, 21000 Novi Sad
Serbia
{sashat,radacha,mira}@im.ns.ac.yu

Summary. Nowadays, a large number of planning systems exist. The majority of them are either applicable only to specific domains (domain dependent) or support only plan generation, but not plan execution and supervision. This paper presents Agent Planning Package (APP), a new domain independent planner which facilitates the creation of intelligent agents. Through the use of APP, agents can generate plans within their operating environment, as well as execute them and supervise the execution. Furthermore, APP supports online changes to the planning problem and domain, as well as creation of multiple plans and plans which contain alternative paths of execution.

1 Introduction

Modern mainstream application development tools offer many features which simplify and quicken the development process. Developers can easily create and incorporate databases into their projects, design user interfaces, enable network communication of programs, use extensive libraries which implement standard data structures and algorithms, etc. Development tools also support the break-down of larger problems into smaller ones, and merging of small problems into larger, but these operations have to be performed manually by the programmer. None of the standard development tools offers *automatic* combining of small functional parts, in the sense of the tool itself generating their order of execution to solve a larger problem. Furthermore, the tools provide little control of pre- and post-conditions for the execution of a part of a program, except for debugging purposes, which again requires the problem solution to be specified by hand. The craft of manual problem solving is one of the most demanding and error-prone parts of the application development cycle.

On the other hand, *planning systems* solve complex problems by combining atomic actions which can be performed. They use a declarative description of actions and the goal, based on which they create a plan, i.e. a sequence of actions which need to be executed in order to take the system from the initial state to the state with the desired properties.

Saša Tošić et al.: *APP: Agent Planning Package*, Studies in Computational Intelligence (SCI) **78**, 217–226 (2008)
www.springerlink.com

When specifying actions one needs to specify their parameters, the preconditions which must hold before the action can be executed, and the effects of the actions. A large number of planning systems also support specifying of the cost of action execution, and the probabilities that actions will be performed successfully.

Although the integration of concepts from planning systems into mainstream application development tools seems far away, the coupling of automatic planning with *agent development systems* is a reality. This paper presents Agent Planning Package (APP), a new domain independent planner which facilitates the creation of intelligent agents. Through use of APP, agents can generate plans within their operating environment, as well as execute them and supervise the execution. Furthermore, APP supports online changes to the planning problem and domain, as well as creation of multiple plans and plans which contain alternative paths of execution. The main purpose of APP is not to present a new deliberative agent architecture (although it may be possible to treat it as such), but rather to enable the extension of existing architectures with planning capability.

The rest of the paper is organized as follows. Related work and major existing planning systems are reviewed in Section 2. APP is introduced in Section 3, through a discussion of its basic anatomy and the plan creation process. Section 4 describes the APP planning cycle in more detail, focusing on the agents' needs and point of view. Section 5 summarizes experimental results which show that APP is among the medium-powered planners in terms of speed, and among the most efficient in terms of generating the shortest existing plan if one exists. Finally, Section 6 concludes and outlines the directions for future work.

2 Related Work

Research on planning systems (more precisely, automated problem solving) started in the late 1950s. One of the first programs in this field was GPS (General Problem Solver) by Newell, Shaw and Simon [15]. In 1971, Fikes and Nillson developed the STRIPS (STanford Research Institute Problem Solver) formalism [6], which was the basis for most research in the field until Pednault presented ADL (Action Description Language), mixing the STRIPS approach with the situation calculus [16]. The ADL formalism provided inspiration to many researchers, and soon other formalisms like UCPOP and UMCP emerged. In 1998, PDDL (Planning Domain Definition Language) was introduced [12], attempting to unify the most important aspects of all prior formalisms. PDDL 2.1 added the possibility to specify durations of actions, and integrated the systems for planning and scheduling [7]. In 2002, this version was further extended to PDDL 2.2 by introducing derived predicates and timed initial literals [5]. Year 2005 witnessed the introduction of PDDL 3.0, which enabled the definition of strong and soft constraints on plan trajectories, as well as strong and soft problem goals [8].

In parallel with the development of planning formalisms, planning systems based on them were being written. Some of the well known systems include SHOP [14, 13], GraphPlan [2], STAN [11], FF [9], HSP [3], SAPA [4] and AltAlt [17].

3 The Agent Planning Package

APP is a planning system written in Java in form of a Java package. Java was chosen as the implementation language for a number of reasons, including its portability and the existence of numerous Java-based agent systems, e.g. JADE, Mole, etc.

The basic purpose of APP is to facilitate the creation of intelligent agents through its integration with existing agent development tools. Thus, APP is designed to be simple to use and to offer vast functionality to an agent, such as the creation of planning instances, their modification, creation of different types of plans and their execution in the agent's environment.

3.1 The Basic Anatomy of APP

The structure of a multi-agent system in which agents utilize APP is shown in Fig. 1(a). Each agent in the environment may use an arbitrary number of instances of the *planner*, according to its needs. As illustrated in Fig. 1(b), one instance of the planner may be used to describe and initiate several *planning instances* which use the same *planning domain*, but may differ in the concrete *planning problem*.

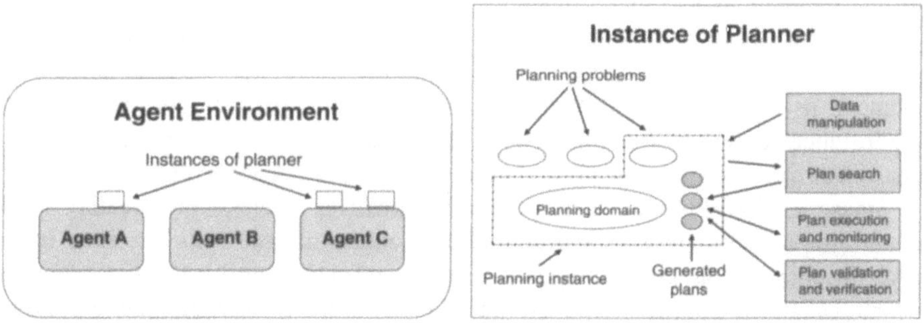

(a) (b)

Fig. 1. Structure of a multi-agent environment in which agents utilize APP

The planning problem describes the objects existing in the environment, declares initial values of functions, defines the relations that exist in the environment and describes the goal. This approach insures that there is no duplication of data characteristic of the planning domain, e.g. definitions of types, predicates, functions, constants and actions.

The core of APP resides in class Planner, which forms the base class for creating instances of the planner. At the instance creation phase, a reference to an agent is passed to the planner; this reference is used at the plan execution stage.

The class Planner contains the field domain, which represents the planning domain, and a list of planning instances. Creation of planning instances is done via a call to the addInstance() method, which returns the planning instance's ordinal number. The instance itself is accessible through method getInstance(num). Besides creating planning instances, it is possible to delete them, get the number of instances and the ordinal number of the last instance, and delete the planning domain if there are no planning instances left.

3.2 Plan Creation in APP

Plan creation is done on the basis of the current state of the planning instance prior to the invocation of one of the plan creation methods. The used algorithm is a mixture of several standard algorithms which utilize a planning graph. The base algorithm is the one used in the GraphPlan system, and is described in [2], while some of the employed modifications stem from [7, 9, 10].

During the creation of the planning graph, every level of the graph is represented by two sublevels, one representing facts (instances of predicates), and fluents (instances of functions), and the other representing instances of actions. The zeroth level of the graph is initialized with facts and fluents from the internal data structure which represents the planning problem[1].

In contrast to many existing planners, APP does not perform *grounding* (creating instances of facts and actions) prior to the execution of the planning algorithm – instances of facts and actions are created when they become needed. The reason behind this approach is that a planning instance represents an agent's model of its environment and may therefore contain a large number of objects. A consequence of possibly having a large number of objects is the expenditure of great amounts of memory, as well as a lot of processing time needed to perform grounding. Creating instances during plan search lessens the memory and time requirements for this phase, but increases the time needed for the creation of deeper levels of the graph.

Phase 1 of APP's plan search algorithm involves successively creating levels of the graph until a level is reached which contains all the facts specified in the goal. In phase 2, the creation of new levels is resumed, until a level is generated which contains no mutually exclusive facts. If any of the two terminating levels from phase 1 or 2 is determined not to be reachable, the algorithm halts.

The search for the plan itself is conducted in phase 3, using a backward-chaining strategy. First, all possible ways of fulfilling the goal are instantiated at the highest (goal) level. This generates all subgoals whose fulfillment on the previous level can lead to the fulfillment of the goal, and also determines the sequences of actions which transfer the system from the subgoal to the goal. The subgoals are then sorted according to a heuristic[2], and their fulfillment is respectively attempted at the previous level. If the root level of the graph is reached, a plan is generated according to the actions chosen during the search. If the root level is unreachable, the subgoal is inserted into a goal tree data structure for its level. During subsequent descents into lower levels of the graph, new subgoals are tested for presence in the goal tree, and if the outcome is positive the search is aborted for that particular subgoal.

[1] The zeroth level of the graph does not contain any action instances.

[2] The currently used heuristic is the number of no-op actions, i.e. the number of matching facts in the prefixes of the goal and subgoal.

The goal tree is a data structure in APP which is filled with goals that were not satisfied at a particular level of the graph. Since every goal is represented by a (sorted) list of numbers denoting facts, the goal tree is implemented as a *multiway trie*, more precisely an *existence trie*, described in [18] (Chapter 15). Each node of the tree contains a single fact, and every path from root to a leaf represents a goal. At goal insertion, the first fact of the goal is compared with the roots of all subtrees of the (empty) root node of the tree, and if a match is found the element of the goal is skipped and the rest of the goal recursively inserted into the appropriate subtree. If no match is found, a new subtree is created and the remaining facts inserted in a straight path down to the leaf.

Compared with the approach of storing all unsatisfied goals in an array and accessing them using a hashing function or some other type of tree, the goal tree conserves space to a certain extent. The worst-case time complexity of goal retrieval in the current implementation is $O(lf)$, where l is goal length, and f is the number of distinct facts instantiated by the system. Although $f \gg l$ for typical planning problems, the chosen data structure functions efficiently in the plan creation setting since it was empirically observed that the tree usually remains sparse, with the number of children of a node never really nearing f. Attempts to optimize retrieval by performing binary search on children, thus turning the complexity estimate into $O(l \log(f))$, resulted in no practical speedup. On average, only several percent of time taken for plan generation is expended on interaction with the goal tree.

On the other hand, the space requirements of the goal tree may become prohibitively high for large planning problems. This calls for further investigation of into methods for (sub)goal compression, as well as search space pruning.

4 The Planning Cycle

APP enables an agent to impose full control over the planning domain and problem. The internal structure of the planning domain and problem in APP are modeled according to the PDDL 2.1 standard [7]. This particular standard was chosen because APP is still in development, and PDDL 2.1 allows easy extension of the system to newer PDDL versions (unlike PDDL 1.x, which is structured somewhat differently).

4.1 Data Manipulation

Data input into a planning instance may be done in two ways: by loading the planning domain and problem from PDDL files, or "manually" from Java code by creating instances of appropriate classes and inserting them into the planning instance.

Input of the domain and problem from files is done via methods `loadDomain(filepath, filename)` and `loadProblem(filepath, filename)`. This way it is possible to load the entire domain and/or problem, and later modify them from Java if necessary. Since the domain independent of the problem(s), domain input is possible not only on the planning instance level, but also for the whole instance of the planner.

During its life cycle, an agent may obtain information about changes in its environment. As it is of crucial importance for the planning instance to be up to date with the state of the agent's surroundings, the agent is given the capability to insert, change or remove data from a planning instance. This is achieved by making available over 100 classes, the structure of which closely mimics the BNF structure of the domain and planning problem described in [7]. Some of these classes represent basic planning concepts (action, predicate, function, constant, object, etc.), while the majority of classes representing some form of expression used for defining preconditions and effects of actions, as well as defining the goal.

For example, if an agent concludes that the value of some function is no longer valid, it is given the opportunity to change the value by accessing the structure containing the planning instance.

4.2 Types of APP Plans

APP can generate various types of plans:

- SingleAction is a plan containing only one instance of an action;
- SerialPlan contains a series of subplans, where the execution sequence of the plans is strictly defined;
- ParallelPlan consists of several subplans, where the execution sequence of the plans is not defined, and subplans may be executed in parallel;
- AlternativePlan is a plan represented as a set of plans. If any subplan is valid, the whole plan is;
- MultiPlan represents a set of plans, where each plan is valid in terms of successfully reaching the goal. MultiPlan was introduced to enable an agent to choose from a set of plans based on its own additional criteria.

Planning instances are provided with methods which initiate the creation of an appropriate type of plan, optionally limited by a specified amount of time allowed for the plan generation process.

4.3 Plan Execution

From the point of view of practical application, one of the main weaknesses of standard planning systems is that after generating the plan they consider the job done. These plans are for the most part not ever executed in a real, possibly changing environment. This was one of the main motivations behind the design of APP.

Every APP plan possesses the method execute() which initiates plan execution. The method returns a boolean value indicating the success or failure of reaching the goal according to the plan. In a static setting the method always returns true for a properly generated plan. But in dynamic environments, changes during the process of plan creation, execution, or in between, may render the plan no longer valid.

SerialPlans are executed starting with the first plan in the sequence, continuing with the next plan after the execution of its predecessor successfully terminates. If the any subplan fails, execution of the whole plan is aborted. With ParallelPlans all subplans are executed simultaneously in their own threads, and the plan is successful if all subplans terminate with success.

During execution of AlternativePlans, if a subplan is not executed successfully, the next alternative is tried. An AlternativePlan is successfully executed if at least one of its subplans is successful. This type of plan may be used within an integrated version of a MultiPlan.

If MultiPlan M contains two plans $[A, B]$ and $[A, C]$, and $[A, B]$ is executing, during the execution of subplan A an unexpected event could arise causing the subsequent failure of subplan B. If plan $[A, C]$ is then tried, it could also fail simply because plan A had already been executed and the agent's environment had changed. However, if M is transformed into a SerialPlan $[A, \{B, C\}]$, then after the execution of A the AlternativePlan $\{B, C\}$ is left. If plan B does not end in success, plan C may freely be tried.

Every APP plan is, at its core, composed of atomic actions represented by a SingleAction plan. To enable the execution of an action in the actual environment of an agent, every action from a planning domain is assigned an instance of class ActionCode, which includes the method run(action, agent). This method is invoked at the execution of the SingleAction plan.

ActionCode is an abstract class which is intended to be inherited and its method run(action, agent) overridden by a method containing code to be executed in an agent's environment. After inheriting the class and creating an instance, it is necessary to assign this instance to an appropriate action within the domain. This is done via planning domain's method insertActionCode(name, actionCodeInstance), where name denotes the name of the action. Within the run(action, agent) method an agent may execute arbitrary Java code and do whatever is in line with performing the action in his practical environment. Parameter action may be used to access various properties of an action, e.g. its name and arguments. Parameter agent allows access to the resources of an agent.

APP automatically insures that the planning instance is consistent with the state of the environment. After every executed action, APP updates the planning problem according to the effects of the action, enabling the addition of new, as well as change and deletion of existing data.

4.4 Properties of APP Plans

Besides plan execution, APP offers some other plan-related functionality. On a standard level, it is possible to check the number of actions in a plan, copy plans, check if two plans are equal and convert a plan into text for easy output. In addition, it is possible to convert plans from one form into another. For example, method mergePlans(multiPlan) transforms the set of independent subplans within the multiPlan into plans which may contain AlternativePlans.

APP allows access to every part of a plan, that way enabling enquiry into certain properties of plans which can not be specified as constraints in the planning domain. The creation of MultiPlans facilitates this functionality by enabling and agent to generate a large number of plans and then filter them according to additional criteria. Such criteria may include the least use of some action which is too costly for the system, avoiding the use of a particular error prone resource, discouragement of executing plans similar to other plans which had failed in the past, etc.

5 Experimental Results

Although the initial objectives behind APP did not include competing with the state-of-the-art planners, APP has proven to be more efficient than anticipated. In order to evaluate its efficiency, APP was tested on examples from the planning competition of the Artificial Intelligence Planning Systems Conference (AIPS) in 2000. This particular benchmark was chosen because APP supports a subset of PDDL 2.1, which was the official language of the AIPS 2000 competition.

Figure 2 shows the results of testing in the *blocks* domain. Figure 2(a) depicts the number of solved problems for various planners, while Fig. 2(b) shows the time in milliseconds taken by each planner to solve a particular problem. The results for other planners were obtained from the official Web site of the conference [1], while APP was tested on a closely reproduced computer configuration, with the same time limit of 30 minutes for each problem.

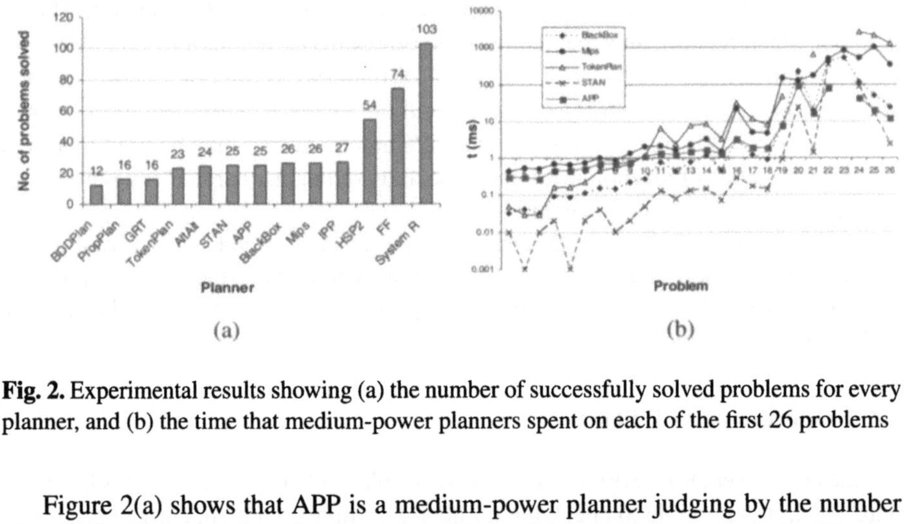

(a) (b)

Fig. 2. Experimental results showing (a) the number of successfully solved problems for every planner, and (b) the time that medium-power planners spent on each of the first 26 problems

Figure 2(a) shows that APP is a medium-power planner judging by the number of solved problems. From the first 26 problems APP could not solve only "blocks-11-1" which was problematic for all planners except the three most efficient. IPP solved two problems more than APP, but the time expended on the two was an order of magnitude longer than the time IPP typically took to solve the other 25 problems.

Figure 2(b) focuses on the time spent by the medium-power planners on the first 26 problems in the test. The fastest planner was STAN which is well known for its efficient implementation, although APP did show a better result on "blocks-11-0" and "blocks-11-2" problems. On simpler problems, APP exhibited about the same efficiency as Mips, while on more difficult problems APP proved somewhat superior. TokenPlan and BlackBox achieved better results on simpler problems, but as problems grew more difficult APP was more up to the task.

Such behavior of APP can be explained by the fact that for smaller problems more time is spent on the creation of the planning graph, while larger problems take up more time on graph searching. As APP creates instances of actions during graph creation, this could be expected. Planning systems which first create all instances of actions, and than search for a plan, achieve better results in these situations. However, real world systems, for which APP is designed, usually contain a large number of objects and actions, punishing this approach to action instantiation. The reason behind APP's better results for larger problems in the benchmark lies a great deal in the use of the goal tree data structure. The goal tree speeds up the check for prior attempts to satisfy a goal, which is important for larger graphs.

It is important to note that unlike most other planners, APP does not use heuristics to prune unpromising paths during the search of the graph. A good consequence of this is that APP is guaranteed to find the shortest possible plan for any given problem, which puts APP among the most efficient systems judging by this criterion. The bad consequence is that search time is greatly increased.

6 Conclusions and Future Work

This paper presented the APP Java package which enables agents to expand their capabilities with automatic planning. We have shown how APP can be integrated into a multi-agent system, and how its communication with an agent may be achieved.

The basic services APP offers to an agent include creating multiple planning instances within one instance of a planner, and accessing the data within a planning instance in order for existing data to be read, changed or deleted, and new data to be inserted. In contrast to APP, most of the other planners allow only one planning instance, and are not able to modify it after the plan had been created. Furthermore, because of its ease of data manipulation, APP may be utilized to implement agents' representations of the current state of the world.

Besides plan creation, explicit plan *execution* is also supported, facilitating the use of APP in dynamic environments. APP also provides services for checking various properties of plans, which is a feature not supported by the majority of planners.

APP uses a modified algorithm for plan creation which employs a planning graph and a goal tree structure, supporting the generation of many different types of plans. For example, a combination of SerialPlans and ParallelPlans may be used to obtain partial order plans (POP), which are supported by the majority of planners, while AlternativePlans allow execution of multiple subplans until one terminates with success. AlternativePlans may be combined with SerialPlans and ParallelPlans to derive plans which have increased chances for behaving robustly in dynamic environments.

Currently, APP supports a subset of PDDL 2.1 which includes :strips, :typing and :fluents requirements. Upgrading to the full :adl set is underway, as well as extension to the third and fourth level of PDDL which incorporate actions with duration. In order to provide faster plan generation, research into heuristics will be done to enable reliable and efficient pruning of unpromising paths during graph search.

Since the main purpose of APP is to serve as a library for creation of intelligent agents, we intend to add a number of new methods for investigating various properties of plans. One example is a method which simulates plan execution without actually firing any actions in the agent environment, in order to check the validity of a plan before actually executing it.

References

1. Aips-00 planning competition. `http://www.cs.toronto.edu/aips2000/`.
2. A. L. Blum and M. L. Furst. Fast planning through planning graph analysis. *Artificial Intelligence*, 90:281–300, 1997.
3. B. Bonet and H. Geffner. Heuristic search planner 2.0. *AI Magazine*, 22(3):77–80, 2001.
4. M. Do and S. Kambhampati. SAPA: A multi-objective metric temporal planner. *Journal of Artificial Intelligence Research*, 20:155–194, 2003.
5. S. Edelkamp and J. Hoffmann. PDDL2.2: The language for the classical part of the 4th international planning competition. Technical Report 195, Institute of Computer Science, University of Freiburg, 2004.
6. R. Fikes and N. Nilsson. STRIPS: A new approach to the application of theorem proving to problem solving. *Artificial Intelligence*, 2:189–208, 1971.
7. M. Fox and D. Long. PDDL2.1: An extension to PDDL for expressing temporal planning domains. *Journal of Artificial Intelligence Research*, 20:61–124, 2003.
8. A. Gerevini and D. Long. Plan constraints and preferences in PDDL3. Technical report, Department of Electronics for Automation, University of Brescia, 2005.
9. J. Hoffmann and B. Nebel. The FF planning system: Fast plan generation through heuristic search. *Journal of Artificial Intelligence Research*, 14:253–302, 2001.
10. S. Kambhampati, E. Parker, and E. Lambrecht. Understanding and extending Graphplan. In *Proc. ECP'97, 4th European Conference on Planning*, LNCS 1348, pages 260–272, 1997.
11. D. Long and M. Fox. Efficient implementation of the plan graph in STAN. *Journal of Artificial Intelligence Research*, 10:87–115, 1999.
12. M. Ghallab et al. PDDL – the planning domain definition language. Technical Report CVC TR-98-003, Yale Center for Computational Vision and Control, 1998.
13. D. S. Nau, T. C. Au, O. Ilghami, U. Kuter, J. W. Murdock, D. Wu, and F. Yaman. SHOP2: An HTN planning system. *Journal of Artificial Intelligence Research*, 20:379–404, 2003.
14. D. S. Nau, Y. Cao, A. Lotem, and H. Muñoz Avila. SHOP: Simple hierarchical ordered planner. In *Proc. IJCAI'99, 16th International Joint Conference on Artificial Intelligence*, pages 968–973, 1999.
15. A. Newell, J. C. Shaw, and H. A. Simon. Report on a general problem-solving program. In *Proc. International Conference on Information Processing*, pages 256–264, 1959.
16. E. P. D. Pednault. ADL: Exploring the middle ground between STRIPS and the situation calculus. In *Proc. KR'89, First International Conference on Principles of Knowledge Representation and Reasoning*, pages 324–332, 1989.
17. R. Sanchez-Nigenda, X. Nguyen, and S. Kambhampati. AltAlt: Combining the advantages of Graphplan and heuristic state search. In *Proc. KBCS'00, 3rd International Conference on Knowledge-based Systems*, 2000.
18. R. Sedgewick. *Algorithms in Java, Parts 1–4*. Addison Wesley, 3rd edition, 2002.

Converting Programs into Constraint Satisfaction Problems*

Franz Wotawa and Mihai Nica[1]

Technische Universität Graz, Institute for Software Technology, Inffeldgasse 16b/2, 8010 Graz, Austria {wotawa,mihai.nica}@ist.tugraz.at

Summary. In this paper we introduce the basic methodology for analyzing and debugging programs. We first convert programs into their loop-free equivalents and from this into the static single assignment form. From the static single assignment form we derive a corresponding constraint satisfaction problem. The constraint representation can be directly used for debugging. From the corresponding hyper-tree representation of the constraint satisfaction problem we compute the hyper-tree width which characterizes the complexity of finding a solution for the constraint satisfaction problem. Since constraint satisfaction can be effectively used for diagnosis the conversion can be used for debugging and the obtained hyper-tree width is an indicator of the debugging complexity.

1 Introduction

Ideally intelligent systems should provide self-reasoning and reflection capabilities in order to react on internal faults as well as on unexpected interaction with their environment. Reflection capabilities are highly recommended for systems with strong robustness constraints, like space exploration probes or even mobile robots. A scenario for example is a robot where with a broken engine which should reach a certain position. Without self-reasoning or reflection such a robot would simple fail to reach its goal. Another example would be a robot where the software fails because of a bug. In this situation a robot should recover and ideally repair itself. Note that even exhaustive testing does not prevent a program from containing bugs which might cause an unexpected behavior in certain situations.

In this paper we do not focus on whole systems which comprise hardware and software. Instead we are discussing how to represent programs to allow for reflection which can be used for enhancing the system with debugging

* This work has been supported by the FIT-IT research project *Self Properties in Autonomous Systems project (SEPIAS)* which is funded by the Austrian Federal Ministry of Transport, Innovation and Technology and the FFG.

Franz Wotawa and Mihai Nica: *Converting Programs into Constraint Satisfaction Problems*, Studies in Computational Intelligence (SCI) **78**, 227–236 (2008)
www.springerlink.com

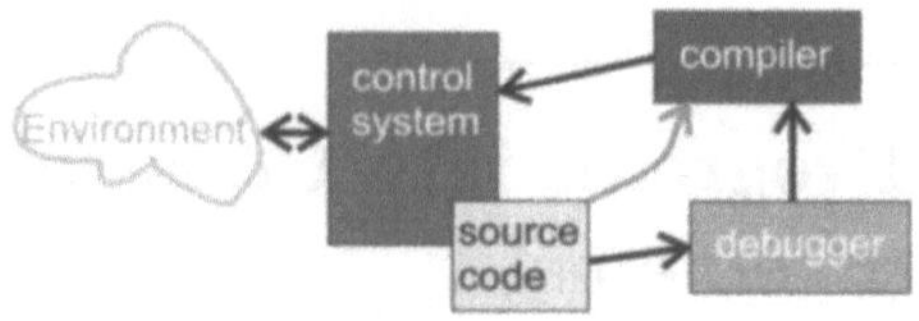

Fig. 1. Interaction between the control and the debugging system

functionality. In the context of this paper debugging is defined as fault localization given a certain test-case. We do not take care of verification and test-case generation which is used for fault detection and repair. In order to compute the fault location we follow the model-based diagnosis approach [14] but do not rely on logical models but use constraints instead for representing programs. The obtained constraint representation can be directly used for computing diagnosis, e.g., by using specialized diagnosis algorithms like the one described in [5, 17, 18].

Although, reflection and debugging capabilities are a desired functionality of a system they provoke additional computational complexity which can hardly be handled by the system directly because of lack of computational power. Note that model-based diagnosis is NP complete. Hence, a distributed architecture would be required which separates the running control program from the debugging capabilities. In Figure 1 we depict the proposed architecture. The debugging module takes the source code of the original system which is in this case a control system and a test-case to localize and repair the fault. The changed source code is compiled and transferred back to the original system.

In the proposed architecture there are several open issues which have to be solved. The first is regarding the test case. In particular someone is interest in the origin of the test case. One way would be to have a monitoring system which checks the internal state of the control system. In case of a faulty behavior, the given inputs coming from the sensors, the internal state and the computed output together with the information regarding the expected output is used as a test case. The second issue is related to fault localization and repair. There is literature explaining fault localization and repair which can be used, e.g., [6, 16, 9, 15]. In this paper we focus on modeling for fault localization, i.e., for identifying the root cause of a detected misbehavior. This does not necessarily directly lead to good repair suggestions. Repair definitely requires knowledge about further specification details and incorporating different test-cases should help a lot. A detailed discussion of repair is outside the scope of this paper and still an important topic of research. The third and last issue is a more technical one. After recompiling the new program has to replace the old one on the fly which might cause additional problems because of ensuring integrity and consistency of the system's behavior.

In the rest of the paper, we first introduce the basic idea behind debugging which is followed by a discussion of the conversion process of the program's source code. This conversion starts with an elimination of all loops using nested if-statements. Afterwards, we show how programs can be converted into their SSA equivalent. We present how to use the SSA form to derive the corresponding constraint satisfaction problem (CSP). Finally, we discuss related research and conclude the paper.

2 Diagnosis/debugging

The idea behind our approach can be best explained using the following tiny program which should compute the area and circumference of a circle given the diameter.

```
1.   r = d/2;
2.   c = r*pi;
3.   a = r*r*pi;
```

Obviously the statement in line 2 is faulty. A possible repair would be `c = 2*r*pi;` or `c = d*pi;`. However, before talking about fault localization and repair we have to detect the fault. In programming this is done using test-cases or properties together with formal verification methods. In our case we rely on test-cases. One test-case which allow to detect the faulty behavior would be setting `d` to 2 and requiring `c` to be 2π and `a` to be π. The program computes the value π for both variables `c` and `a` which contradicts the test-case.

This contradiction can be used to locate the fault. One way would be to have a look at the first definition of variable `c` which has assigned a wrong value and trace back using the dependencies between variables. In this case `r` is defined in line 1 and used in line 2. Hence, line 1 and 2 are possible fault candidates. This approach uses only structural information coded in the program and might overestimate the number of diagnosis candidates. Another approach would be to assume some line of the program to be faulty. In this case the statement corresponding to the line does not provide any known functionality. For example, when assuming line 2 to be faulty, we do not know how to compute a value for `c`. Therefore, we do not get any contradicting information and the assumption is consistent with the given test-case. Unfortunately, this is also the case when assuming line 1 to be faulty and only considering the computation of values like specified in the semantics of the programming language, i.e., from the begin of the program to its end.

The situation changes when considering the statement as equations where no direction of information flow is given. An equation like $r = d/2$ is a constraint which specifies a relationship between r and d. When we now assume the constraint corresponding to line 1 to be faulty we can compute a value for r using the given test-case. From the constraint corresponding to line 3 $a = r^2\pi$

and $a = \pi$ we derive $r = 1$ and finally using $c = r\pi$ leads to $c = \pi$ which contradicts the given expected value for c which is 2π. Hence, the assumption in this case contradicts the observations and cannot be a single-fault candidate. Only line 2 remains as single fault. The reason for this improvement is that when using constraints we have the capabilities for reasoning backwards.

Summarizing the above discussion, a solution to the debugging problem would be to convert programs into their constraint representation and use it for debugging. However, there are some problems we have to overcome. First, variables might be defined more than once in a program. Every definition has to be considered separately. Programs comprises conditional and loop statements. How to handle them? For the first problem and the conditionals we propose the use of the static single assignment (SSA) form of programs which is used in compiler construction. In order to represent loops we make use of the following observation. Loops can be represented by nested if-statements where the nesting depth is infinite. When restricting the nesting depth to a finite value the nested if-statements still behave the same as the loop statement when considering only inputs which do not cause the number of loop-iterations to exceed the nesting depth. Hence, under this assumption the nested if-statements are good enough to represent loops and we have reduced the problem of handling loops to the problem of handling conditionals.

```
1.   if x < y {
2.     min = x;
     } else {
3.     min = y;
}
```

Fig. 2. A program computing the minimum of two numbers

```
1.   i = 0;
2.   r = 0;
3.   while (i < x) {
4.     r = r + y;
5.     i = i + 1;
     }
```

Fig. 3. A program for computing the product of two natural numbers

3 Conversion

In this section we describe the conversion process of programs into their equivalent CSP representation in detail. For more information regarding CSPs we refer the reader to Dechter [4]. We start with converting programs into their loop-free equivalents which are used as basis for the conversion into the SSA form. Finally, we show how to extract the CSP from the SSA form. The complexity of the proposed compilation approach is composed of the complexity of each conversion step. While SSA construction and CSP extraction can be both handled in polynomial time computing a loop-free equivalent depends on the number of necessary iterations. This number depends on the program's complexity. Throughout this section we use the programs given in Figure 2 and 3 as running examples. We further assume without restricting generality of the approach that the programs to be converted have a Java like syntax (ignoring object-oriented elements).

3.1 Loop-free programs

When executing while-statements they behave like a conditional statement in one step. If the condition is fulfilled the statements in the block are executed and the while-statement is executed again afterwards. Otherwise, the while-statement is not executed. Hence, it is semantically correct to represent while-statements using an infinite number of nested if-statements, i.e., `while (`C`) {` B `}` is equivalent to `if (` C `) {` B `if (` C `) {` B `if (` C `)` B `if ... } } }` where C is the condition and B the list of statements in the block of the while statement.

Of course it is not possible in practice to compile while-statements into an infinite number of conditionals. Instead we assume that the number of iterations of the while-statement never exceeds a certain limit say n. We argue that faults can be detected using test-cases which cause a small number of iteration and that it is therefore – for the purpose of debugging – not necessary to consider larger values of n. Moreover, we might introduce a procedure which is called whenever the limit is reached. This information would give us back additional information which we might use for increasing n in a further step. To set a small bound for the number of iterations is also used in a different context. Jackson [10] uses a similar idea which he called *small scope hypothesis* in his Alloy system for verification.

We formalize the bounded conversion from while-statements into nested if-statements by introducing the function $\Gamma : PL \times \mathbb{N} \mapsto PL$ where PL represents the programming language.

$$\Gamma(\texttt{while (} C \texttt{) \{ } B \texttt{ \}}, n) = \\ = \begin{cases} \texttt{if (} C \texttt{) \{ } B\ \Gamma(\texttt{while (} C \texttt{) \{ } B \texttt{ \}}, n-1) \texttt{ \}} & \text{if } n > 0 \\ \texttt{if (} C \texttt{) \{ too\_many\_iterations(); \}} & \text{otherwise} \end{cases}$$

Considering the above discussion it is obvious that the following theorem which states the equivalence of the program and its loop-free variant with respect to their behavior has to hold.

Theorem 1. *Given a program $\Pi \in PL$ written in a programming language PL and a number $n \in \mathbb{N}$. Π behaves equivalent to $\Gamma(\Pi, n)$ for all inputs I iff the run of $\Gamma(\Pi, n)$ does not call the* `too_many_iterations()` *function.*

We now use the program from Figure 3 and $n = 2$ to show the application of Γ which leads to the following program:

```
i = 0;
r = 0;
if (i < x) {
  r = r + y;
  i = i + 1;
  if (i < x) {
    r = r + y;
    i = i + 1;
    if (i < x) {
      too_many_iterations();
    } } }
```

The loop-free variant can be used in all cases where $x = 0$ or $x = 1$ without causing a different behavior to the original program. It is interesting to note that the time complexity of a program which is corresponds to the number of iterations depending on the size of the inputs is now represented by the size of the converted program. Hence, we easily can give an estimate for the size fo the converted programs (and also the number of iterations) when limiting the size of the input.

3.2 Static single assignment form

The SSA form is an intermediate representation of a program, which has the property that no two left-side variables share the same name, i.e., each left-side variable has a unique name. Because of this reason the SSA form is of great importance when building the CSP. Since all variables are only defined once the SSA form allows for a clear representation of the dependencies that are established between different variables inside the corresponding program. The SSA representation of a program is sometime also an intermediate step in the compiling process; basically before compiling a java file, first a transformation is undertaken and the SSA form is obtained. The obtained form will be the form used as input for the compiler.

In order to compile loop-free programs into SSA we have to analyze the program and rename all variables such that each variable is only defined once without changing the behavior of the program. Basically, this compilation is done by adding an unique index to every variable which is defined in a statement. Every use of this variable in the succeeding statement is indexed using the same value. This is done until a new definition of the same variable occur.

For example, the following program fragment on the left side can be converted into its SSA representation on the right.

```
x = a + b;          x_1 = a_0 + b_0;
x = x - c;          x_2 = x_1 - c_0;
```

Although, converting programs comprising only assignment statements is easy, it is more difficult for programs with loop or conditional statements. In our case we only need to consider conditional statements. The idea behind the conversion of conditional statement is as follows: The value of the condition is stored in a new unique variable. The if- and the else-block are converted separately. In both cases the conversion starts using the indices of the variables already computed. Both conversions deliver back new indices of variables. In order to get a value for a variable we have to select the last definition of a variable from the if- and else-part depending on the condition. This selection is done using a function Φ. Hence, for every variable which is defined in the if- or the else-branch we have to introduce a selecting assignment statement which calls the Φ function.

For example, corresponding SSA form of the program fragment

```
if (C) {.. x = .. } else { .. x = .. }
```

is given as follows:

```
var_C = C; ... x_i = ... x_j = ... x_k = Φ(x_i,x_j,var_C);
```

where the function Φ can be defined as follows: $\Phi(x,y,b) = \begin{cases} x \text{ if } b \\ y \text{ otherwise} \end{cases}$

For algorithms for computing the SSA form and more information regarding the Φ function we refer the reader to [3, 1, 19]. The SSA representation for the programs from Figure 2 and 3 are depicted in Figure 4 and 5 respectively. Note that in this conversion we ignored the procedure call to `too_many_iterations()` and represent the Φ function by a call to the procedure `phi`.

```
var_1 = ( x_0 < y_0 );
min_1 = x_0;
min_2 = y_0;
min_3 = phi(min_1,min_2,var_1);
```

Fig. 4. The SSA form of the program from Fig. 2

```
i_1 = 0;
r_1 = 0;
var_3 = (i_1 < x_0);
r_2 = r_1 + y_0;
i_2 = i_1 + 1;
var_6 = (i_2 < x_0);
r_3 = r_2 + y_0;
i_3 = i_2 + 1;
r_4 = phi(r_3,r_2,var_6);
i_4 = phi(i_3,i_2,var_6);
r_5 = phi(r_4,r_1,var_3);
i_5 = phi(i_4,i_1,var_3);
```

Fig. 5. The SSA form of the loop-free variant of the program from Fig. 3 where procedure calls are ignored.

3.3 Constraint representation

Constraint satisfaction problems are effectively used in Artificial Intelligence as general knowledge representation paradigm. A CSP (V, D, CO) is characterized by a set of variables V, each variable having a domain D, and a set of constraints CO which define a relation between variables. The variables in a relation $r \in CO$ are called the scope of the relation. There are very effective reasoning algorithms available for CSP. In particular there are diagnosis algorithms [5, 17] available which can be used for debugging when representing programs as CSPs. More information about CSPs can be found in Rina Dechter's book on constraints [4].

The extraction of the constraint representation for a program can be easily performed on its SSA representation. The algorithm for constructing the CSP is pretty straight forward and only implies a simple analyze of the SSA representation. Each line of the SSA representation has to be interpreted as a constraint that is established between the left side and the right side of an assignment. Hence, the SSA form of a program as given in this paper is equivalent to its CSP representation when interpreting an assignment as an equation.

For example, the program from Figure 2 can be expressed as CSP when using its SSA form which is depicted in Figure 4 as follows:

$$\textbf{Variables: } V = \{var\_1, x\_0, y\_0, min\_1, min\_2, min\_3\}$$
$$\textbf{Domains: } D = \{D(x) = \mathbb{N} | x \in V\}$$
$$\textbf{Constraints: } CO = \left\{ \begin{array}{c} var\_1 = (x\_0 < y\_0), min\_1 = x\_0, \\ min\_2 = y\_0, min\_3 = phi(min\_1, min\_2, var\_1) \end{array} \right\}$$

CSPs can be themself represented as hyper-graphs where the variables represent vertices and the constraint scopes the hyper-edges. Thus hyper-edges connect possible more than one vertex. Hyper-graphs can be used to classify CSPs regarding their complexity of computing a solution. For example, solutions for CSPs with corresponding acyclic hyper-graphs can be computed in polynomial time. Such hyper-graphs can be directly represented as hyper-trees. However, general hyper-graphs can be converted into hyper-trees when joining constraints. Such resulting hyper-trees can be again effectively used for computing solutions. However, because of the join operator the resulting system becomes larger. Hence, someone only gains computational advantages when the number of constraints to be joined is as minimal as possible. The number of constraints that has to be joined is referred to as hyper-tree width. More information regarding hyper-graphs and hyper-tree composition which allows to convert hyper-graphs into hyper-trees can be found in [7, 8]. The hyper-graph and its corresponding hyper-tree for the CSP introduced above is depicted in Figure 6. The hyper-tree width for this example is 2.

4 Related research and Conclusion

Collavizza and Ruehner [2] discussed the conversion of programs into constraint systems and their use in software verification. They also make use of the SSA form but there are differences in the description of the corresponding

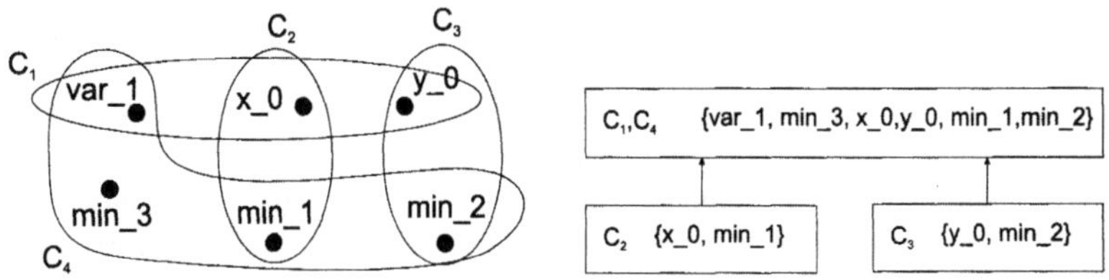

Fig. 6. Hyper-graph (left) and corresponding hyper-tree (right) for the SSA form given in Fig. 4

CSP. Moreover, we focus on debugging and program analysis rather than on verification.

Various authors, e.g., [6, 16, 12, 13], have described models to be used for fault localization using model-based diagnosis. Almost all of the work makes assumptions regarding the program's structure, uses abstractions which lead to the computation of too many diagnosis candidates, or does not handle all possible behaviors at once. The latter models, for example, consider one execution trace which prevent the diagnosis engine of remove some diagnosis candidates. In our representation we consider all possible behaviors up to a given boundary. This should lead to a reduction of the number of diagnosis candidates. In [11] Köb and Wotawa discussed the use of Hoare logic for model-based debugging which requires a Hoare logic calculus for computing diagnosis. In our case we rely on CSPs which are a natural representation of assignment statements. Moreover, from the structure of the CSP, i.e., its hyper-tree representation, we obtain the hyper-tree width which is an indicator for the complexity of computing diagnoses.

In this paper we introduced a methodology for compiling programs into their equivalent CSP representation. We argued that the CSP representation can be effectively used for debugging and, moreover, allows for computing a metrics for the complexity of debugging, i.e., the hyper-tree width. Because of the worst-case case complexity of debugging using the CSP representation which is NP complete a distributed approach where the system to be debugged and the debug system are separated is recommended. The system sends its source code and a test-case to the debugger and receives a new (compiled) version as a result. We currently have implemented the compilation of Java-like programs into their CSP representation. Debugging tests and the relationship between program complexity and hyper-tree width will be the next steps.

References

1. Marc M. Brandis and H. Mössenböck. Single-pass generation of static assignment form for structured languages. *ACM TOPLAS*, 16(6):1684–1698, 1994.

2. H. Collavizza and M. Rueher. Exploration of the capabilities of constraint programming for software verification. In *Tools and Algorithms for the Construction and Analysis of Systems (TACAS)*, pages 182–196, Vienna, Austria, 2006.
3. Ron Cytron, Jeanne Ferrante, Barry K. Rosen, Mark N. Wegman, and F. Kenneth Zadeck. Efficiently computing static single assignment form and the control dependence graph. *ACM TOPLAS*, 13(4):451–490, 1991.
4. Rina Dechter. *Constraint Processing*. Morgan Kaufmann, 2003.
5. Yousri El Fattah and Rina Dechter. Diagnosing tree-decomposable circuits. In *Proceedings* 14th *International Joint Conf. on Artificial Intelligence*, pages 1742 – 1748, 1995.
6. Gerhard Friedrich, Markus Stumptner, and Franz Wotawa. Model-based diagnosis of hardware designs. *Artificial Intelligence*, 111(2):3–39, July 1999.
7. Georg Gottlob, Nicola Leone, and Francesco Scarcello. On Tractable Queries and Constraints. In *Proc. 12th International Conference on Database and Expert Systems Applications DEXA 2001*, Florence, Italy, 1999.
8. G. Gottlob, N. Leone, and F. Scarcello. A comparison of structural CSP decomposition methods. *Artificial Intelligence*, 124(2):243–282, 2000.
9. A. Griesmayer, R. Bloem, and B. Cook. Repair of boolean programs with an application to C. In *Proc. 18th Conference on Computer Aided Verification (CAV'06)*, pages 358–371, Seattle, Washington, USA, August 2006.
10. Daniel Jackson. *Software abstractions: logic, language, and analysis*. MIT Press, 2006.
11. Daniel Köb and Franz Wotawa. Fundamentals of debugging using a resolution calculus. In Luciano Baresi and Reiko Heckel, editors, *Fundamental Approaches to Software Engineering (FASE'06)*, volume 3922 of *Lecture Notes in Computer Science*, pages 278–292, Vienna, Austria, March 2006. Springer.
12. Cristinel Mateis, Markus Stumptner, and Franz Wotawa. Modeling Java Programs for Diagnosis. In *Proceedings of the European Conference on Artificial Intelligence (ECAI)*, Berlin, Germany, August 2000.
13. Wolfgang Mayer, Markus Stumptner, Dominik Wieland, and Franz Wotawa. Can ai help to improve debugging substantially? debugging experiences with value-based models. In *Proceedings of the European Conference on Artificial Intelligence*, pages 417–421, Lyon, France, 2002.
14. Raymond Reiter. A theory of diagnosis from first principles. *Artificial Intelligence*, 32(1):57–95, 1987.
15. S. Staber, B. Jobstmann, and R. Bloem. Finding and fixing faults. In *Proc. 13th Conference on Correct Hardware Design and Verification Methods*, pages 35–49. Springer-Verlag, 2005. LNCS 3725.
16. Markus Stumptner and Franz Wotawa. Debugging Functional Programs. In *Proceedings* 16th *International Joint Conf. on Artificial Intelligence*, pages 1074–1079, Stockholm, Sweden, August 1999.
17. Markus Stumptner and Franz Wotawa. Diagnosing tree-structured systems. *Artificial Intelligence*, 127(1):1–29, 2001.
18. M. Stumptner and F. Wotawa. Coupling CSP decomposition methods and diagnosis algorithms for tree-structured systems. In *Proc.* 18th *International Joint Conf. on Artificial Intelligence*, pages 388–393, Acapulco, Mexico, 2003.
19. M.N. Wegman and F.K. Zadek. Constant propagation with conditional branches. *ACM Transactions on Programming Languages and Systems*, 13(2), April 1991.

Part III

Short Papers

Models for Temporal and Spatial Terrain Coverage

Harry Bullen and Ronaldo Menezes

Florida Institute of Technology, Computer Sciences
Computer Vision and Bio-Inspired Computing Laboratory
Melbourne, Florida, USA
hbullen@fit.edu rmenezes@cs.it.edu

Summary. The world we live in is becoming increasingly more complex, requiring non-trivial, creative solutions. One important problem is the one of continuous monitoring, reconnaissance, or terrain coverage. Here the terrain may be constantly changing and agents are required to frequently return to previously visited locations. We call this temporal terrain coverage. This paper describes several models of terrain coverage and compare them wrt. their ability to deal with temporal and spatial coverage.

1 Introduction

Terrain coverage applications abound: military, robotic automation, and many others (see Section 2.1). A gap in the current literature of terrain coverage is on metrics to evaluate the proposed approaches. When the term terrain coverage is used, it does not convey what task need to performed until a metric is defined such as the minimization of unvisited cells in the terrain, the minimization of the time between visits to a particular cell, or the pattern of the coverage (when using multiple agents). In addition, the literature has neglected one variation of the terrain coverage characterized by agents that are required to continuously cover a terrain due to the fact that terrain itself may be dynamic and constantly changing: we call it *temporal terrain coverage*. Hence the purpose of this paper is two-fold: first we provide a survey of various approaches for terrain coverage and how (and if) they apply to the temporal case; secondly, we describe various metrics for evaluating a good model for continuous dynamic terrain coverage and provide comparative results for all the coverage approaches. In order to place temporal coverage in context, we briefly discuss a more standard approach to coverage: spatial coverage.

2 Terrain Coverage

As pointed out by VanderHeide and Rao [6] there is quite a bit of confusion about the definition of terrain coverage since it relates to path planning and terrain model

Harry Bullen and Ronaldo Menezes: *Models for Temporal and Spatial Terrain Coverage*, Studies in Computational Intelligence (SCI) **78**, 239–244 (2008)
www.springerlink.com

acquisition. In a nutshell, one can say that terrain coverage refers to the problem of covering as much surface area in a defined terrain as possible. In this paper we focus on two aspects of terrain coverage: temporal and spatial. A good temporal coverage relates mostly to the amount of time between cell visits—the standard deviation of the time between visits in each cell should be low.

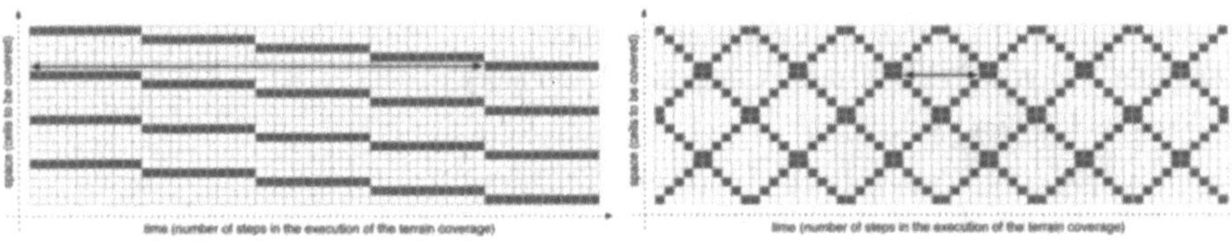

Fig. 1. Distinct scenarios of temporal coverage.

In Figure 1 we depict two distinct scenarios of temporal terrain coverage. Each row represents the evolution of visits to a cell in the terrain. Hence one entire column represents the status of all the cells in the terrain at a time t. We mark the maximum time a cell remains unvisited using an arrow. Note that in Figure 1(left) cells remain unvisited for a long time while in (right) the time is shorter (more desirable).

Spatial coverage is quite different and relates only to the number of times a cell is visited, independently of the time-distribution of the visits. Using the examples in Figure 1 we can say that spatially, both coverages are similar since all cells are, at the end of the coverage execution, visited the same number of times (12 times).

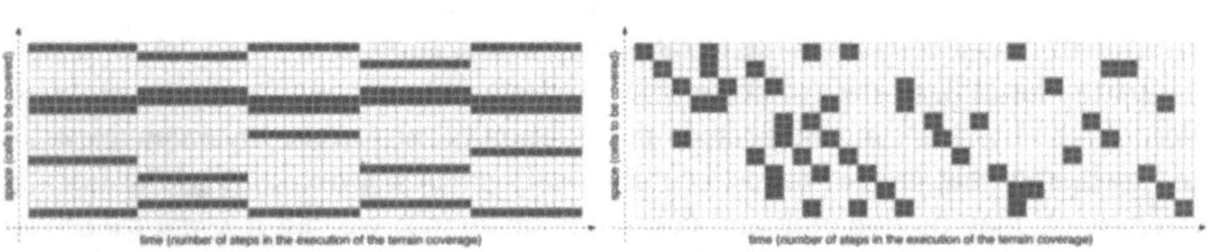

Fig. 2. Examples of distinct configurations of spatial coverages.

Figure 2 shows two other possibilities of spatial coverage differing significantly. Figure 2(right) is better than the other (on the left) when it comes to spatial coverage. The reason is that in the example of Figure 2(right) all cells are visited 10 times at the end of the coverage, whereas in Figure 2(left) some cells are visited 50 times while others are never visited.

2.1 Applications of Terrain Coverage

Terrain coverage has many well-known applications beyond the obvious *cover every position of a terrain*. The term is commonly used whenever the goal is to interact with all locations of a given terrain. We refer to this approach as spatial terrain coverage. There are many situations where spatial terrain coverage is useful, for example checking a document for spelling or grammatical errors would be a trivial example of spatial coverage. Other typical examples include vacuum cleaning, painting, and harvesting. These are all examples of spatial coverage because, to complete the task, each location in the environment has to be visited a fixed number of times independently of the periodicity of the visits.

There are many other situations where it is important to have a space visited again and again during a long period of time—we refer to this here as temporal coverage. Temporal coverage should be thought of as pertaining to situations where it would be desirable (though impractical) to have an agent at each location of the terrain in every time interval—hence the existing agents need to be moving constantly to maximize their value. The most common situation is surveillance; while it might be possible to cover a building with cameras, it would be very difficult to cover a city. In this situation autonomous robots could move through the city providing information about the location they visit. In this surveillance example, one could have the robots feeding a database with information about what is taking place in their location; the information could then be used to distribute the police force in the city. Note that the dynamism of the city requires robots to return to previously visited locations to re-check the status of that location. The temporal coverage would aim at minimizing the time between re-visits so that areas are not left unvisited for a long period of time.

Similar to surveillance would be exploration, perhaps of disaster areas such as the one that took place in New Orleans, USA with the hurricane Katrina (in 2005), where many small rovers could work to collect data about the destruction. Given the dynamism of the area, rovers would be required to return to previously visited areas. Another interesting application of temporal coverage is crawling the Internet, where the agents (spiders) retrieve web pages for use elsewhere, typically searching. In this case, it is important for the search engine to have an up-to-date copy of every page.

3 Models of Terrain Coverage

Node counting has been suggested as a simple, intuitive, and effective model. It is also used as starting point for more complex models. Here, each time an agent visits a cell it increases a counter that records the number of times the cell has been visited. The agents then deterministically move to the neighboring cell that has been visited the fewest times. This has been shown to quickly cover an area [5, 1]. This model can be modified to work probabilistically. Another possible modification is to cause the count of visits in each state to "decay" over time. This makes the model have preference for states that have not been visited recently over ones that have.

Recency is a model where the decision of where to move is based on how recently the cells were visited. Note that this is like Node Counting but rather than counting spatially, the counting is temporal. This can be implemented setting a value to cells that are visited. Over time this value decays at a fixed rate so the agents can tell how long ago a vertice was visited. Note that this is very similar to pheromone-based models such as the Alarm model—the decaying of the value relates to pheromone evaporation. This model was used by Thrun [5] to test exploration in reinforcement learning and he found it performed well against node counting.

The Learning Real Time A* (LRTA*) [2] is a derivative of the A* search designed for online searching. This model works like the A* search, by using a non-underestimating heuristic to speed up searching. Like A*, each vertex has a heuristic estimating the distance to the goal, however, the LRTA* search improves these heuristics by updating them with new information during the search. A heuristic can be updated by changing it to the value of the next vertex plus the cost to travel there if this gives a higher, and thus more accurate, value. LRTA* was adapted for use in terrain coverage by observing that when searching a terrain for a goal with an unknown location, the search agent would in the worst case have to cover the terrain. So, to adapt this LRTA* to perform terrain coverage, the goal cell was simply removed and the agents allowed to run as normal.

Thrun [5] has proposed a probabilistic model for online mapping (henceforth called OMA). The model can be considered a sort of a cross between LRTA* and node counting. It uses the LRTA* update rule, but always increases the value of the pheromone (count of visits) by at least one. This ensures that every time the cell is visited, it becomes less attractive to other agents. As such, it is expected that this model will have qualities of both LRTA* and node counting. The OMA presented here is an adaptation of the original proposed by Thrun.

The Alarm model [4] is inspired by ant-colony optimization, where ants search for an optimal solution by sending many agents through a terrain who lay down pheromone based on the quality of the solution they found. The other agents then chose their path based on the pheromones left behind on the terrain. This results in all the agents converging to one of the shortest paths. In the Alarm model the pheromone repels the agents in order to spread them out; this mimics the alarm pheromone behavior common in many species of ants [3].

4 Performance Evaluation

The first metric in our evaluation is the distribution of coverage, or the number of times each cell was visited—a metric of spatial coverage. This is applicable to areas where the number of visits to each cell is important such as painting and harvesting. The goal is to have all the cells covered with the same number of visits.

Figure 3 show the results for spatial coverage. One can clearly see from the results of the standard deviation of the number of visits that both Alarm and Node Counting perform significantly better than the other models. More importantly, the result demonstrates the ability of a probabilistic terrain coverage model (Alarm) to perform as well as a deterministic approach (Node Counting). From the same result we see that the LRTA* model does not do well in spatial terrain coverage, which relates to the fact that LRTA* was designed for target location rather than coverage.

The primary metric used in temporal coverage is the time between visits and the length of time since each vertex has been visited. These two are closely related.

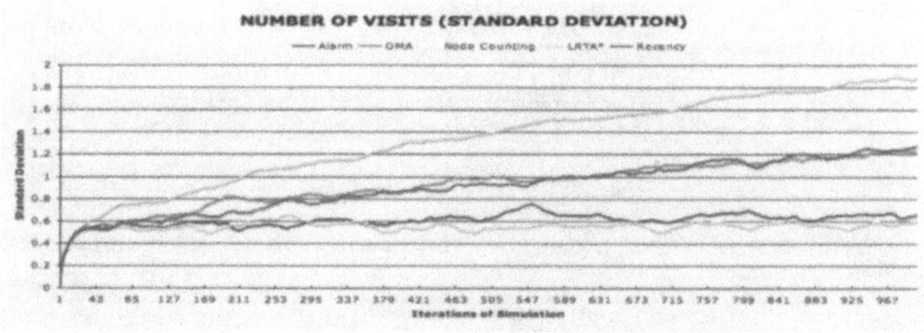

Fig. 3. Results for spatial coverage. Graphs relate to the total number of visits in each cell.

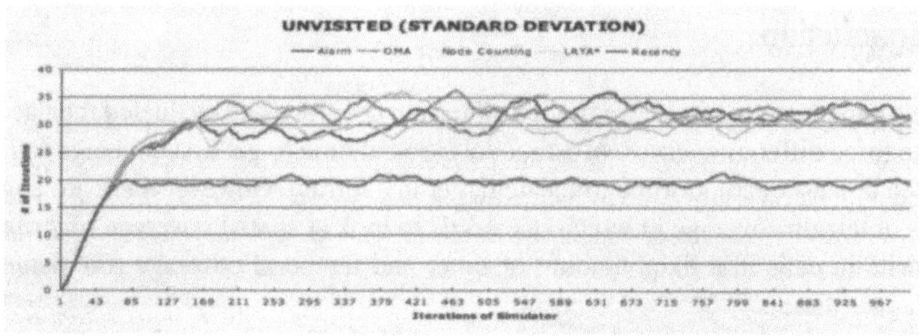

Fig. 4. Results describing the performance of the models for temporal coverage. Here the time since the last visit is depicted showing the efficiency of the Recency model.

First, the time since a vertice has been visited is called *unvisited*, and it is the main metric used in the Recency model. This measurement should be as low as possible indicating all locations are getting attention frequently. Also, the value of the cell with the longest (max) amount of time since it has been visited, tells valuable information about the uniformity of the agents.[1] Figure 4 shows the performance of the models with regards to time since the cells have been visited (the amount of time a cell remains unvisited).

The Recency model is clearly better than all the others. Recency is a model designed to solve this problem. All the other models perform at the same level with a slight advantage to Alarm and Node Counting over OMA and LTRA*.

The last metric we used to evaluate the models is the time between visits or *revisits*. In theory the revisits time should approximate the number of cells divided by the number of agents. This is slightly different from unvisits because it evaluates the ability of agents to return to previously visited cells—the sensitivity to cells that have been visited. Figure 5 confirms the superiority of the Recency model.

We have performed many other experiments which are not shown here due to space limitations. In summary, our studies show that Node Counting and Alarm perform very well in spatial terrain coverage while Recency is the clear winner in terms of temporal coverage.

[1] this result is not displayed due to space restrictions.

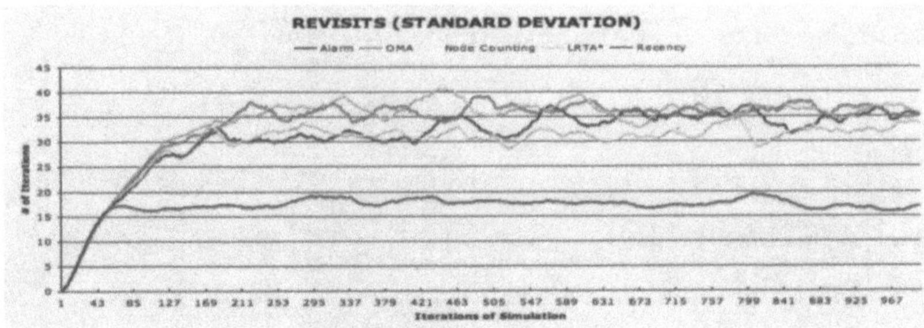

Fig. 5. Time between visits.

5 Conclusion

In this paper we described five models of terrain coverage and evaluated then according to three different metrics. Most models look at time to get first coverage (all cells visited at least once), or time to locate target in a terrain. However there are applications of terrain coverage in which one needs to look at spatial coverage (distribution of visits in cells in a fixed amount of time) and temporal coverage (the pattern of visits over time).

Our results demonstrated that Alarm and Node Counting perform very well in spatial coverage while the Recency model is the best at temporal coverage. Other more specialized models (originally designed with the purpose of locating targets in terrains) such as OMA and LTRA* did not perform well in these settings.

References

1. S. Koenig and Y. Liu. Terrain coverage with ant robots: A simulation study. In *AGENTS '01: Proceedings of the fifth international conference on Autonomous agents*, pages 600–607, New York, NY, USA, 2001. ACM Press.
2. R. Korf. Real-time heuristic search. *Artificial Intelligence*, 42(2-3):189–211, 1990.
3. C. Lloyd. The alarm pheromones of social insects: A review. Technical report, Colorado State University, 2003.
4. R. Menezes, F. Martins, F. E. Vieira, R. Silva, and M. Braga. A model for terrain coverage inspired by ant's alarm pheronome. In *SAC '07: Proceedings of the 2007 ACM Symposium on Applied Computing*, pages 728–732, New York, NY, USA, 2007. ACM Press.
5. S. Thrun. A probabilistic on-line mapping algorithm for teams of mobile robots. *International Journal of Robotics Research*, 20(5):335–363, 2001.
6. J. R. VanderHeide and N. S. V. Rao. Terrain coverage of an unknown room by an autonomous mobile robot. Technical report, Oak Ridge National Laboratory, Tennessee, United States, 1995.
7. U. Wilensky. Netlogo. Technical report, Center for Connected Learning and Computer-Based Modeling, Northwestern University, 1999. http://ccl.northwestern.edu/netlogo.

Using Genetic Algorithms to Optimise Rough Set Partition Sizes for HIV Data Analysis

Bodie Crossingham and Tshilidzi Marwala

School of Electrical and Information Engineering, University of the Witwatersrand
bodie@lutrin.co.za

Summary. In this paper, we present a method to optimise rough set partition sizes, to which rule extraction is performed on HIV (Human Immunodeficiency Virus) data. The genetic algorithm optimisation technique is used to determine the partition sizes of a rough set in order to maximise the rough sets prediction accuracy. The proposed method is tested on a set of six demographic properties of individuals obtained from the South African antenatal survey, with the outcome or decision being either HIV positive or negative. Rough set theory is chosen based on the fact that it is easy to interpret the extracted rules. The prediction accuracy of equal width bin partitioning is 69.8% while the accuracy achieved after optimising the partitions is 87.5%.

1 Introduction

In the last 20 years, over 60 million people have been infected with HIV (Human Immunodeficiency Virus), and of those cases, 95% are in developing countries [1]. During this year, AIDS (Acquired Immune Deficiency Syndrome) claimed an estimated 2.9 million lives [2]. HIV has been identified as the cause of AIDS. It is thus evident that the analysis of HIV is of the utmost importance. By correctly forecasting HIV, the causal interpretations of a patients being seropositive (infected by HIV) is made much easier. Poundstone *et al* related demographic properties to the spread of HIV [3]. In their work they justified the use of demographic properties to create a model to predict HIV from a given database, as is done in this study. RST (rough set theory) uses the social and demographic factors to predict HIV status, this in turn provides insight into which variables are most sensitive in determining HIV status. Rough sets have been used in many applications. Rowland *et al* compared the use of RST and neural networks for the prediction of ambulation spinal cord injury [4], and although the neural network method produced more accurate results, its "black box" nature makes it impractical for the use of rule extraction problems. RST compromises accuracy over rule interpretability but for HIV it can be argued that interpretability of the data

Bodie Crossingham and Tshilidzi Marwala: *Using Genetic Algorithms to Optimise Rough Set Partition Sizes for HIV Data Analysis*, Studies in Computational Intelligence (SCI) **78**, 245–250 (2008)
www.springerlink.com

is of more importance than just prediction. In order to achieve the best accuracy, the rough set partitions or discretisation process needs to be optimised. The optimisation process is done using genetic algorithm (GA), where the fitness function aims to achieve the highest accuracy produced by the rough set. Literature reviews have shown that limited work has been done on the optimisation of rough set partition sizes.

In the following sections, the background of the topic is stated, followed by a summarised explanation of RST. Section 4 explains how the genetic algorithm is used to optimise the rough set partitions, and in section 5 the results are given and compared.

2 Background

Rough set theory was introduced by Zdzislaw Pawlak in the early 1980s [5]. RST is a mathematical tool which deals with vagueness and uncertainty. Rough sets are useful in the analysis of decisions in which there are inconsistencies. To cope with these inconsistencies, lower and upper approximations of decision classes are defined [6]. One of the advantages of RST is that it does not require *a priori* knowledge about the data set, and it is for this reason that statistical methods are not sufficient for determining the relationship between the demographic variables and their respective outcomes.

The data set used in this paper was obtained from the South African antenatal sero-prevalence survey of 2001. The data was obtained through questionnaires completed by pregnant women attending selected public clinics and was conducted concurrently across all nine provinces in South Africa [7].

The six demographic variables considered are: *race, age of mother, education, gravidity, parity* and, *age of father*, with the outcome or decision being either HIV positive or negative.

The HIV status is the decision represented in binary form as either a 0 or 1, with a 0 representing HIV negative and a 1 representing HIV positive. The input data is discretised into four partitions and represented numerically as either 1, 2, 3 or 4. This number is chosen as it gives a good balance between computational efficiency and accuracy.

3 Rough Set Theory and Rough Set Formulation

Rough set theory deals with the approximation of sets that are difficult to describe with the available information [8]. Some concepts that are fundamental to RST are mentioned briefly but for a complete explanation refer to [5]. The data is represented using rows and columns in an information table. Once the table is obtained, the data is discretised into four partitions as mentioned earlier. Each case is evaluated and cases that are conflicting are referred to as an indiscernibility relation and this is the main concept of rough set theory, (indiscernibility meaning indistinguishable from one another). RST offers a tool to deal with indiscernibility and the way in which it works is, for each concept/decision X, the greatest definable set containing X and the least definable set containing X are computed. These two sets are called the lower and upper approximation respectively.

It is through these lower and upper approximations that any rough set is defined. It must be noted that for most cases in RST, reducts are generated to enable us to discard functionally redundant information [5]. And although reducts are one of the main advantages of RST, they are ignored for the purpose of this paper, i.e. the optimisation of discretised partitions.

A membership function then determines the plausibility of which any object x belongs a particular set X. The results are illustrated using a receiver operating characteristic (ROC) curve. The ROC curve displays the relationship between sensitivity (true-positive rate) and 1-specificity (false-positive rate) across all possible threshold values that define the positivity of being infected with HIV. The area under curve (AUC) of the respective ROC curves is used as the performance criterion. The higher the AUC, the better the classification accuracy is. The AUC is the accuracy used to optimise the rough set partition sizes. The process of modelling rough sets can be broken down into five stages and can be summarised as follows:

Stage one would be to select the data. Stage two to pre-process and discretise the data. If reducts are considered, the third stage would be to use the cleaned data to generate reducts. A reduct is the most concise way in which we can discern object classes [9]. To cope with inconsistencies, lower and upper approximations of decision classes are defined [5, 6, 9]. Stage four would be to extract the generated rules and these rules can be presented in an *if* CONDITION(S)-*then* DECISION format, e.g.

- **If** Race = African **and** Mothers Age = 23 **and** Education = 4 **and** Gravidity = 2 **and** Parity = 1 **and** Fathers Age = 20 **Then** HIV = Positive with plausibility = 0.33333

The final stage would be to test the newly created rules on a test set. The accuracy achieved is then sent back to the GA to optimise the partition sizes.

4 Genetic Algorithm

A genetic algorithm (GA) is a stochastic search procedure for combinatorial optimisation problems based on the mechanism of natural selection [10]. GAs are popular and widely used due to their ease of implementation, intuitiveness and the their ability to solve highly nonlinear optimisation problems. Genetic algorithms are a particular class of evolutionary algorithms that use techniques inspired by evolutionary biology such as inheritance, mutation, selection, and crossover. The fitness/evaluation function is the only part of the GA that has any knowledge about the problem. The fitness function tries to maximise the AUC. The GA represents the design variables of each individual design with binary strings of 0's and 1's, these are referred to as chromosomes. The variables are limited to the predefined upper and lower bound values. These limits are coded into the GA encoding feature. They perform the task of optimisation, the GA employs three main operators to propagate its population from one generation to the next.

The first operator is *selection*. During each successive generation, a proportion of the population is selected to breed a new generation. This is done on the basis of "survival of the fittest". The fitness of each solution is evaluated using a fitness function, and it is this function that maximises the AUC. After each generation, the AUC is evaluated and sent to the rough set for evaluation, this process continues until the termination criteria is reached. Several selection functions are available and in this paper tournament selection is used. Tournament selection is whereby a "tournament" is run on a few individuals chosen at random and the strongest individual (winner) is chosen for the process of crossover.

The second operator is *crossover* which mimics reproduction or mating in biological populations. The crossover technique used is uniform crossover; in this technique, two parents are combined to produce two new offspring. The way in which it works is individual bits in a string of two parents are compared, and then swapped with a probability of 0.5.

The third operator used is *mutation*, this is whereby an arbitrary bit in a generic sequence or string will be mutated or changed. This is analogous to biological mutation. The reason for implementing this operator is to promote diversity from one generation of chromosomes to another, this prevents the GA from getting stuck at a local optimum but rather a global optimum. Boundary mutation is chosen, this is whereby a variable is randomly selected and is set to either the upper or lower bound depending on a randomly generated uniform number. It must be noted that the chosen operators are case specific, dependent on the nature of the optimisation, different operators would have to be explored.

The pseudo-code algorithm for genetic algorithms is given below;

1. Initialise a population of chromosomes
2. Evaluate each chromosome (individual) in the population
   a) Create new chromosomes by mating chromosomes in the current population (using crossover and mutation)
   b) Delete members of the existing population to make way for the new members
   c) Evaluate the new members and insert them into the population
3. Repeat stage 2 until some termination condition is reached, in this case until 100 generations are reached.
4. Return the best chromosome as the solution

An initial population of 20 individuals is chosen. As mentioned in order to prevent premature convergence to a local minima, the mutation diversification mechanism is implemented. Other diversification mechanisms such as elitism can also be implemented in an attempt to improve the accuracy.

5 Results Obtained

First the accuracy of the rough set is computed using equal width bin (EWB) partitioning and the resulting accuracy is 69.8%. Next the GA discretisation method is run with GA using a tournament selection function, a boundary mutation and uniform crossover. An initial population of 20 members is selected and the termination function is 100 generations. The GA version produced an accuracy of 87.5%.

Tabulated below are the number of rules (No. Rules), AUC and ROC curves obtained using equal width bin partitioning (EWB) and as well when using a genetic algorithm (GA) to discretise the partitions.

	No. Rules	AUC
EWB	488	0.698
GA	49	0.875

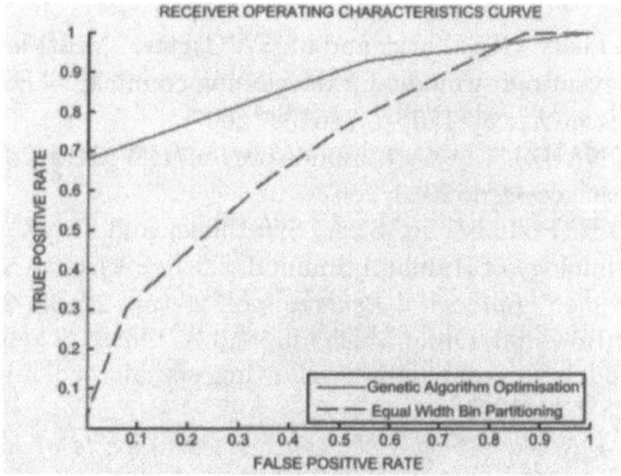

Fig. 1: Tabulated results and ROC curves obtained for EWB and GA optimised partitioning

The ROC curve is chosen to illustrate the results as they allow the performance of the classifiers to be evaluated on how well they predict. The area under curve (AUC) of the respective ROC curves is used as the performance criterion.

Once RST is applied to the HIV data, for the first case of equal width bin partitioning, the 283 cases of the lower approximation are rules that always hold, or are definite cases. The 205 cases of the upper approximation can only be stated with a certain plausibility. For the second case of GA optimised partitions, there are 26 unique discernible cases (in the lower approximation) and 28 indiscernible cases (in the upper approximation). It can be seen that a lower amount of rules resulted in a higher accuracy.

6 Conclusion

A genetic algorithm is successfully applied to RST on the HIV data set. As a result of implementing RST on the data set, the rules extracted are explicit and easily interpreted. RST will however compromise accuracy over rule interpretability, and this is brought about in the discretisation process where the granularity of the variables are decreased. An accuracy of 69.8% is produced by the rough set when applied to the HIV data set for equal width partitioning, an improved accuracy of 87.5% is achieved for the genetic algorithm optimised partitions. Recommendations for future work include the application of other optimisation techniques such as particle swarm optimisation (PSO). PSO is advantageous over GAs as it is easy to implement and there are fewer parameters to adjust.

References

1. A. Lasry, G. S. Zaric, and M. W. Carter. "Multi-level resource allocation for HIV prevention: A model for developing countries." *European Journal of Operational Research*, vol. 180, p. 786799, 2007.
2. "UNAIDS." www.unaids.org/en/HIV_data/2006GlobalReport/default.asp/. Last accessed: 20/3/2007.
3. K. E. Poundstone, S. A. Strathdee, and D. D. Celentano. "The Social Epidemiology of Human Immunodeficiency Virus/Acquired Immunodeficiency Syndrome." *Epidemiol Reviews*, vol. 26, pp. 22–35, 2004.
4. T. Rowland, Ohno-Machado, and A. Ohrn. "Comparison of multiple prediction models for ambulation following spinal cord injury." *In Chute*, vol. 31, pp. 528–532, 1998.
5. Z. Pawlak. *Rough Sets, Theoretical Aspects of Reasoning about Data*, chap. 3, p. 33. Kluwer Academic Publishers, 1991.
6. M. Inuiguchi and T. Miyajima. "Rough set based rule induction from two decision tables." *European Journal of Operational Research*, vol. In Press, Corrected Proof, 2006.
7. R. Department of Health. "National HIV and Syphilis Sero-Prevalence Survey of Women Attending Public Antenatal Clinics in South Africa." http://www.info.gov.za/otherdocs/2002/hivsurvey01.pdf, 2001.
8. A. Ohrn and T. Rowland. "Rough Sets: A Knowledge Discovery Technique for Multifactorial Medical Outcomes." *American Journal of Physical Medicine and Rehabilitation*, vol. 79, pp. 100–108, 2000.
9. F. Witlox and H. Tindemans. "The application of rough sets analysis in activity-based modelling. Opportunities and constraints." *Expert Systems with Applications*, vol. 27, p. 585592, 2004.
10. S. Malve and R. Uzsoy. "A genetic algorithm for minimizing maximum lateness on parallel identical batch processing machines with dynamic job arrivals and incompatible job families." *Computers and Operations Research*, vol. 34, p. 30163028, 2007.

1

Semantic Web and Rule Reasoning inside of E-Learning Systems

Ion-Mircea Diaconescu, Sergey Lukichev and Adrian Giurca

Department of Internet Technology
Institute of Informatics
Brandenburg Technical University at Cottbus, Germany
{M.Diaconescu, lukichev, giurca}@tu-cottbus.de

Summary. This paper describes how the Semantic Web and the JenaRules can work together in order to empower e-Learning platform Moodle with rich reports and flexible courses management. We discuss aspects of using Semantic Web technologies in e-learning in general and the JenaRules in particular. The main focus is towards using rules for reasoning on top of existing Moodle content.

Key words: e-Learning, Moodle, Semantic Web, Jena, RDF, RDFS, OWL, SPARQL.

1.1 Introduction

Learning is an important aspect of our live. Online learning enables previously disqualified groups such as single parents and full time workers to learn outside the time constraints of conventional campus structures. Free software has delivered tools such as Moodle, Sakai, Atutor that fulfill this any time functionality with feature-rich sets and dynamically active group awareness. These online systems are also known as e-Learning Systems. Such systems provide an easy way for tutors to expose their courses, to interact with students, and to track student interest and activities over the presented domain. In addition, they have forums, where students and tutors can exchange opinions or receive answers to some questions.

The use of Semantic Web technologies in e-Learning ([1]) may bring benefits to tutors and students. For example, a course tutor may want to know which students are active in a course or how many students use a recommended documentation for the course. A traditional solution to such problem is relational databases, but moving to RDF [2] is a shift to the open world with many distributed resources, identified by URIs as a mechanism for referring to global constants on which there exists some agreement among multiple data providers. Queries can be performed not only over a single database,

Ion-Mircea Diaconescu et al.: *Semantic Web and Rule Reasoning inside of E-Learning Systems*, Studies in Computational Intelligence (SCI) **78**, 251–256 (2008)
www.springerlink.com

but over the content of several distributed educational systems, including resources, which are externally available on the Web. In addition, we may list the following advantages of using RDF and rules:

- Data can be shared using common vocabulary;
- New rules can be easily added or modified at any time without making changes in the system code;
- Modules can be easily adapted to new requests from tutors and students.

In our position statement [3] we define a project for empowering Moodle with rules and semantics. The project consists of the following steps:

1. RDF-izing existing Moodle content;
2. Automatic rules mining on top of RDF fact base, obtained in the previous step;
3. Extending automatically generated rules from the previous step manually.

In this paper we focus on the use of the JenaRules for implementing rules, which are used for advanced reporting. The JenaRules is discussed and implementation of some rule examples is provided.

1.2 Rule Reasoners for the Semantic Web

Moodle uses MySQL database to store logs. In order to get benefits of using Semantic Web technologies, this data should be structured in semantically meaningful way, i.e. RDF-ized. The Moodle information model for representing logs is discussed in [3]. Having an RDF fact base, we may start creating rules on top of it. These rules are further executed by a reasoning engine, which may run as an external service. There are several Semantic Web reasoners:

- Notation3, or *N3*[1] as it is more commonly known, is a shorthand non-XML serialization of RDF, designed with human-readability in mind. N3 has a reasoning engine, CWM[2], written in Python and part of SWAP, a Semantic Web Application Platform. It is open source under the W3C software license.
- Semantic Web Rule Language (SWRL)[3] is a combination between the OWL DL and OWL Lite and the Unary/Binary Datalog RuleML[4] sublanguages of the Rule Markup Language.
- Racer (Renamed ABox and Concept Expression Reasoner)[5] is an OWL based reasoner. RacerPro (the commercial version) allows users to describe

[1] N3 Specification: `http://www.w3.org/DesignIssues/Notation3`
[2] CWM, `http://www.w3.org/2000/10/swap/doc/Processing`
[3] `http://www.w3.org/Submission/SWRL/`
[4] `http://www.ruleml.org`
[5] `http://www.racer-systems.com/products/racerpro/index.phtml`

their data and benefit from powerful ontology-based query answering systems. It has been extended with support for applying SWRL rules to set of OWL facts or to RDF descriptions.

- JenaRules[6] is an extension of the standard Jena reasoner[7]. The system provides a rule language which is somehow similar with N3 notation. A Java API for JenaRules and SPARQL is provided.
- Mandarax[8] is an open source java class library for deduction rules. It provides an infrastructure for defining, managing and querying rule bases. It is based on backward reasoning and is pure OO.

We consider that JenaRules is the best choice at this moment, because it is specially designed for the Semantic Web. There are several possibilities to RDF-izing Moodle content. The first option is a script, which periodically passes through the Moodle log and updates the RDF fact base with new data. The second option is an update of the RDF fact base at the same time when the log is updated. This solution requires some modifications into existing Moodle code. And the third solution is to use Jena RDF front-end to MySQL. We are in favor of this solution since it does not require too much work and is natural way of accessing databases without a need to duplicate existing log data. Another advantage of this approach is that current version of Jena implementation loads the full fact base from the RDF source, which is memory-consuming in case of real life Moodle system. Using the front-end to MySQL solves this problem.

1.3 Inside JenaRules

Jena[9] is a Java framework for building Semantic Web applications. It provides a programmatic environment for `RDF`[10], `RDFS`, `OWL`[11] and `SPARQL`[12] and includes a rule-based inference engine. Jena includes a reasoner to reason with rules. Rules can be created using a text file, or in applications using a String data type.

1.3.1 Jena Rules

An informal description of the simplified text rule syntax can be found at : `http://jena.sourceforge.net/inference/index.html#RULEsyntax`. A sample file containing Jena rules is:

[6] `http://jena.sourceforge.net/inference`
[7] `http://jena.sourceforge.net`
[8] `http://mandarax.sourceforge.net/`
[9] Jena Semantic Web Framework: `http://jena.sourceforge.net/`
[10] RDF specification and documentation: `http://www.w3.org/RDF`
[11] OWL specification and documentation: `http://www.w3.org/2004/OWL`
[12] SPARQL specification: `http://www.w3.org/2001/sw/DataAccess`

```
// Prefixes definition
@prefix moodle: http://www.moodle.org/#
@prefix userdef: http://www.moodle.org/annotations#
@prefix lecture: http://www.moodle.org/lectures#
@prefix xs: http://www.w3.org/2001/XMLSchema#
// We may include rules from another file.
// @include http://www.example.org/rules/relations.rules

// If a student has between 25 and 49 posts in a specific forum,
// then the student rank is 2 for that forum.
[forum_rank: (?student, userdef:rank '2'^^xs:integer)
            <-
            (?student rdf:type moodle:Student)
            (?student userdef:numberOfPosts ?postsNumber)
            ge(?postsNumber, "25"^^xs:integer)
            le(?postsNumber, "49"^^xs:integer)]
```

A rule consists of two parts :

- *conditions*, for example `(?student userdef:numberOfPosts ?postsNumber)`
- *conclusions*, i.e. `(?student, userdef:rank '2'^^xs:integer)`.

Atoms in conditions and conclusions are implicitly connected by conjunction. JenaRules do not allows disjunction nor negation. If conditions are satisfied, then the conclusion part is performed. In the conclusion part some actions may be performed or some other operations can be executed. In our example, `forum_rank` is the name of the rule. Head disjunction can be simulated using many rules and in many cases negation can be simulated using `noValue` built-in. Jena offers support for user-defined built-ins and functors to create operations or actions. For example, in the previous rule, `ge` and `le` are built-ins, which verify the math inequalities *greater than or equal* and *less than or equal*.

1.3.2 Jena Java API

Jena rules are executed on top of RDF triples. As a result of the rules execution the RDF fact base is changed (new triples are added, triples are updated and/or deleted). Consider the following RDF triple:

```
<moodle:Student>
 <userdef:numberOfPosts rdf:datatype="xs:integer">28</userdef:numberOfPosts>
</moodle:Student>
```

This triple satisfies all conditions of the rule `forum_rank` therefore the rule is executed and, as a consequence, a new fact is created. After execution the working memory contains:

```
<moodle:Student>
 <userdef:numberOfPosts rdf:datatype="xs:integer">28</userdef:numberOfPosts>
 <userdef:rank rdf:datatype="xs:integer">2</userdef:rank>
</moodle:Student>
```

The RDF triples are loaded using the method `loadModel` offered by the class `FileManager` of the Java API. The RDF fact base is used to create the RDF

Model (an instance of the Java class `Model`), then an instance of the `InfModel` is used to create the connection between the rules and the fact base.

1.4 Using Rule-based Reasoning to generate Moodle reports

In this section we give example of Moodle report, which can be implemented using Jena. Facts are represented using RDF. Reports are generated using SPARQL query language.
Consider that we need to view all *diligent students* for a specific course. First, we define the concept of a *diligent student* for a specific course by means of a derivation rule:

> *A student, who takes the Semantic Web course, completes all assignments and is active into the forum of this course is a diligent student.*

It corresponds in JenaRules to the following rule:

```
[diligent_student: (?student userdef:diligentFor lecture:SemanticWeb)
                   <-
                   (?student rdf:type moodle:Student)
                   (?student userdef:takeCourse lecture:SemanticWeb)
                   (?student userdef:assignmentsCompleted 'true'^^xs:boolean)
                   (?student userdef:activeInForum lecture:SemanticWeb)]
```

In the condition part we have four triples :

- `(?student rdf:type moodle:Student)` - binds `student` variable to an instance of `moodle:Student` from the memory. According with the RETE algorithm, the conditions will be checked for every instance from memory of the `moodle:Student` class.
- the next three triples checks if all conditions for the rule are satisfied, respectively if the student is enrolled to the Semantic Web course, has completed all assignments and is active in the forum of this course.

If all conditions are satisfied for that student, then, in the conclusion part we have a `diligent student` for the `SemanticWeb` course:

```
(?student userdef:diligentFor lecture:SemanticWeb)
```

Considering the following RDF facts:

```
<moodle:Student>
 <moodle:name>John Smith</moodle:name>
 <userdef:takeCourse rdf:resource="lecture:SemanticWeb"/>
 <userdef:assignmentsCompleted rdf:datatype="xs:boolean">
  true
 </userdef:assignmentsCompleted>
 <userdef:activeInForum rdf:resource="lecture:SemanticWeb"/>
</moodle:Student>

<moodle:Student>
 <moodle:name>John Doe</moodle:name>
```

```
  <userdef:takeCourse rdf:resource="lecture:SemanticWeb"/>
  <userdef:assignmentsCompleted rdf:datatype="xs:boolean">
   true
  </userdef:assignmentsCompleted>
</moodle:Student>
```

The rule concludes that the student *John Smith* is diligent for the *Semantic Web* course, but *John Doe* is not because it is not active in the Semantic Web forum.

```
<moodle:Student>
  <moodle:name>John Smith</moodle:name>
  <!-- ... -->
  <userdef:diligentFor rdf:resource="lecture:SemanticWeb"/>
</moodle:Student>
```

In order to generate a report for all student names, which are *diligent* for the *Semantic Web* course we use a query language for RDF called SPARQL, which is also supported by Jena.

```
SELECT ?studentName
WHERE {
       ?student rdf:type moodle:Student .
       ?student moodle:name ?studentName .
       ?student userdef:diligentFor lecture:SemanticWeb
      };
```

1.5 Conclusions and future work

Nowadays, e-Learning systems are an important part of the learning process. Students and tutors can collaborate, using a virtual system for optimal learning performance. We have seen in this paper how we can represent information using Semantic Web technologies like RDF and JenaRules, and how we can add reasoning and generate advanced reports. Our further work on this project is towards practical implementation of a Moodle module, which will help tutors to have a flexible course management and reporting in Moodle e-Learning system.

References

1. K. Mller, U. Bojars and J. G. Breslin. Using Semantics to Enhance the Blogging Experience. In 3rd European Semantic Web Conference (ESWC2006), Budva, Montenegro, June 2006.
2. G. Klyne and J.J. Caroll (Eds.), Resource Description Framework (RDF): Concepts and Abstract Syntax, W3C, 2004
3. Lukichev S., Diaconescu M., Giurca A.: Empowering Moodle with Rules and Semantics. 3rd Workshop on Scripting for the Semantic Web, Innsbruck, Austria, June 2007, to appear.

Web-based Remote Digital Signal Processing (DSP) Laboratory Using the Integrated Learning Methodology (ILM)

Zohar Dvir

Afeka College of Engineering, Electrical and Electronics Engineering Department
Bney Efraim 218, Tel Aviv, 69107, Israel
zoard@afeka.ac.il

Summary. This paper presents an approach to conducting Digital Signal Processing (DSP) laboratory by using a new learning methodology. The suggested methodology is based on integration of the on-site and distance learning environments. The pilot of the proposed approach was implemented at AFEKA - Tel Aviv College of Engineering as part of undergraduate students' curriculum in Electrical and Electronic Engineering during the spring semester 2007, and this paper is accompanied with a description of pilot results.

1 Introduction

In engineering teaching programs, practical hand-on experimentation of the academic disciplines is essential. During recent years we have witnessed a significant progress in the development of web-based methodologies that support pedagogic and didactic efforts. This has been done in wide range of engineering disciplines, by remote experimentation technological tools in the virtual environment [1, 2]. Most impressive is the progress in the use of web-based tools for experimentation in Digital Signal Processing (DSP) labs in the virtual and remote-controlled environment [3].

On the other hand, physical experiments in the real on-site laboratory room constitute an important ingredient in order to carry out hands-on learning activities. This approach states that students must actually "be there" for gaining the true knowledge based on actual experience of the described discipline.Thus, the main disadvantage in using distant learning technology tools is that students lack the benefits of learning in a community along with other students within a laboratory.

The Integrated Learning Methodology (ILM) offers an answer to the disadvantages of the on-site and web-based (virtual or remote-access) learning methods when used separately, by combining their advantages.

Zohar Dvir: *Web-based Remote Digital Signal Processing (DSP) Laboratory Using the Integrated Learning Methodology (ILM)*, Studies in Computational Intelligence (SCI) **78**, 257–262 (2008)
www.springerlink.com

2 The Integrated Learning Methodology (ILM)

2.1 Traditional learning

The introductory DSP laboratory course in Afeka Tel-Aviv College of Engineering is of 14 weeks duration (typically).

During the DSP laboratory course the students are required to perform 10 experiments, and in addition, to design and implement a small-scale engineering project (4 weeks long, including a project exam) that evaluates their academic achievements. At the end of each experiment, students hand in an experiment for evaluation. The final grade of the course is based on the scores of the 10 experiments and the exam.

Although course materials are available online for registered students, access to the laboratory equipment and the experimentation requires the physical presence of the student in the lab. There are two lab rooms of 10 stations each, equipped for teaching DSP laboratory course where students work in pairs. Each experiment lasts 2 hours and scheduling of the laboratory classroom imposes tight time management by the college administration. For this reason, students encounter serious difficulties when they are looking for additional study time in the laboratory.

2.2 ILM-based learning

The main idea of ILM is integration of distance learning with classroom meetings, where in part of the course the student uses remote-access experimentation methodology in web-based environment, while in the other part he participates on-site. The schedule of the suggested ILM-based introductory DSP (iDSP) laboratory course is presented in Table 1.

Table 1. The schedule of the suggested iDSP lab

Week	Subject	Learning Method
1	Introduction to EZ-KIT Lite	**On-Site Classroom**
2	Using EZ-KIT Lite simulator	Remote Experimentation
3	Fixed-point v. Floating point	Remote Experimentation
4	Introduction to computing units	Remote Experimentation
5	Writing of the first application	**On-Site Classroom**
6	Branches and loops	Remote Experimentation
7	Interrupts, timer and Serial/Parallel I/O ports	Remote Experimentation
8	Parallel instructions, quantization error, basic algorithms	Remote Experimentation
9	Implementation of real-time FIR filter	Remote Experimentation
10	Implementation of real-time IIR filter	Remote Experimentation
11	Design of small-scale DSP project	**On-Site Classroom**
12	Implementation of small-scale DSP project	Remote Experimentation
13	Implementation of small-scale DSP project	Remote Experimentation
14	Project exam	**On-Site Classroom**

Although many improvements have been developed in the area of virtual and remote-access experimentation [4], not much attention has been paid to the combination of those methods with classical on-site learning [5]. A major weak point of the "virtual environment (VE)" approach is that a student has no familiarity with "real" equipment, components and tools that he will be dealing with in the "real world" as a professional electronics engineer. However, the "on-site environment" approach has also drawbacks that were described previously.

The main advantages and disadvantages of on-site, virtual and remote laboratories are summarized in Table 2 [6].

Table 2. Comparison of different laboratory methodologies

Lab type	Advantages	Disadvantages
On-site	Realistic data. Interaction with real equipment. Collaborative work. Interaction with supervisor.	Time and place restrictions. Requires scheduling. Expensive. Supervision required.
Virtual	Good for concept explanation. No time and place restrictions. Interactive medium. No interaction with real equipment.	Idealized data. Lack of collaboration. No interaction with real equipment.
Remote	Interaction with real equipment. Realistic data. No time and place restrictions. Medium cost.	Only "virtual presence" in the lab.

3 The ILM-based laboratory infrastructure

3.1 System technology

In order to implement remotely controllable experiments, we use the waveform generator and the oscilloscope that are network-enabled via TCP/IP protocol. The Tavor Electronics WW2572 waveform generator is used to trigger the Analog Devices EZ-KIT LITE board. Connectivity to the board from the distant location is provided via RS232 to LAN adapter in order to allow the student "inserting" experiment software program into the EZ-KIT LITE board. The output results of the experiment is detected by the Yokagawa Electric Corporation DL1740 oscilloscope.

The EZ-KIT LITE board is based on ADSP-2106x SHARC (Super Harvard Architecture Computer) chip of Analog Devices. The SHARC builds on the ADSP-21000 Family DSP core to form a complete system-on-a-chip, adding a dual-ported on-chip SRAM and integrated I/O peripherals supported by a dedicated I/O bus. The network connectivity diagram is shown in Figure 1.

During the hardware configuration designing process, special attention was paid to overcome the network connectivity problems. The latter are primarily influences

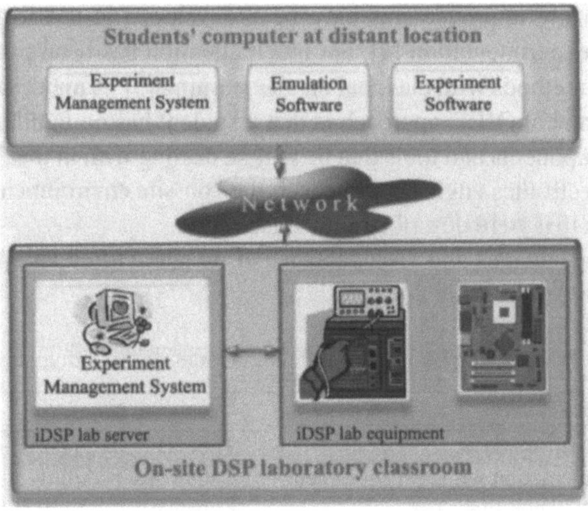

Fig. 1. The network connectivity diagram

of the "firewall" on the accessibility of the distant located students to the laboratory equipment.

Graphical User Interface (GUI) of the emulation software provides connectivity to the oscilloscope and waveform generator via instrument applets in order to interact with the laboratory instruments.

Changing instrument settings on the GUI at distant location actually changes instrument settings in the on-site laboratory classroom. Thus, it allows remote-access experimentation. The DSP Lab Server uses a gateway to the computer-based learning environment, and includes all the hardware and software components to allow successful remote-access experimentation from the distant location.

The proposed hardware configuration allows performing the remote-controlled experimentation at the on-site laboratory classroom from the distant location.

3.2 Access management and system administration

The laboratory administrator creates accounts and establishes time quotas for each experiment. All users, including instructors, gain access to the system via DSP lab server. The distance learning management system includes a Login Page that allows inserting the User ID and Password in order to obtain an authentication, and clearance to start the session. After clearance is granted, and the student has requested particular experiment, he is obliged to submit a preliminary questionnaire of the experiment, which is intended for checking that the student has sufficient knowledge for the experiment. The preliminary report is checked by the responsible instructor. If it is successful, the student is allowed to allocate a time quota for the experiment. After performing the experiment the student submits the final report and gets a score and remarks for the experiment. The distance learning management system consists (in addition to the Login Page) of several dialog boxes for creating and deleting accounts, for establishing individual time quotas, for defining time slots, for analyzing the logging messages and for managing student-instructor related communication, such as submitting of reports, getting remarks, scores etc.

4 The ILM-based laboratory pilot implementation

The pilot was implemented with a group of three third year B.Sc. students enrolled in the Digital Signal Processing (DSP) course at AFEKA Tel-Aviv Academic College of Engineering during the spring semester 2007.

Students who participated in the distance learning research group gained access to the ILM resources and performed iDSP laboratory course activities as described in Table 1. They were obliged to physical presence in the on-site laboratory classroom for experiment sessions four times during the semester as shown in the Table 3.

Table 3. On-site sessions scheduling

Time-line	Objective
Week one	Introduction to the DSP lab. Introduction to the teacher and classmates. Giving the students a first opportunity to "being there". Implementation of the first experiment.
Week five	Midterm meeting. Writing of the first application. Implementation of convolution algorithm. Midterm exam.
Week eleven	Getting directives for small-scale DSP project.
Week fourteen	Conclusive meeting. Examination. Final questionnaire.

The other students in the class performed traditional on-site DSP laboratory.

Learning outcomes for the research was assessed by on-line survey questionnaires and the midterm (week 5) and final (week 14) exams of the semester. The survey questionnaires pay attention to the following topics [7]:

- Effectiveness of the ILM-based learning compared to traditional lab in providing applications of course concepts to real world systems.
- Specific aspects of the labs as to their value in promoting understanding of the course's concepts, such as: preliminary homework report, final report, team work, physical presence in lab etc.
- Level of students' satisfaction with the ILM-based learning.

5 Pilot findings

The preliminary analysis of the pilot findings shows that ILM-based iDSP laboratory course offers a following benefits:

- Flexible laboratory scheduling and more efficient time management.
- Enjoying a learning in community along with distance learning advantages.
- Activating a real equipment and manipulating a realistic data.

From the other hand, some significant remarks has been raised:

- Adding a video camera for experiments board's status lights tracking.
- Providing hardware control to the "Reset" button on the experiments board.
- Providing hardware control to the "Flag" switch on the experiments board.

6 Concluding Remarks

In this note a new approach to the DSP laboratory implementation was presented. The new methodology is based on the idea of combining on-site and distance learning approaches. The integrated course offers additional way to contend with practical experimentations of the academic disciplines. As future work we intend to expand ILM-based DSP lab course to more participants and to continue a meticulous research.

References

1. Schmid, C.: Remote experimentation in control engineering. In: *Proc. 145th International Scientific Colloquium* (IWK2000), Ilmenau, Germany (2000).
2. Leleve, A., Benmohamed, H., Prevot, P., Meyer, C.: Remote laboratory toward an integrated training system. In: *Proc. ITHET 03*, Marrakech, Morocco, July 2003.
3. Boukadoum, M., Bari, M.: Remote-access to a digital signal processing laboratory using a client-server architecture. In: *ITHET 04*, Istanbul, Turky, June 2004.
4. Bahring, H., Keller, J. Schiffmann, W.: A combined virtual and remotely accessible microprocessor laboratory. In Proc. 11th Workshop on Computer Architecture Education (WCAE 2004), pages 136-141, June 2004, In http://www4.ncsu.edu/ efg/wcae/2004.
5. Huang, W., Luce, T.: Proposing an Effective Teaching Pedagogical Mode for Online MBA Education: An Exploratory Empirical Investigation. Issues in Information Systems, V(1). 2004.
6. Nedic, Z., Machotka, J., Nafalski, A., :Remote Laboratories versus Virtual and Real Laboratories. In Proc. ASEE/IEEE Frontiers in Education Conference, Boulder, CO, pp. T3E-1 - T3E-6., 2003
7. Huang, W., Luce, T. :A conceptual research framework for investigating the effects of teaching pedagogy and technology on learning performance in online education. In Proc. Annual Conference of International Association of Computer Information Systems (IACIS), Las Vegas, USA, Oct. 2003

Reasoning on Activity-centered Semantics in Request-Service Matchmaking

Alexandra Galatescu, Taisia Greceanu, Cristian Neicu, and Doina Enache

National Institute for R & D in Informatics
8-10 Averescu Avenue, 011455, Bucharest, Romania
{agal, gresta, ncristi, doinae}@ici.ro

Summary. Web service and Semantic Web technologies impose the restructuring of the organization processes and the resolution of specific problems. Still a problem is the request-service matchmaking in a service-oriented process. The paper proposes a solution for optimizing it, based on activity-centered semantics for the service definition/ annotation and for the request composition. It proposes the use of ontology-based templates semantically compatible for activities, services and requests and their reuse by similar categories of providers.

1 Introduction

Future Web-based organizations will rely on service-oriented (intra and inter—organization) processes able to support the enterprise activities that involve internal or external service providers (SPs) (e.g. technological, organizational, marketing, etc, processes). Although most processes are predefined, they should be adaptable to the technological, economical and social changes. The activities that compose them should be interoperable and, for each activity, the system should be able to discover the best SP (and service) to execute it, according to a specific request.

For solving the activity and service *technological interoperability*, Web service technology is going to change the enterprise processes. Also, for the service *semantic interoperability* and for their conceptual discovery and integration, the benefits from the Semantic Web technology and from ontologies are already known [1] - [10]. Generic ontologies like OWL-S [6], WSMO [7] are proposed for service modeling and domain ontologies enrich the service definitions by their post-publishing annotation. However, for the service semantic discovery and integration, the resolution of the *request-service matchmaking* (service dynamic discovery based on the requestor's preferences) has not yet spectacular results. Main drawbacks are: (1) annotation (usually, manual) is after and in addition to the service definition and publishing; (2) there is no or less concern for the explicit semantic correlation between the concepts the service requestors (SRs) instantiate in their requests and the concepts the SPs use for annotating the service definitions. Without this correspondence, the request-service matchmaking needs complicated algorithms, not always

Alexandra Galatescu et al.: *Reasoning on Activity-centered Semantics in Request-Service Matchmaking*, Studies in Computational Intelligence (SCI) **78**, 263–268 (2008)
www.springerlink.com

successful; (3) there is no criterion to select the services that should be annotated and the concepts in ontology that should be used for the service annotation.

For the execution of an activity, a solution is the definition/ annotation of the services with concepts *semantically compatible* (i.e. *equivalent* (same or synonymous type) or *nearly equivalent* (subtype/ super-type at any distance in the ontological hierarchy)) with the concepts that compose the requests for that type of activity. This solution requires a criterion for the concept selection (during the service annotation or definition from scratch), for the service semantic correlation with the requests and for the request-service semantic comparison. Now, this criterion exists in the process/ service designers' minds and is not explicit and automated.

The paper proposes an explicit criterion and reasoning in the context of a predefined service-oriented process, dynamically generated by the dynamic (at run-time) discovery and binding of the services needed for the execution of the process activities. It is meant to enhance the service semantics and to help for their automatic matching with the requests. It consists in: (1) the use of activity (ontology-based) templates as a conceptual common base for the service definition and request composition; (2) the customization of the activity templates, resulting in service and request templates, semantically compatible with the activity template; (3) the reasoning on the activity-centered semantics of the requests and services, for the verification of their semantic completeness, correctness and compatibility.

Section 2 motivates the use of ontologies in defining semantic templates for activities, services and requests and in their customization. Section 3 gives the steps of an algorithm for the request-service matchmaking based on the activity-centered semantics. For limited space reason, the algorithm is not detailed in this paper.

2 Activity Semantic Templates for Services and Requests

For the semantics-based definition of services from scratch, for semantically enhancing, by annotation, the already published services and for the request composition, this paper proposes the use of *ontology-based templates for the signatures of the activities*. First reason is that the service discovery is needed before the execution of each activity in a process and the activity conceptual signature (input/ output concepts) helps to semantically identify the candidate services for its execution. Services are discovered according to their own signatures that should be composed of concepts semantically compatible with concepts in the activity signature.

Second reason is to avoid the post-publishing annotation of the services defined from scratch. Service definition relies on *service templates*, initially identical to the templates of the corresponding activities (see Fig. 1). Experienced designers may build *variants* of the service initial templates, by specializing concepts. Inexperienced designers can only instantiate concepts in existing templates.

According to this proposal, the concepts selected for the annotation of the already published services that can execute an activity, should be semantically compatible with the concepts in the activity signature. Also, during the *request composition* for each process, the requestor is asked to instantiate concepts semantically compatible with the mandatory input concepts in the activity signatures. This semantic compatibility minimizes unsuccessful request-service matchmaking, when the request composition and service definitions do not rely on semantically compatible concepts.

For service and request semantic definition, the *reference ontology* unifies the following basic concepts (along with the rules for their correlation and constrain): *Service* that can be atomic or complex; *Process* that describes a complex service and is composed of concepts of type Process (sub-processes in the upper-level process) or of type Activity; *Activity* which is an atomic operation executed by an atomic service; *Agent* that has two roles (SP and SR) and can be human or automated; *Object* that describes an activity, any atomic service that can executes it and the requests for the activity execution; *Object Attribute* correlated to an object in the definition of an activity/ service/ request; *Quality Attribute* correlated to objects, activities, services and requests in their definition.

The concepts in the *domain ontology* specialize concepts in the reference ontology. Fig. 1 exemplifies taxonomies for objects, object attributes and quality attributes, used in the examples of activity, service and request templates. A particular feature of the domain ontology is that the relationships between the basic types of concepts are not statically defined, but dynamically established during the activity/ service definition and request composition (e.g. relationships between objects and their attributes, between activities/ services/ objects and their quality attributes).

Although, lately, the ontologies are used for modeling Web service-oriented processes (e.g. in [9], [10]), most ontology editors and management tools are not process-oriented and do not have capabilities for the explicit representation, decomposition and interpretation of the process semantics. The designers devise their own concepts and relationships, usually complying with a process-oriented language (e.g. PSL, BPML). An alternative solution is the annotation of the activities signatures. Annotation can be explicit, after the process definition or implicit, during the process definition, by composing activity templates implicitly annotated with concepts in ontology. This paper proposes the second solution with benefits for service definition, request composition and for the request-service matchmaking.

Fig. 1 exemplifies the service customization according to the semantics of the activity 'Find supplier' with the target: to buy a car for organization. In order to be general, the activity description refers to both products and services and the object attributes and quality attributes are not specifically imposed. Fig. 1 suggests different types of SPs: chambers of commerce (that can find producers of cars in their databases), import companies and also producers of cars. Each type of SP defines differently the service it provides, by customizing the activity semantic template.

Fig. 1 also suggests the actions for the service definition based on the activity semantics. The dialog between the system and service designer is as follows:

1. ***System***: Asks the designer to select/ specify an existing/ new activity type.
2. ***Service designer*** has two alternatives: (2.1) Selects an *existing activity type* A_i and, implicitly, its semantic template composed of the activity mandatory signature (input/ output objects and attributes, dragged from the domain ontology),

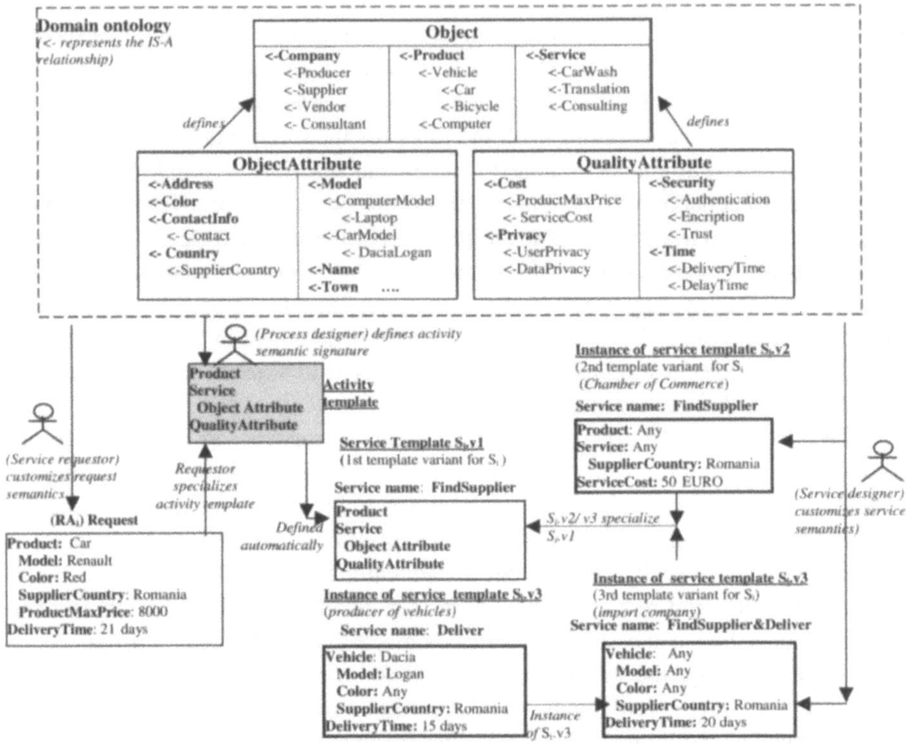

Fig. 1. Semantic customization of the service and request templates based on the semantics of the activity Find supplier

needed for the discovery of the best service for A_i; or (2.2) Defines a *new activity type* A_i and its semantic template.

3. ***System***: (3.1) For a new activity type, automatically creates the first template variant $S_i.v_1$ for the service S_i (type of service compatible with A_i) and opens it; (3.2) For an existing activity type, displays a list of service templates semantically compatible with the activity template.
4. ***Service designer***: (4.1) In the case of an existing activity, selects a template for the service S_i; (4.2) On the selected/ first template: (a) either instantiates the template for S_i; or, customizes the template for S_i and creates a new variant of template for S_i by specializing concepts (by dragging subtypes of objects and attributes from ontology, possibly adding in ontology new concepts and rules) and instantiates the new template. E.g. in Fig. 1, a Chamber of Commerce creates $S_i.v_2$, a new variant of $S_i.v_1$, obtained by the specialization of the initial template. Another chamber can then instantiate the template $S_i.v_2$. An import company of vehicles can also specialize the initial template $S_i.v_1$ and build the template $S_i.v_3$. A company that produces vehicles only instantiate the template

$S_i.v_3$. Then, (4.3) Attaches references to the service specific code/ application; (4.4) if necessary, defines a service-specific request template (composed of concepts and relationships specific to a certain service and SP).

5. ***System***: Publishes the service in a public registry. If the SPs have already published (in other registries) services semantically compatible with A_i, the SPs should enhance their semantics by annotating them with concepts compatible with the concepts in the activity template. In order to execute an activity A_i in the process P, the Process Manager tool should collect from the requestor values for concepts semantically compatible with the input concepts in the activity signature. E.g. the type of product the requestor needs, the model, color, maximum accepted price etc. This information is necessary for the discovery of the service matching the request for the activity A_i and for its execution. Fig. 1 suggests that the requestor is allowed not only to instantiate concepts but, also, to semantically enhance the request by specializing concepts in the activity template.

3 Request-Service Matchmaking Based on Activity Semantics

First phase of the service discovery is according to the service functional and domain-specific semantics. In our case, it relies on the activity-centered semantics. For an activity type Ai in a process, the system should discover the best service having the type S_i, semantically compatible with A_i and satisfying the request RA_i. The semantic compatibility of the services and requests relies on the compatibility of the concepts they involve (in service/ request structure and instances). *Second phase* consists in filtering the services resulted from the first phase according to the required quality. For space reason, the steps of the request-service matchmaking are only sketched. However, they suggest the matchmaking complexity.

The *steps for request-service matchmaking* based on activity semantics are:

1. ***Before the service discovery***
   a) *Semantic verification of the services* (after the service definition from scratch or their annotation): (a) Check on the semantic completeness of the service template: each mandatory concept in the signature of A_i should have a semantically compatible concept in the service template; (b) Check on the semantic correctness of the defined/ annotated service, i.e. check on the compatibility of the concepts types with their concept-like instances.
   b) *Semantic verification of the requests*: (a) Check on the semantic completeness of the request (versus the activity template); (b) Check on the semantic correctness of the request (similar to the check on the service correctness).
2. ***During the service discovery***
   a) *Semantics-based selection of the candidate services and request-service comparison*: (a) Select the service types semantically matching the activity type Ai; (b) Check their semantic compatibility with the request, i.e. the compatibility of the concept types and instances in the request and services; (c) Check the request-service matching results and errors (unknown or incompatible concepts in services and request).

Checking the request and service completeness and correctness is necessary because the requestor/ provider is allowed to semantically vary the inputs in the activity template, in order to match their vocabularies and needs. The semantic compatibility of the services and request relies on the semantic compatibility between the concepts they involve. The algorithm checks not only the concept types, but, also, their concept-like instances. Also, the algorithm implicitly ranks the services like in [4], [8], [11] and gives the explicit causes of the mismatching.

4 Conclusions

The paper has motivated the ontology-based activity, service and request templates for optimizing the request-service matchmaking during the generation of a service-oriented process. Service and request templates are obtained by the semantic adaptation of the activity templates. The activity semantics is used in the selection of the services and during the verification of the service and request semantic correctness and completeness. In a next version of the system under development, the algorithm will be enhanced with an additional check on the semantic compatibility of the dependencies between objects and between attributes in the service and request definition.

References

1. Tsalgatidou, A., Athanasopoulos, G. et al.: Semantically Enhanced Unified Service Discovery. In: *Proc WS on Frameworks for Semantics in Web Services*, Insbruck (2005)
2. Benatallah, V., Hacid, M-S, Leger, A., Rey, C., Toumani F.: On automating Web service discovery. In: *VLDB Journal 14* (2005)
3. Srinivasan, N., Paolucci, M., Sycara, K.: Semantic Web Service Discovery in the OWL-S IDE. In: *Proc 39th Annual Hawaii Intl. Conf. on System Sciences* (2006)
4. Klusch, M., Fries, B., Sycara, K.: Automated Semantic Web Service Discovery with OWLS-MX. In: *Proc 5th Intl. Conf. on Autonomous Agents and Multi-Agent Systems (AAMAS)*, Hakodake, Japan (2006)
5. Maximilien, M., Singh, M.: A Framework and Ontology for Dynamic Web Services Selection. In: *IEEE Internet Computing*, Sept-Oct. (2004)
6. Martin, D., Paolucci, M. et al.: Bringing Semantics to Web Services: OWL-S Approach. In: *Proc. WS on Semantic Web Services and Web Process Composition*, San Diego, (2004)
7. W3C: Web Service Modeling Ontology. In: *www.w3.org/Submission/WSMO/#lang* (2005)
8. Jaeger, M., Rojec-Goldmann, G. et al.: Ranked Matching for Service Descriptions Using OWL-S. In: *Kommunikation in Verteilten Systemen, Kaiserslautern*, Germany, Feb. (2005)
9. Brambilla, M., Ceri, S., Fraternali, P., Manolescu, I.: Process Modeling in Web Applications, In: *ACM Journal, No. 20* (2006)
10. Sivashanmugam, K., Miller, J., Sheth, A., Verma, K.: Framework for Semantic Web process Composition, In: *TR 03-008, LSDIS, Univ. of Georgia*, USA (2003)
11. Tsetsos, V., Anagnostopoulos C. et al.: On the Evaluation of Semantic Web Service Matchmaking Systems, In: *Proc. European Conf. on Web Services*, Zurich (2006)

Marked-Controlled Rewriting EFPN for Dynamic Knowledge Inference

Emilian Guţuleac[1] and Mihai Mocanu[2]

[1] Technical University of Moldova, Computer Science Department,
168 Bd. Ştefan cel Mare, Chişinău, MD-2004, Republic of Moldova
egutuleac@mail.utm.md

[2] University of Craiova, Software Engineering Department
Bvd.Decebal 107, Craiova, RO-200440, Romania
mocanu@software.ucv.ro

Summary. In this paper, we introduce the descriptive marked-controlled rewriting Extended Fuzzy Petri Nets (EFPN) that can at run-time modify their own structure by rewriting some of their descriptive expression components where the enabling of a rule also depends on the net marking. They can then be used in the design of "dynamic knowledge inference" frameworks, which are adjustable according to knowledge variation as human cognition and thinking. This is very important for some domains, such as expert systems.

Keywords: descriptive fuzzy Petri nets, knowledge inference, rewriting rules.

1 Introduction

Recently, Fuzzy Petri Nets (FPN) have been widely used for modeling systems that involve approximate reasoning and uncertainty knowledge inference. Expert systems may be regarded as such dynamic systems in which knowledge is vague, but updated or modified frequently. Suitable models for them should have the ability to adjust themselves according to the systems' changes. However, the lack of dynamically adjustment mechanism in FPNs can not cope with potential structure changes of actual systems [1, 2, 7]. Some adjustable FPNs were proposed. A generalized FPN model (GFPN) which can be transformed into neural networks with OR/AND logic neurons, in which parameters of the corresponding neural networks can be learned (trained), was proposed in [6]. Just as other fuzzy Petri net (FPN) models, adaptive Fuzzy Petri Nets (AFPN) with adjustable weights, and also the learning ability of neural networks, can be used for knowledge representation and reasoning, and, more important, for dynamic knowledge inference - but they need first to be transformed into neural networks. Based on AFPN transition firing rule, modified back propagation learning algorithms may be developed to assure weights convergence [5].

In this paper, we introduce an alternative model: the descriptive marked-controlled rewriting Extended Fuzzy Petri Nets (EFPN) for knowledge inference that can at

Emilian Guţuleac and Mihai Mocanu: *Marked-Controlled Rewriting EFPN for Dynamic Knowledge Inference*, Studies in Computational Intelligence (SCI) **78**, 269–276 (2008)
www.springerlink.com

run-time modify their own structure by rewriting some of their descriptive expression components. We aim to prove how easily and directly EFPN models can change their structure dynamically. We also try to illustrate the concepts described by giving a small example at the end of the paper.

2 Extended Fuzzy Petri Nets

Various extensions have been made to the framework of FPN since its introduction and early works in [1, 2, 7]. In this section, we describe a extended variant of FPN.

Definition 1. *An **extended FPN** is a 13-tuple structure:*
$F\Gamma =< P, T, D, I, O, I^H, I^T, W, G, Pri, \alpha, \beta, Th >$, *where:*

- *P is the finite set of places (a place is graphically represented as a single circle);*
- *T is a finite set of transitions (a discrete transition is drawn as a black bar);*
- *D is a finite set of propositions,* $P \cap T \cap D = \emptyset, |P| = |D|$;
- *I,* I^T *and* I^H : $P \times T \to Bag(P)$ *are forward flow, test and inhibition functions, which define a mapping from transitions to bags of places, respectively.*
- $O : P \times T \to Bag(P)$ *is the backward flow function in the multisets of P. These functions define the set of arcs A and describe the marking-dependent weights* $W = (W_I \cup W_O \cup W_h \cup W_t) \times Bag(P) \to [0, 1]$ *of of normal arcs* A_n, *inhibitory arcs* A_h *and test arcs* A_t, *respectively.*
- $G : T \times Bag(P) \to \{true, false\}$ *is the guard function defined for each transition. For* $t \in T$ *a guard function g(t, M) will be evaluated in marking M;*
- $\alpha : T \times Bag(P) \to \{0, 1\}$ *is an association function which assigns a real value between zero to one to each place;*
- $\beta : P \to \{0, 1\}$ *is a bijective mapping between the proposition and place label for each node* $P \cup T$, $P \cap T \cap D = \emptyset$, $|P| = |D|$;
- $Th : T \times Bag(P) \to \{0, 1\}$ *is the function which assigns a threshold value* λ_j *from zero to one to transition* t_j.

Defined in this way, the structure of EFPN is static. The dynamics of a net structure is specified by defining its initial marking and its marking evolution rule.

Definition 2. *A **marked EFPN** is a pair* $FN =< F\Gamma, M_0 >$, *where* $F\Gamma$ *is an EPPN structure and* M_0 *is the initial marking of the net.* $M : P \to \{0, 1\}$ *is the current marking of the net which is described by a vector-column* $M = (m_i > 0, \forall p_i \in P)$, *where* m_i *is the marking of a place defined as the **certainty factor** of the token in it.*

Enabling and Firing of Transitions. Given a $t_j \in T$, we denote ${}^\bullet t_j$ and $t_j^\bullet$ the input and output set places of transition t_j over normal arcs, and with ${}^\circ t_j$ and ${}^* t_j$we denote the inhibitor and test set places of transition t_j, respectively. We say that a transition t_j is enabled in current marking M of net FN if the following enabling condition (logic expression) $ec(t_j, M)$ is verified:

$$ec(t_j, M) = (\widehat{\forall_{p_i \in {}^\bullet t_j}}(m_i \cdot I(t_j, p_i) > \lambda_j)) \wedge (\widehat{\forall_{p_l \in {}^\circ t_j}}(m_l \cdot I^H(t_j, p_l) < \lambda_j)) \\ \wedge(\widehat{\forall_{p_n \in {}^\circ t_j}}(m_n \cdot I^T(t_j, p_n) > \lambda_j)) \wedge g(t_j, M) \quad (1)$$

Let $T(M)$ be the set of transitions enabled by the current marking M. When t_j is enabled and fires, it produces a new certainty factor:

$$CF(t_j) = \begin{cases} x_j, \ x_j > \lambda(t_j) \\ 0, \ x_j < \lambda(t_j) \end{cases}, \quad \text{where } x_j = \sum_{\forall p_i \in {}^\bullet t_j} m_i \cdot w^I_{i,j} \tag{2}$$

According to the enabling condition (1), a transition is enabled if all its input places have tokens, and if the certainty factor produced by it is greater than its threshold. A transition $t_j \in T(M)$ *fires* if no other enabled transition $t_k \in T(M)$ has higher priority. If $x_j > \lambda(t_j)$ transition t_j fires and at the same time, token transmission takes place:
1)If only a place p_k has one input transition t_j, $|{}^\bullet t_j| = 1$, a new token with certainty factor $w^O_{j,k} \cdot CF(t_j)$ is put in each output place $p_k \in t_j^\bullet$ and all tokens in $p_i \in^\bullet t_j$ are removed.
2)If a place p_k has more than one input transitions, i.e. $|{}^\bullet p_k| > 1$, (as it will be the case in the example illustrated by figure 3) and more than one of them fires, then the new certainty factor α_k of p_k is decided by the center of gravity of the fired transitions (this rule is also consistent with [5]):

$$m_k = (\sum_{\forall p_k \in t_j^\bullet} w^O_{j,k} \cdot CF(t_j))/(\sum_{\forall p_k \in t_j^\bullet} w^O_{j,k}) \tag{3}$$

3)If the enabled transitions $t_j \in p_i^\bullet$, $p_i^\bullet = \{\forall t_j \in T : O(t_j, p_i) > 0\}$ are in structural conflict, i.e. $|p_i^\bullet| > 1$, and they are scheduled to fire in marking M, then the new certainty factor of $p_k \in t_j^\bullet$ is decided by the following relation:

$$m_k = w^O_{j,k} \cdot CF(t_j) \cdot \lambda_j / \sum_{\forall t_n \in {}^\bullet p_i} \lambda_n \tag{4}$$

Thus, through firing transitions, certainty factors can be reasoned from a set of known antecedent propositions to a set of consequent propositions step by step.

3 Descriptive expressions of EFPN

We use the concept of *descriptive expressions* (*DE*) for analytical representation and compositional construction of EPPN. Due to the space restrictions, we will give only a brief overview to this topic and refer the reader to [3, 4] and the references therein. In the following, if not mentioned apart, we assume that $Pri(t_j) = 0$ and $g(t_j, M) = 1$ and therefore we can omit $Pri(t_j)$ and $g(t_j, M)$ in the formal definitions of *DE* components of compositional operations and its *FN* translations.
A basic DE element (*bDE*) for a basic *FN* net (*bFN*) is: $bDE = \left|^{\lambda_j}_{t_j} m^0_i p_i[W^+_i, W^-_i]\right|^{\lambda_k}_{t_k}$.
The translation of this *bFN* is shown in figure 1a, where t_j is the input transition with the threshold value λ_j and t_k is the output transition of the place p_i with initial marking $m^0_i = M(p_i)$. The flow relation functions $W^+_i = O(t_j, p_i)$ and $W^-_i = I(t_k, p_i)$, return the multiplicities of the input and output arcs of p_i. The derivative elements of *bDE* are as in figure 1b:

- for $p_i^\bullet = \phi$, $W_i^- = 0$ is $\Big|_{t_j}^{\lambda_j} m_i^0 \cdot p_i[W_i]$ with final place p_i of t_j,
- for ${}^\bullet p_i = \phi$ is $m_i^0 \cdot p_i[W_i]\Big|_{t_k}^{\lambda_k}$ with entry place p_i of t_k.

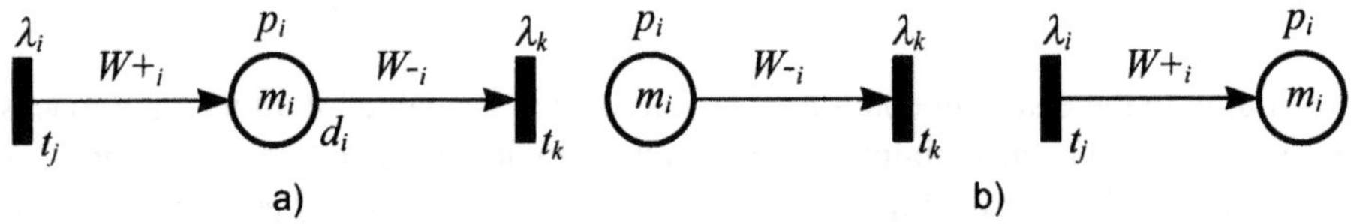

Fig. 1. Translation in *bFN* of *bDE* (a) and its derivatives (b)

If the initial marking m_i^0 of place p_i is zero tokens, we can omit $m_i^0 = 0$ in *bDE*. Also, by default, if $W_i^+ = W_i^- = 1$, we present *bDE* and its derivatives as follows:

$$\Big|_{t_j}^{\lambda_j} m_i^0 p_i\Big|_{t_k}^{\lambda_k}, \; m_i^0 p_i\Big|_{t_k}^{\lambda_k}, \text{ or } \Big|_{t_j}^{\lambda_j} m_i^0 p_i$$

From a *bDE* we can build more complex *DE* of *FN* components by using composition operations. The DE is either a bDE or a composition of DE: $DE ::= bDE/DE * DE/ \circ DE$, where $*$ represents any binary composition operation and $\circ$ denotes any unary composition operation.

Also, the unary *inhibition* "$-$" ($\bar{p}_i$) or *test* "$\sim$" ($\tilde{p}_i$) operations describe the inhibitory or test arc, respectively. The composition operations are reflected at the level of the *DE* components of *FN* models by fusion of places, fusion of transitions with the same name (label) or sharing of as subnets.

The binary *place-sequential* operation "|" determines the logics of an interaction between two local states: pre-condition and post-condition. The specified conditions are always fulfilled. This operation is *not commutative*.

The binary *synchronization* operation " $\bullet$ " or " $\wedge$ " d escribes the rendez-vous synchronization (the transition t_j) of two or more conditions represented respectively by place $p_i \in^\bullet t_j, i = \overline{1,n}$, i.e. it indicates that all preceding conditions of occurrence actions must have been completed.

The binary *split* operation "$\diamond$" determines the causal relations between t_j and its post-conditions. After completion of the preceding action of t_j concomitantly several other post-condition can occur in parallel.

As an example, in figure 2 are shown the mappings of the three type of Weighted Fuzzy Production Rule (WFPR) into FN_i with the $DE_i, i = 1,2,3$. These types of WFPR may be represented as follows:

- *Type 1* - a **Simple** *WFPR*: IF d_i THEN d_k, $Th(t_j) = \lambda_j$, $W_O(t_j, p_k) = w_k$, $W_I(t_j, p_i) = w_i$ is described by DE_1. The $DE_1 = m_i^0 p_i[W_i]\Big|_{t_j}^{\lambda_j} m_k^0 p_k[W_k] \neq m_k^0 p_k[W_k]\Big|_{t_j}^{\lambda_j} m_i^0 p_i[W_i]$ means that the specified conditions associated with place-symbol p_i are always fulfilled and happen before the occurrence of the conditions associated with p_k by means of the transition t_j. The translation of DE1 in FN1 is shown in figure 2a.

- *Type 2* - a **Composite Conjunctive** $WFPR$: IF $(d_1\&d_2\&d_3\&\overline{d_4})$ THEN $(d_4\&d_5)$, $Th(t_j) = \lambda_j$, $W_O(t_j,p_3) = w_3$, $W_O(t_j,p_5) = w_5$, $W_I(t_j,p_i) = w_i$, $i = 1,2$, $W_I(t_j,p_3) = w_3$, $W_{I^H}(t_j,p_4) = w_4$ is described by DE_2. The translation of DE_2 in FN_2 is shown in figure 2b, $DE_2 = (1p_1[w_1]\cdot 1p_2[w_2]\cdot 1\tilde{p_3}[w_1]\cdot \overline{p_1}[w_1])|_{t_j}^{\lambda_j} p_5[w_5]$
- *Type 3* - a **Conjunctive-Fork** $WFPR$: IF d_i THEN $(d_k\&d_l\&d_n)$, $Th(t_j) = \lambda_j$, $W_O(t_j,p_k) = w_k$, $W_O(t_j,p_l) = w_l$, $W_O(t_j,p_n) = w_n$, $W_I(t_j,p_i) = w_i$ is described by DE_3. The translation of DE_3 in FN_3 is shown in figure 2c, and $DE_3 = 1p_i[w_i]|_{t_j}^{\lambda_j}(p_k[w_k]\Diamond p_l[w_l]\Diamond p_n[w_n])$.

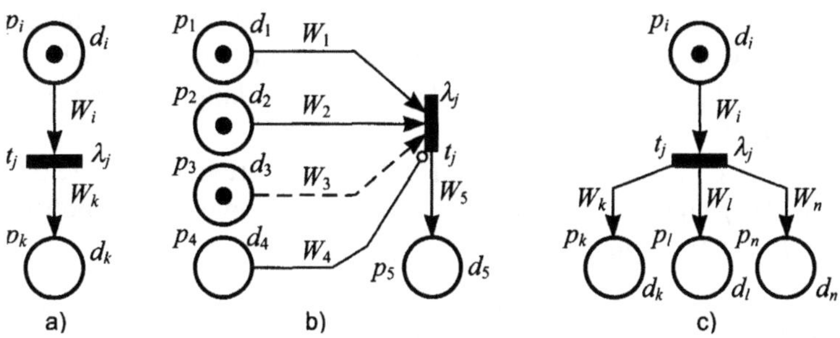

Fig. 2. Translation in DE_1 in FN_1 (a), DE_2 in FN_2 (b) and DE_31 in FN_3 (c)

The compositional binary *competing parallelism* operation " $\vee$ " means that it can be applied over two FN nets: FN_A with $DE_A = A$, and FNB with $DE_B = B$. The resulting net FNH_R with $DE_R = R$ can be represented by $DE_R = R = A \vee B$. The fused nodes, with the same names, will inherit the incidence arcs of the marking-controlled nodes of descriptive expressions A and B.

The *precedence relations* between the operations in the *DE* are the following:

1. the evaluation of operations in DE are applied left-to-right;
2. an unary operation binds stronger than a binary one;
3. the "$\bullet$" operation is superior to "|" and "$\Diamond$", and in turn, these are superior to the " $\vee$ " operation.

4 Dynamic Rewriting EFPN

We are now ready to introduce in this section the model of *descriptive dynamic EFPN rewriting systems* and prove its usefulness in dynamic knowledge inference.

In the following, if not stated otherwise, we will use some widespread denotations, that is, if $X\rho Y$ is a binary relation, then the *domain* of ρ is $Dom(\rho) = \rho Y$ and the *codomain* of ρ is $Cod(\rho) = X\rho$. Also, let $A =< I,O,I^H,I^T >$ be a set of arcs belonging to net $F\Gamma =< P,T,D,I,O,I^H,I^T,W,G,\alpha,\beta,Th,l >$.

Definition 3. *A **dynamic rewriting labeled EFPN** is a system*

$$RF =< F\Gamma, R, \phi, G_{tr}, G_r, M >, \text{where}:$$

- $R = r_1, ..., r_k$ *is a finite set of rewriting rules about the run-time structural modification of a net, so that* $P \cap T \cap R = \emptyset$*;*
- $\phi : E \rightarrow \{T, R\}$ *is a function which indicates the type of event that can occur, where* $E = T_D \cup R$ *denotes the set of events of the net, for every rewriting rule;*
- $G_{tr} : R \times Bag(P) \rightarrow \{true, false\}$ *is the transition rule guard function associated with* $r \in R$*;*
- $G_r : R \times Bag(P) \rightarrow \{true, false\}$ *is the rewriting rule guard function defined for each rule of* $r \in R$*, respectively.*

In the graphical representation, the R rule is drawn as two embedded empty rectangles. For $\forall r \in R$, the functions $g_{tr}(M) \in G_{tr}$ and $g_r(M) \in G_r$ will be evaluated in each marking, and if they are evaluated to *true*, the rewriting rule r may be enabled, otherwise it is disabled. The default value of $g_{tr}(M) \in G_{tr}$ in marking M is *true*, and for $g_r(M) \in G_r$ it is *false*.

Let $R\Gamma =< FN, R, \phi, G_{tr}, G_r >$ and $RN =< R\Gamma, M >$ be represented by the descriptive expression $DE_{R\Gamma}$ and DE_{RN}, respectively. A **dynamic rewriting structure modifying rule** $r \in R$ of RN is a map $r : DE_L \triangleright DE_W$, where:

- the *codomain* of the $\triangleright$ rewriting operator is a fixed descriptive expression DE_L of a subnet RN_L of current net RN, with $RN_L \subseteq RN$, $P_L \subseteq P$, $E_L \subseteq E$ and the set of arcs $A_L \subseteq A$;
- the *domain* of the $\triangleright$ is a descriptive expression DE_W of a new RN_W, with $P_W \subseteq P$, $E_W \subseteq E$ and the set of arcs $A_W \subseteq A$.

The rewriting operator $\triangleright$ represents the binary operation which produces a *structure change* in the DE_{RN} and the net RN by replacing (rewriting) the fixed current DE_L of the subnet RN_L(DE_L and RN_L are dissolved) with the new DE_W of the subnet RN_W, now belonging to the new modified resulting $DE_{RN'}$ of the net $RN' = (RN \backslash RN_L) \cup RN_W$, with $P' = (P \backslash P_L) \cup P_W$ and $E' = (E \backslash E_L) \cup E_W$, the meaning of $\backslash$ and $\cup$ operations being net or set removal or adding, and $A' = (A - A_L) + A_W$. In the new net RN', obtained by firing of enabled rewriting rule $r \in R$, places and events with the same attributes which belong to RN' are fused. By default, the rewriting rules $r : DE_L \triangleright \phi$ or $r : \phi \triangleright DE_W$ describe the rewriting rule holding $RN' = RN \backslash RN_L$ or $RN' = RN \cup RN_W$. A state configuration of a net RN is a pair $(R\Gamma, s)$, where $R\Gamma$ is the current structure of net together with a current state $s = (M, \beta(M))$. The $(R\Gamma_0, s_0)$ with $P_0 \subseteq P$, $E_0 \subseteq E$ and state s_0 is called the initial state configuration of a net.

Enabling and Firing of Events. The rewriting rule $r_j \in R$, that changes the structure of net RN, is enabled in current marking M if equation (1) and $g_{tr}(r_j, M)$ are verified [3]. Also, the discrete rewriting rule $r_j \in R$, that changes the structure of RN, is enabled in current marking M if the $ec(e_j, M)$ and $g_{tr}(r_j, M)$ equations are verified.

Let $T_D(M)$ and $R(M)$, $T_D(M) \cap R(M) = \phi$, be the sets of enabled discrete transitions and enabled rewriting rules in current marking M, respectively. We denote the set of enabled events in a current marking M by $E(M) = T_D(M) \cup R(M)$. The event $e_j \in E(M)$ fires if no other event $e_k \in E(M)$ with higher priority is enabled. Hence, for each e_j if $((\phi_j = t_j) \vee (\phi_j = r_j) \wedge (g_{tr}(r_j, M) = False))$ then the firing of transition $t_j \in T_D(M)$ or rewriting rule $r_j \in R(M)$ changes only the current marking: $(R\Gamma, s) \xrightarrow{e_j} (R\Gamma, s')$. Also, for event e_j IF $((\phi_j = r_j) \wedge (g_r(r_j, M) = True)$ THEN the event e_j occurs to firing of rewriting rule r_j and it changes the configuration and marking of the current net in: $(R\Gamma, s) \xrightarrow{r_j} (R\Gamma', s')$.

The accessible state graph of a $RN =< R\Gamma, M >$ net is the labeled directed graph whose nodes are the states and whose arcs which are labeled with events or rewriting rules of RN are of two kinds: (a) *firing* of an enabled event $e_j \in E(M)$ determines an arc from state $(R\Gamma, s)$ to state $(R\Gamma, s')$ which is labeled with event e_j when this event can fire in the net configuration $R\Gamma$ at marking M and leads to a new state s'; (b) *change configuration*: arcs from state $(R\Gamma, s)$ to state $(R\Gamma', s')$ labeled with the rewriting rule $r_j \in R$, so that $r_j : (R\Gamma_L, M_L) \triangleright (R\Gamma_W, M_W)$ which represent the change configuration of the current RN net. Also, for rewriting rule r_j it is required to identify if the RN_L net belongs to $R\Gamma$. Upon firing, the enabled events or rewriting rule modify the current marking.
As an example, we considered in figure 3 the RN1 net given by the following DE_{RN1}:

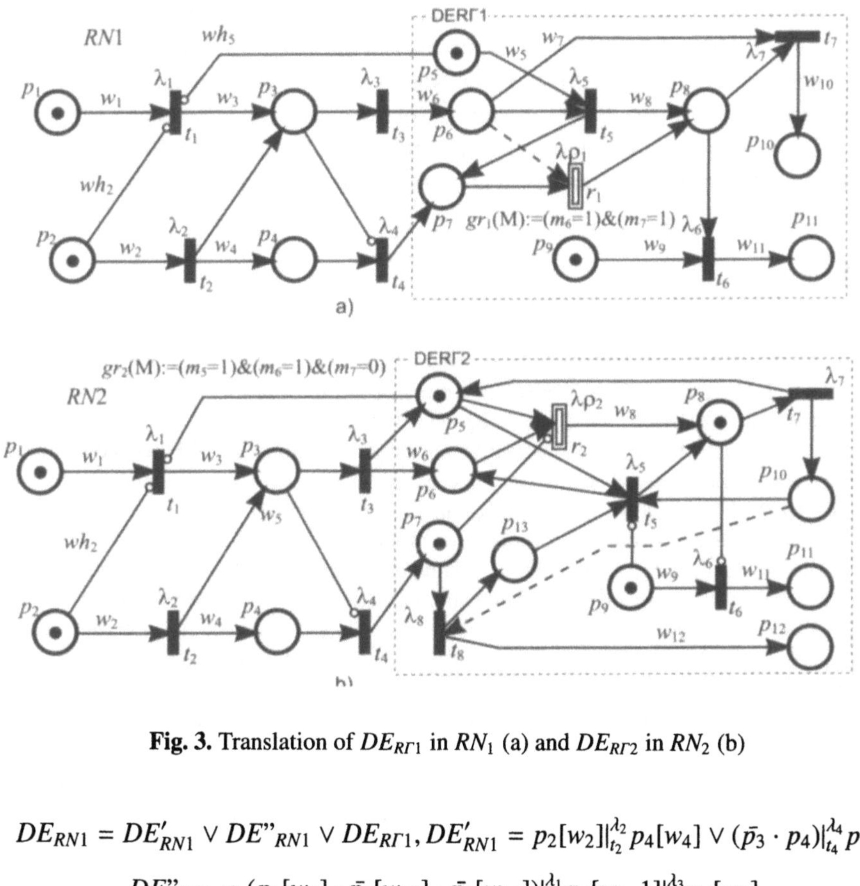

Fig. 3. Translation of $DE_{R\Gamma 1}$ in RN_1 (a) and $DE_{R\Gamma 2}$ in RN_2 (b)

$$DE_{RN1} = DE'_{RN1} \vee DE''_{RN1} \vee DE_{R\Gamma 1}, DE'_{RN1} = p_2[w_2]|_{t_2}^{\lambda_2} p_4[w_4] \vee (\bar{p}_3 \cdot p_4)|_{t_4}^{\lambda_4} p_7,$$

$$DE''_{RN1} = (p_1[w_1] \cdot \bar{p}_2[w_{h2}] \cdot \bar{p}_5[w_{h5}])|_{t_1}^{\lambda_1} p_3[w_3, 1]|_{t_3}^{\lambda_3} p_6[w_6],$$

$$DE_{R\Gamma1} = (p_5[w_5] \cdot p_6)|_{t_5}^{\lambda_5}(p_7 \Diamond p_8[w_8]) \vee (\tilde{p}_6 \cdot p_7)|_{r_1}^{\lambda\rho_1} p_8 \vee DE'_{R\Gamma1},$$

$$DE'_{R\Gamma1} = (p_9[w_9] \cdot \bar{p}_8)|_{t_6}^{\lambda_6} p_{11}[w_5] \vee (p_6[w_7] \cdot p_8)|_{t_7}^{\lambda_7} p_{10},$$

$$M_0 = (m_1, m_2, m_5, m_9), g_{r1}(M) := (m_6 = 1)\&(m_7 = 1), r_1 : DE_{R\Gamma1} \triangleright DE_{R\Gamma2}.$$

Resulting the firing of r_1 we obtain the $DE_{RN2} = DE'_{RN1} \vee DE''_{RN1} \vee DE_{R\Gamma2}$, where

$$DE_{R\Gamma2} = (p_5 \cdot \bar{p}_9 \cdot p_{10} \cdot p_{13})|_{t_5}^{\lambda_5}(p_6 \Diamond p_8) \vee (p_7 \cdot \tilde{p_{10}})|_{t_8}^{\lambda_8}(p_{12} \Diamond p_{12}) \vee DE'_{R\Gamma2},$$

$$DE'_{R\Gamma2} = \vee(p_5 \cdot p_6 \cdot \tilde{p_7})|_{r_2}^{\lambda\rho_2}(p_8[w_8, 1]]|_{t_7}^{\lambda_7}(p_5 \Diamond p_{10}) \vee (\bar{p}_8 \cdot p_9[w_9])|_{t_6}^{\lambda_6} p_{11},$$

$$M = (m_7, m_8, m_9), g_{r2}(M) := (m_5 = 1)\&(m_6 = 1)\&(m_7 = 0), r_2 : DE_{R\Gamma2} \triangleright DE_{R\Gamma1}.$$

The translation of $DE_{R\Gamma1}$ in RN_1 is shown in figure 3a, and that of $DE_{R\Gamma2}$ in RN_2 is shown in figure 3b.

5 Concluding Remarks

In this paper we have used a set of descriptive composition operations and expressions for the creation of marked-controlled rewriting $EFPN$ models. We have shown how easily and directly can an $EFPN$ model change its structure dynamically. We proved how these models are able to support basic inference and suggested how they can be subsequently used for fuzzy knowledge inference in expert systems.

In order to prove the $EFPN$ potential for modeling uncertain knowledge and inference, we are currently developing a software visual simulator with a friendly interface for the verification and evaluation of dynamic rewriting $EFPN$.

References

1. T. Cao and A. C. Sanderson, Representation and analysis of uncertainty using fuzzy Petri nets, In: J. Intell. Fuzzy Syst., vol. 3, pp. 3-19, 1995.
2. S. Chen, J. Ke, and J. Chang, Knowledge representation using fuzzy Petri nets, In: IEEE Trans. Knowl. Data Eng., vol. 2, no. 3, pp. 311-319, 1990.
3. E. Guţuleac and M. Mocanu, Descriptive Dynamic Rewriting GSPN-based Performance Modeling of Computer Systems, In: Proceedings of the 15th International Conference CSCS15, 25-27 May 2005, Bucureşti, Romania, pp. 656-66, 2005
4. E. Guţuleac, Descriptive Compositional Construction of Generalized Stochastic Petri Net Models for Performance Evaluation of Computer Systems, Bul. Inst. Politehnic, Tomul L (LIV), Fasc. 1-4, Automatica şi Calculatoare, Iaşi, Romania, pp. 143-159, 2004
5. X. Li, Wen Yu, and F. L. Rosano, Dynamic Knowledge Inference and Learning under Adaptive Fuzzy Petri Net Framework, In: IEEE Trans. Syst., Man and Cybernetics - Part C: Applications and Reviews, vol. 30, no.4, pp. 442-450, November 2000.
6. W. Pedrycz and F. Gomide, A generalized fuzzy Petri net model, In: IEEE Trans. Fuzzy Syst., vol. 2, no. 4, pp. 295-301, 1994.
7. D. S. Yeung and E. C. C. Tsang, A multilevel weighted fuzzy reasoning algorithm for expert systems, In: IEEE Trans. Syst., Man and Cybernetics - Part A, vol. 28, no. 2, pp. 149-158, 1998.

1

Off-Line Signature Verification using Feature Based Image Registration

Zsolt Kertész[1] and Bence Kővári[2]

[1] Budapest University of Technology and Economics, Department of Control Engineering and Information Technology kertesz@iit.bme.hu
[2] Budapest University of Technology and Economics, Department of Automation and Applied Informatics beny@aut.bme.hu

Summary. This paper discusses the issues of off-line signature verification and shows that signature verification methods can and should be improved by applying feature based image registration methods. First the different aspects and problems of signature verification are discussed in conjunction with off-line analysis methods. It is assumed that area based registration methods fail to improve their results because the high level of abstraction they use. To overcome this limitations a feature based image registration method is introduced, altered step by step to get a better matching between the endpoints of two signatures.

1.1 Introduction

Signature recognition is probably the oldest biometrical identification method, with a high legal acceptance. The most accurate systems almost always take advantage of dynamic features like acceleration, velocity and the difference between up and down strokes. This class of solutions is called on-line signature verification. However in the most common real-world scenarios, this information is not available, because it requires the observation and recording off the signing process. This is the main reason, why off-line signature analysis is still in focus of many researchers. These methods can only operate on static image data; therefore they often try to compare global features like size of the signature or similarities of the contour [12] [13] [14]. To get a tractable abstraction of the two dimensional images, these methods often involve some image transformation, like the Hough or Radon transformations [6] or work on the density models of the signatures [11]. These methods seem to give a good representation of the signature, allowing the researchers to reach Equal Error Rates (EER) between 10% and 15% [9]. The drawback of this methodology is that loosing the semantic information makes it almost impossible to improve the algorithm or to explain the results in detail. Jose L. Camino et al. take an other approach [4] they try to guess the pen movements during the

Zsolt Kertész and Bence Kővári: *Off-Line Signature Verification using Feature Based Image Registration*, Studies in Computational Intelligence (SCI) **78**, 277–282 (2008)
www.springerlink.com

signing by starting at the left and bottom most line-end and then following it. There are also other approaches trying to reconstruct the signing process. In [7] stroke, and sub-stroke properties are extracted and used as a basis for the comparison. Based on own experience, these latter approaches seem to be the most promising, because their results can be explained (and therefore improved) in a semantically meaningful way. There is also a wide variety of classifiers used to compare the results: hidden Markov models [4] [5], Support Vector Machines [6], multi-layer perceptron [3], genetic algorithms, and neural networks [15] are the most widely used solutions.

1.1.1 Preprocessing

In our method signatures are scanned with a resolution of 600dpi, resulting in an average image size of 1000*250 pixels. This resolution has shown to be necessary to correctly interpret the line crossings. Signature preprocessing is a necessary step to improve the accuracy of the latter algorithms, and to reduce their computational needs. The following steps are done:

1. A noise filter is applied to remove the noise caused by the scanner.
2. The image is cropped, to the bounding rectangle of the signature.
3. Transformation from color to grayscale, and finally to black and white.
4. Thinning the black and white image [10].

1.1.2 Feature extraction

Once, the preprocessing has finished, the endpoints of strokes are identified. To help improve the robustness of the latter processing algorithms, endpoints are also characterized by the direction of their corresponding strokes. Direction vectors are calculated from the first 10 stroke pixels next to the attached endpoint (Fig. 1.2). Connections are defined in a similar way except that they have 3 directions associated with them (Fig. 1.1). There are two side effects of the thinning algorithm which should be noted here.

- Rapid changes in pen angle often result in falsely detected endpoints. This reduces the quality of our "abstraction" but it does not reduce the quality of our matching algorithm because such changes in angle are as characteristic as real stroke endpoints.
- Connection points will always have three branches. This property only simplifies our mathematical model.

1.2 Matching of signatures

In this section the mutual correspondences of the signatures is discussed. Assumed, that a set of original signature images is available as reference, we

Fig. 1.1. Connections are represented by their position and 3 directions

Fig. 1.2. Endpoints are represented by their location and direction

have to decide from a new image, whether it has the same characteristics or not. This image correspondence process is called image registration in the literature. A large number of different applications use the image registration, from the medical imaging to the remotely sensed data processing. There are two major categories of the image registration [1], [2], the area-based and the feature based methods. The area-based methods are correlation-like techniques. These algorithms use Fourier properties and other means of structural analysis. There are some limitations of these methods, e.g. these often work with a rectangular window and if the images are deformed by more complex transformation, the window is not able to find the same parts of the scene. Since the signatures appear very variously, the area-based methods are less applicable for off-line signature analysis. The feature-based methods assume that two sets of features have been detected. These find the pair wise correspondence between reference and examined image using their spatial relations or various feature descriptors: lines, curves, points, line intersections, boundaries etc. The simplest procedure for giving the correspondence is the Iterative Closest Point algorithm [16], which matches two clouds of points. It estimates the transformation (translation, rotation) between the coordinates of points on two images iteratively.

1. Initialize: Define point P set it as P_0 and set k iteration variable to 0
2. Iteration: Find the new point set Q_k as the closest points on the image from the P_k Find the optimal rigid transformation T from P_k to Q_k Calculate P_{k+1}=$T(P_k)$
3. If the distance metric d_k-d_{k+1} $< \tau$, then terminate (τ is a predefined threshold value)

As an extension of the spatial relations methods, the correspondence of the features can be estimated using the properties above. It can be done through the following metrics:

- absolute distance from an appropriate reference point

Fig. 1.3. Correspondence between 'a' letters

- orientation of the above distance vector
- and the own properties of the features (at the endpoints this is an angle compared to horizontal axe, at junctions there are three angles compared to the horizontal direction, etc.)

The algorithm runs cyclically (like the Iterative Closest Point algorithm), it calculates the weighted distances of above metrics and changes the reference point in every loop. For the first iteration, the appropriate reference point can be simply the center of gravity (COG) in both images. In the next loops new reference points are selected, as the pair of points with the smallest distance.

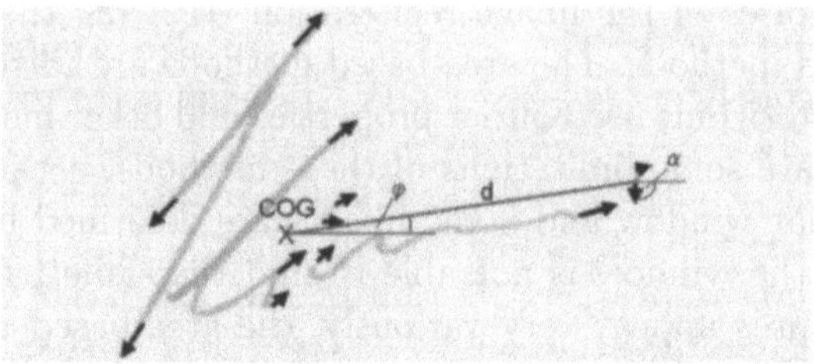

Fig. 1.4. Properties of the features (the arrows represent the position and orientation of endpoints in a signature).

1.3 Experimental results

In our experiments 16 original signatures were taken from 50 different signers. Then some of the signers were requested to stay and create forgeries of the original signatures, 16 pieces for every person. 640 signatures (320 originals and 320 forgeries) from 20 signers were used to tune our algorithm, and the other 960 signatures were reserved for benchmarking purposes. The effectiveness of a system is most commonly described with its "false rejection rate" (FRR, Type I error), its "false acceptance rate" (FAR, Type II error), and the "equal error rate" (EER). The false rejection rate is the percentage of original signatures the system rejects. The false acceptance rate is the percentage of forgeries the system accepts as original, and the equal error rate is the point at which the two factors intercept.

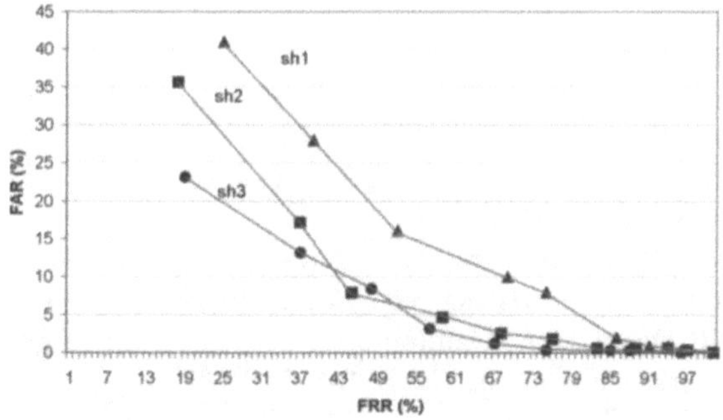

Fig. 1.5. Signature verification results measured on the "Test" database with different threshold configurations

Our results are demonstrated on Fig. 1.5. In the first case (sh1) only spatial information is taken into consideration. Better results (sh2) can be achieved by calculating with properties of the features. But there is still a weak point in the algorithm. If in any iteration a wrong (very distant) pair of points is chosen as reference, the matching will fail. In the third case only the first ten best pairs are used as reference points to minimize the probability of such errors (sh3).

1.4 Conclusions

A novel signature verification method was introduced, which is able to match features in signatures and thereby calculate a distance measure for the signatures. The differences between experiments sh1 and sh2 show, that not only the position of endpoints but also their direction has a significant impact on the accuracy of our system. This also implies that these measures can be effectively used in signature recognition systems. Experimental results are promising, even now, when the algorithm uses only a fragment of the available semantic information. Further benefits of this approach are, that each decision made by the algorithm, can be directly interpreted by humans, which facilitates the improvement and debugging of the whole system. Therefore future development will mainly focus on the extraction and comparison of further features.

1.5 Acknowledgements

This project was supported in part by the Innovation and Knowledge Centre of Information Technology BME (IT^2), the National Office for Research and Technology (NKTH) and the Agency for Research Fund Management and Research Exploitation (KPI).

References

1. Jan Flusser Barbara Zitov. Image registration methods: a survey. In *Image and Vision Computing 21*, pages 977–1000, 2003.
2. L. G. Brown. A survey of image registration techniques. In *ACM Computing Surveys 24*, pages 326–376, 1992.
3. R.C.L. Cajote R.D., Guevara. Global word shape processing using polar-radii graphs for off-line handwriting recognition. In *IEEE*, 2004.
4. Morales C.R. Ferrer M.A. Camino J.L., Travieso C.M. Signature classification by hidden markov model. In *IEEE, no. 5.*, 1999.
5. R. Sabourin C.y. Suen El-Yacoubi, M. Gilloux. An hmm-based approach for off-line unconstrained handwritten word modeling and recognition. In *IEEE*, 1999.
6. Najoua Ben Amara H A Sofien Touj. Global feature extraction of off-line arabic handwriting. In *IEEE SMC, vol. 2, no. 4.*, 2003.
7. Azriel Rosenfeld Jinhong K. Guo, David Doermann. Off-line skilled forgery detection using stroke and sub-stroke properties. In *Proceedings of the International Conference on Pattern Recognition (ICPR'00)*, 2000.
8. Bence Kővári. Extraction of dynamic features for off-line signature analysis. In *Automation and Applied Computer Science Workshop (AACS)*, Budapest, Hungary, June 2006.
9. Bence Kővári. The development of off-line signature verification methods, comparative study. In *microCAD 2007 International Scientific Conference*, April 2007.
10. Ching Y. Suen Louisa Lam, Seong-Whan Lee. Thinning methodologies - a comprehensive survey. In *IEEE Transactions on Pattern Analysis and Machine Intelligence archive Volume 14, Issue 9*, pages 869–885, 1992.
11. Jalal Mahmud and Chowdhury Mofizur Rahman. On the power of feature analyzer for signature verification. In *Proceedings of the Digital Imaging Computing, Techniques and Applications (DICTA 2005)*, 2005.
12. Alonso J. B. Ferrer M. A. Martinez L. E., Travieso C. M. Parametrization of a forgery handwritten signature verification system using svm. In *IEEE,vol. 2, no. 3*, pages 193–196, 2004.
13. Jesus B Alonso Miguel A Ferrer and Carlos M Travieso. Offline geometric parameters for automatic signature verification using fixed-point arithmetic. In *IEEE Transactions on Pattern Analysis and Machine Intelligence, vol. 27, no. 6*, pages 993–997, June 2005.
14. Etienne Sum Wah Robert Sabourin J-P, Drouhard. Shape matrices as a mixed shape factor for off-line signature verification. In *Fourth International Conference Document Analysis and Recognition (ICDAR'97)*, page 661, 1997.
15. Michael Blumenstein Stephane Armand and Vallipuram Muthukkumarasamy. Off-line signature verification based on the modified direction feature. In *The 18th International Conference on Pattern Recognition (ICPR'06)*, 2006.
16. Z. Zhang. Iterative point matching for registration of free-form curves and surfaces. In *International Journal of Computer Vision, vol. 13, no. 2*, pages 119–152, 1994.

Collaboration of Intelligent, Autonomous Systems: Situation Aware Behavior Change

Willibald Krenn[1] and Franz Wotawa[2]

[1] Institute for Software Technology wkrenn@ist.tugraz.at
[2] Institute for Software Technology fwotawa@ist.tugraz.at

Random encounters of autonomous devices capable of inter-device communication open up the possibility for collaboration. Often times, however, a change of each device's behavior is needed in order to leverage synergies arising from co-location. We present an approach based on a knowledge base that is able to control a device in an isolated situation as well as during co-location with other devices. In both instances, control is done in a fault tolerant way. The algorithm is capable of smooth behavior transitions as well as behavior shifts where the behavior of the device changes abruptly, which can be seen as a change in strategy. The underlying knowledge base stores all possible behaviors of the device. We present an experiment that illustrates shifts in behavior as well as behavior changes in the context of internal faults.

1 Introduction

Consider a Container Tracking Unit (CTU) that, apart from reading GPS data and sending it by GSM to some data terminal, also is equipped with another communication channel that provides medium-distance inter-CTU communication capabilities. Two different scenarios almost instantly spring to mind that show how a co-location of CTUs can be used to improve the over all service quality of the complete system if all CTUs are allowed to talk to each other without restriction. Let us consider in a first scenario that one CTU has incured a hardware defect and that its GSM module is broken. In a second scenario, a set of containers equipped with fault free CTUs is transferred by train. It is obvious that collaboration will improve the system's performance in both scenarios: In the first one, the faulty CTU will use another CTU as a relay so it can send its position data. In the second scenario, the CTUs can save energy by building up a cluster and declaring one CTU to be the "head" speaking for all of them while the rest can have some sleep.

Let us assume that each CTU knows about the goals `BuildCluster`, `SendPositionByRelay`, and `SendPositionDirectly`. Plans on how to reach

Willibald Krenn and Franz Wotawa: *Collaboration of Intelligent, Autonomous Systems: Situation Aware Behavior Change*, Studies in Computational Intelligence (SCI) **78**, 283–288 (2008)
www.springerlink.com

these goals are expressed as rules and stored in a knowledge base. It is easy to see that the importance of reaching a goal is dependent on the current situation: Co-located devices, the internal state of the device (e.g., faults, available energy), and the history of how often a goal has been reached have their influence.

We use a function (γ) to express the user-defined policy of attaining a particular goal: γ specifies how often a certain goal has to be reached within some time frame. Using γ and some additional factors, the system derives a weight for each goal that corresponds to a desirability value and works like a gradient because the closer the "activity" of a goal gets to the value specified by γ, the smaller the weight will be.

This describes the main properties of our approach: We want to reach each goal every time we iterate through the set of goals. Thereby the weight reflects the urgency of reaching a goal and at the same time it also indicates whether a goal has been reached too often. We include a factor in weight calculation that reflects past experiences gained while trying to reach a goal. The system decreases the weight of goals that constantly fail but removes this damping factor when the goal can be reached again. Apart from smooth transitions in behavior caused by the damping factor, changes to γ can lead to "shifts" in behavior. Coming back to our example, a behavior shift most likely will occur when one CTU detects another one co-located and is preparing for collaboration. In that case, the weight of goals `SendPositionByRelay` and `BuildCluster` will be increased, while the weight of `SendPositionDirectly` will be taken back.

This paper is organized as follows: In the next section we give an overview of the rule base our algorithm is using. Thereafter we discuss weight calculation before giving first results gained when running an example rule set on a prototype implementation. Finally, we conclude the paper.

2 Knowledge Base

We use two dimensions when selecting rules from the knowledge base. The first dimension is Boolean and indicates whether a certain rule is runnable at all while the second dimension expresses the importance, or desirability, to choose and run a rule. This section presents the Boolean-based part of our approach, namely the rules within the system knowledge base. We start by defining the knowledge base.

Definition 1 (System Knowledge Base (SKB)). *The system knowledge base (SKB) is a tuple* (P, R, G, γ) *where*

- P *is a set of propositions.*
- R *is a set of definite horn-clauses of the form* $x_1 \wedge \ldots \wedge x_n \rightarrow y$ *where* $x_1, \ldots, x_n, y \in P$. $x_1 \wedge \ldots \wedge x_n$ *is the antecedence and* y *is the consequent.*

- *$G \subseteq P$ is a set of goals. Goals are consequents of R marked as top-level.*
- *$\gamma : R \times \mathcal{R} \mapsto \mathcal{R}$ is a function mapping a real number to a real number for the consequent of a rule in R.*

Note that propositions can represent actions, i.e., activities which might alter the state of the system or the environment, e.g., `TurnOnGps` which turns on the GPS, basic propositions which are used to store certain knowledge, e.g., the current position of the system which has been measured by the GPS, and goals. In order to distinguish the different kind of propositions we assume that actions are represented by `action(X)`, and goals are of the form X:.

Following the given definition, `SendPositionByRelay` can be encoded in the knowledge base as a rule of the form `SendPositionByRelay: <= NewTick & HavePosition & action(NF.SendPosition);`
Thereby `SendPositionByRelay` is the consequent of the rule and because it is top-level (indicated by the colon) the consequent is a goal too. The rule encoding `SendPositionDirectly` looks similar but contains a call to `action(GSM.SendPosition)` instead of `action(NF.SendPosition)`.

While `NewTick` simply is a proposition in memory that is either true or false, `HavePosition` is a proposition that gets computed by another rule not shown here. In other words, `HavePosition` is the consequent of at least one other rule. Depending on the rule set there might be more than one rule defining the or-ed truth value of `HavePosition`.

The truth value of a rule consequent is calculated by assuming all actions can be carried out without error. This is equal to blending out all actions while doing the truth calculation and only checking the remaining propositions. Because action execution is monitored during runtime, the system knows when this assumption is violated, resulting in a cancellation of rule execution.

Due to these unexpected errors and suppose there is more than one rule providing a certain consequent, the system needs to have another quantity in order to choose between rules. Therefore we introduce another ordering on the rules that tells us about the desirability to run the rule and about the history of past runs. This ordering is based on data given by the supplied γ function and is described in the next section. Note that because we allow rules to interact with the γ function at runtime, we are able to force behavior "shifts" on the system.

3 Calculating Weights

Weight calculation is largely based on user supplied information. There are, however, two factors the system has to track, namely the damping and the activity factor. During runtime, as a goal has been reached or failed, these factors get updated accordingly.

The activity factor is a value that describes how often the particular rule has been selected. It is a simple counter that gets aged after a complete run through the knowledge base. The damping factor stores information about failed attempts to run the rule: It is increased whenever an error occurs during rule execution and decreased when the rule executes without problem.

The weight, or desirability, is calculated by a function $\delta : \Gamma \times \mathcal{R} \times \mathcal{R} \mapsto \mathcal{R}$ that returns the result of $\gamma'_n(r, activity) * L * (1 - damping)$. Note that Γ is the set of all possible functions $R \times \mathcal{R} \mapsto \mathcal{R}$ where $\mathcal{R}$ denotes the real numbers and R the rules. Within the presented formula for weight calculation, γ'_n returns a normalized slope for a rule consequent at some point of activity. L is the length from the point of the current activity to the next local maximum on γ. Note that $0 \leq L \leq 1$ and $0 \leq damping \leq 1$.

One important property of weight calculation is that the system tries to minimize all weights and, thus, get the activity factor of each consequent to the value specified by the user. This reflects the idea that the complete system behavior is a superposition of several embedded, smaller behaviors. In our approach, the atoms of "small behavior" are all plans stored inside the knowledge base telling the system how to attain a certain goal. The superposition of behaviors not only depends on whether plans meet their preconditions and are therefore run but also on γ that tells the system how often a goal shall be reached in the first place. The latter property is important for being able to control the system behavior, biasing it according to the situation the device is in. Our approach features two distinct mechanisms that can be used for strategy control: The first mechanism, using the damping factors, is completely automatic and an inherent part of our approach. The second one can be exploited by user-supplied rules and permits a change of γ.

As the system developer is completely free in implementing actions (as long as they can be monitored) collaboration of co-located devices can go as far as transferring γ and/or damping factors, thus allowing quick adaption. Transmission of rules is thinkable too but will need better support from the framework.

4 Experiments

We have conducted first experiments with an algorithm based on the presented ideas. It is a prototype implementation running on a PIC 18F micro-controller that has 128K bytes flash- and about 4K bytes random access memory. On this prototype we loaded a rule set that is able to acquire the current position and either send it by relay or by GSM. The knowledge base also comprises a rule that checks whether a cluster of CTUs is available and, if so, triggers a shift in behavior. Because of the rather limited hardware resources the γ function in our prototype implementation is restricted to straight lines for each rule consequent. However, this does not interfere with our aim of demonstrating the practicability of the approach.

The experiment we discuss in this section demonstrates smooth behavior change as well as behavior shift and is meant to clarify the ideas from this paper. Basically, within the knowledge base there are three rules we are interested in:

```
SendPositionByRelay: <=  NewTick & HavePosition &
      action(NF.SendPosition);
SendPositionDirectly: <=  NewTick & HavePosition &
      action(GSM.SendPosition);
SendPositionDirectly.ChangeGammaMax(1)    :- InCluster;
```

We already know the first two rules. They tell the system that a position reading can either be transmitted directly by GSM, or by using another CTU as relay. The last rule in the given excerpt, however, is new: It tells the system that whenever the proposition `InCluster` is true, it has to change γ(`SendPositionDirectly:`, _) and set the maximum to one. We cannot cover the details of how exactly the system handles this request other than saying that the new value of the γ-maximum most likely won't be one at once, as the system also takes the current maximum (and possibly other change requests) into account.

In our experiment we have simulated 50 runs of the knowledge base. During each run the system has tried to reach all goals. At the beginning of the experiment the γ-function was equally set up for both goals, so the system would have chosen each goal equally often. However, as we simulated a fault within the NF component during the first nine runs of the knowledge base, the system could not reach the first goal (send by relay) during that time. The idea in this situation is that the system will adapt the damping factor and learn to prefer the second goal that is not inhibited by faults. From run number 11 to 41, we set the proposition `InCluster` to true, hence the system was expected to shift its behavior, preferring the goal it first had learned is failing. Runs 42 to 50 then again simulate an error in the NF hardware. In addition the `InCluster` proposition is cleared (set to false) in these runs. Because the rule set does not contain any rule that tells the system to reset the maximum value of γ in case `InCluster` does *not* hold, we were interested in whether the system would be clever enough to select the second goal and send the position directly, even if this had been practically disallowed by the γ change.

Table 1 presents the results of the presented experiment. In summing up, the system behaved as intended and it was possible to send position data during all parts of the experiment. For further experiments we refer the interested reader to [2]. A better overview of our system design along with other experiments is given in [3].

5 Conclusion

There is lot of work based on similar ideas: The rule set can be compared to, e.g., TR-programs [4], the notion of desirability is also used in [5] and a bit

Run	1	2	3	4	5	6	7	8	9	10	11	...	41	42	...	47	48	49	50	51
Rule	G	(R)	G	G	(R)	G	G	G	G	G	R	R	R	(R)	(R)	G	-	-	-	R
Cluster	n	n	n	n	n	n	n	n	n	n	y	y	y	n	n	n	n	n	n	n

Table 1. The experiment shows that the system is able to learn and has the ability for behavior shift. "R" indicates that position data was sent by relay. "(R)" means that an error occurred (as designed) when sending the data. "G" means that the position was transmitted by GSM and the dash indicates that no rule was selected.

differently in [1]. One can also relate the management of the damping factor to machine learning algorithms and the idea of selecting rules and continuously scanning through the knowledge base can be brought in connection with biological systems [6].
In summing up, we present an approach for controlling an autonomous device that is optimistic in a sense that it wants to reach all goals specified by the system developer all the time. Having said that, the exact strategy of how often a goal needs to be attended is given by a user-supplied function (γ) and is also influenced by past experiences. A global change of strategy is supported by changing γ. Collaboration of co-located devices apart from the functional level is possible and can be provided by allowing exchange of knowledge.

Acknowledgments

This work has been supported by the FIT-IT research project *Self Properties in Autonomous Systems (SEPIAS).*

References

1. Cynthia Breazeal (Ferrell) and Brian Scassellati. Infant–like social interactions between a robot and a human caretaker. Technical report, Massachusetts Institute of Technology, 1998.
2. Willibald Krenn and Franz Wotawa. Gradient-based diagnosis. In *DX–07 18th International Workshop on Principles of Diagnosis*, pages 314–321, 2007.
3. Willibald Krenn and Franz Wotawa. Intelligent, fault tolerant control for autonomous systems. In *Proceedings of the Fifth International Workshop on Intelligent Solutions in Embedded Systems*, pages 194–205, 2007.
4. Nils J. Nilsson. Teleo-reactive programs for agent control. *Journal of Artificial Intelligence Research*, (1):139–158, January 1994.
5. A. Saffiotti, K. Konolige, and E. H. Ruspini. A multivalued-logic approach to integrating planning and control. *Artificial Intelligence*, 76(1-2):481–526, 1995.
6. Gregory A. Wray, Matthew W. Hahn, Ehab Abouheif, James P. Balhoff, Margaret Pizer, Matthew V. Rockman, and Laura A. Romano. The evolution of transcriptional regulation in eukaryotes. *Molecular Biology And Evolution*, pages 1377–1419, 2003.

On the Distributed Water Pollution Control Solving with an Agent-Based Approach

Mihaela Oprea[1] and Constantin Nichita[2]

[1] University Petroleum-Gas of Ploiesti, Department of Informatics, Bd. Bucuresti Nr. 39, Ploiesti, 100680, Romania, mihaela@upg-ploiesti.ro
[2] University Petroleum-Gas of Ploiesti, Department of Informatics, Bd. Bucuresti Nr. 39, Ploiesti, 100680, Romania, ccnichita@yahoo.com

Summary. The use of an agent-based approach for distributed problem solving represents a natural solution as it provides many advantages such as parallelism, robustness, and scalability. However, these advantages are contingent of an efficient coordination mechanism. In this paper we present the case study of distributed problem solving by using an agent-based approach in a water pollution monitoring, diagnosis and control system. We have proposed a simple and efficient coordination mechanism for our geographically distributed agent-based system.

1 Introduction

The use of multi-agent systems for distributed problem solving represents a natural solution that was adopted in several real world distributed problems. In [4], the main task is decomposed in subtasks that are performed (in certain cases, periodically) by intelligent agents that are geographically distributed. Our research work proposes the use of an intelligent agent model to automate the water pollution monitoring, diagnosis and control in order to improve the execution of these tasks by using an efficient coordination mechanism. Related work in the field of environmental monitoring is presented for example in [3].

The paper is organized as follows. In section 2 some general information related to water pollution control are introduced. The case study of solving the distributed problem of water pollution control is detailed in section 3. An example of application is given in section 4. Last section concludes the paper.

2 Water Pollution Control

Water is widely used: for industrial processing (as a cooling and heating medium, as an intermediate agent for chemical processes, to generate steam of various qualities, etc.), for civil use (drinking water distribution, firefighting,

Mihaela Oprea and Constantin Nichita: *On the Distributed Water Pollution Control Solving with an Agent-Based Approach*, Studies in Computational Intelligence (SCI) **78**, 289–294 (2008)
www.springerlink.com

etc.), in agriculture, etc. Water is required in widely differing qualities (for instance pharmaceutical and electronics industry require water that is highly purified). Drinking water quality standards are defined in the COUNCIL DIRECTIVE 98/83/EC [2].

As a future member of the European Union, Romania will have to establish monitoring programs to comply with the European legislation in the field: "Member States shall take all measures necessary to ensure that regular monitoring of the quality of water intended for human consumption is carried out, in order to check that the water available to consumers meets the requirements ..." [2].

Tap water testing comes too late to ensure safe drinking water, due to the nature of current microbial sampling and testing, which typically provides results only after water has been distributed and often consumed. Appropriate application of monitoring parameters is needed to ensure safe drinking water through the design of appropriate risk management plans. The monitoring has to take place throughout the water distribution system in order to meet specific information requirements regarding: catchment's protection and assessment, source water quality assessment, assessment of treatment efficiency, monitoring the quality of drinking water leaving the treatment facility and in the distribution system [6].

The work described in this paper addresses the need to develop a predictive system that will warn when a hazardous situation is imminent and will enable timely and cost-effective correction of the situation in the form of water safety plans. The system considered is in fact a source-to-tap system. Triggers of potentially unsafe water include: inadequate performance of drinking water treatment plant, spillage of a hazardous substance into source water, failure of power supply to a critical asset, extreme rainfall in a catchment, detection of unusually high turbidity (source or treated water), unusual taste/odor/appearance of water, disease outbreak for which water is a suspect vector. Table 1 presents a selection of the main parameters (left column) that may be used in the monitoring of water distribution.

The main stages of the monitoring are in fact associated with corresponding steps in the distribution process, denoted in the table by the columns heads 1,2,...,8. In Table 1 these symbols are: 1-source water characterization, 2-groundwater characterization, 3-treatment efficiency (removal), 4-treatment efficiency (disinfection), 5-treated water, 6-distribution system (ingress), 7-distribution system (regrowth), 8-outbreak investigation, S means suitable and SA suitable alternative. Indicators for the safety and for the quality of water can be selected and systematically monitored throughout the water supply chain or cycle. Such indicators should yield information in real time to enable corrective action to prevent unsafe water being supplied. Alert levels can be set, against which to compare observations. Alert levels would typically be just within critical limits of operation, outside of which confidence in water safety would be lost [6]. Pre-determined corrective actions can be implemented once alert levels are exceeded. An example of response to tur-

Table 1. Selection of the main parameters that may be used in the monitoring of water distribution

PARAMETER	1	2	3	4	5	6	7	8
Rainfall events								S
Flow	S			S	S		S	S
Color					S			S
pH				S				S
Solids (Total and dissolved)	S	S						S
Conductivity	S	S						S
Turbidity	S	S	S					S
Particle size analysis			S					S
Microscopic particulate analysis		S		S	S			S
Disinfectant residual						SA		S
Organic matter (TOC, BOD, COD)	S	S					S	S
Ammonia	S	S						S

bidity indicator levels will be presented in the next section. Turning off water supplies can have major economic consequences due to lost production and damaged equipment. Rapid notification procedures need to be practical and available at any time. Systems that enable monitoring of water from source to consumer can assist in better targeting of these types of responses to minimize the extent of their impacts.

3 Distributed problem solving in the case of water pollution control

Upon detection of an indicator parameter exceeding a predetermined allowable level or limit, the monitoring station MA2 in Fig. 1a would react by sending notification to adjacent stations (nodes), MA1 and MA3. These nodes would in turn send notification to the adjacent nodes, and eventually the notification will reach the supervising station (SV). Figure 1b illustrates a detailed description of a monitoring station. The activity of each regular station is controlled by several agents and a measuring device (MD) which corresponds to real life automated online analyzers. The agents are: Monitoring Agent (MA), the Water Pollution Controlling Device (which can be replaced by an interface with a Human Agent–WPCD/HA), the Database Management Agent (DMA) and the Reasoning Agent (RA). The RA agent is responsible for implementing high level of decision support and online decision making in emergent situations. The RA agent can make predictions regarding the location and distribution of hazardous components at future times. The integration of population demographics into the computer model will enable predictions of the number of people potentially affected by a detected disaster. For completing its tasks, the RA agent can use rule-based systems, case-based systems and/or neural

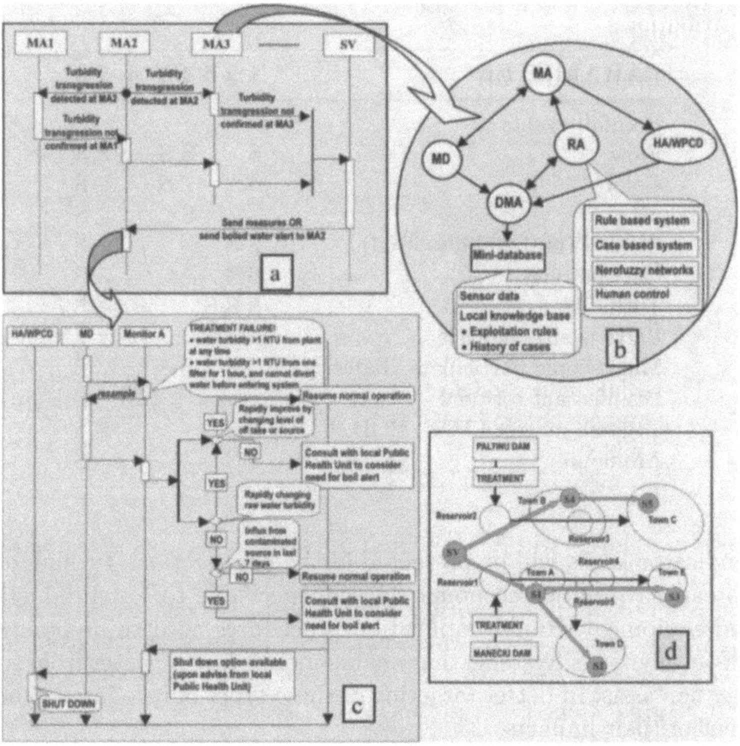

Fig. 1. Multi-Agent System Architecture

networks techniques in addition or to complement the activity of a human expert.

The MA agent will be configured to reconcile differences associated with multiple generated configurations. Also detection at multiple locations can be combined into a single collaboratively developed model. The collaborative action can be in the form of sharing data, receiving data from adjacent nodes, or collaborative working with other similar agents to develop predictive analysis. MA can also asses relevant data available, making a determination as to whether to send data in unprocessed form or to process it before sending the information. The communication between agents of adjacent nodes gives the system the ability to view a disaster from any of several nodes, giving the system different perspectives on the event. The model can be periodically or continuously upgraded as additional information becomes available in order to increase the accuracy of the calculations.

In the case described in Fig. 1a, however the decision-making process can be made locally, since the transgression in the turbidity is not confirmed at the adjacent nodes. Under these circumstances the standard protocol to be followed is described by the collaborative diagram in Fig. 1c. NTU in the diagram stands for Nephelometric Turbidity Units, which are the most commonly used units for turbidity. The turbidity is monitored at several points in the source water supply system. If there is an increase of greater than 5 NTU within three hours then an incident is declared and alternative sources of water should be used. Filtered water turbidity is also monitored continuously and values exceeding 1 NTU are set to trigger warnings. Turning off water can have major economic consequences due to lost production and damaged equipment, and must be exercised only in extreme situations.

4 Application

To develop our multi-agent system we used specific techniques and tools. For the analysis and design phase we used GAIA methodology [7]. Protocols are modeled using the Agent UML sequence diagrams [1]. More details about the implementation of the system are presented elsewhere [5].

The implemented multi-agent system will monitor the local water distribution system in the region Prahova. We identified the principal surface sources of drinking water in Prahova region, namely the dams at Paltinu and Maneciu, the second one serving a population of about 21.670 people. Next we consider a hypothetical situation(shown in Fig. 1d), which better describes the use of the multiagent system . The five towns illustrated in the diagram, have monitoring stations denoted S_i, i=1,5. Each station is equipped with a server, which controls a local wireless network of sensors placed in several key points of the local distribution system. The server is connected to a mini database which stores temporal series of indicator parameters which are monitored continuously/hourly/daily. The database also contains rules of interpretation and operation, comprised of exploitation rules of the system and history of cases which together form the knowledge base. Each station is connected through a high-speed network with the neighboring stations located on the same water distribution line. For instance S3 is connected with S1 and S1 is connected with S3 and also with SV. All stations have the same warning/alarm system software. The equipment of the supervising station SV is similar to that of the regular stations, with the exception of the data acquisition and monitoring system.

The SV station has a high-speed connection line with the zonal Water Directorate (SGA Prahova) and the Public Health Directorate Prahova, and a visual interface with a human dispatcher. The comprehensive database of the supervising station is used to periodically update and correct the knowledge bases for each regular station. Based on the data received from the sensor network, there may appear emergency situations. The rule-based algorithms from the knowledge base may be used to interpret these events as warnings or alerts and to propose solutions for the remedial of the corresponding situation. Through a system of data replication, the supervising station and all stations affected by a common contamination problem will have on display the same information on which they can make action plans based on inference algorithms.

In a first phase, the supervising station will have the decision role, meaning, its decision will be implemented by the station(s) where nonstandard situation is encountered. The monitoring agent corresponding to the station which sent out the alarm, must confirm when the action took place. These requests, approvals, proposals, confirmations should be recorded in the database for future references. In a more advance system, this decision could actually be made locally, through cooperation and negotiation between the monitoring agents of each station.

5 Conclusions and future work

Multi-agent systems represent a natural solution to model distributed systems. We have described the case study of a distributed problem solving by using an agent-based approach in a water pollution monitoring and diagnosis system. Future work will be done on negotiation and dynamic coordination issues between the agents of the multi-agent system. We plan to further investigate the use of alarms that allow room for sensor fuzzy readings, using thus fuzzy logic instead of boolean logic rules, as described in [3].

Acknowledgment. The research work reported in this paper was funded by the Romanian Education and Research Ministry, under the postdoctoral research project CEEX nr. 1533/2006.

References

1. Bauer B, Muller JP, and Odell J (2001) Int J Softw Eng Know, 11(3):207–230
2. COUNCIL DIRECTIVE 98/83/EC on the quality of water intended for human consumption
3. Hatzikos E, Bassiliades N, Asmanis L, and Vlahavas I (2007) Expert Syst 24(3):143–161
4. Lesser V, Ortiz C, Tambe M (2003) Distributed Sensor Networks: A Multiagent Perspective. Kluwer Academic Publishers, Boston
5. Oprea M, Nichita C (2007) An application of agent-based systems in environmental protection. CSCS 16, Bucharest, May 22–26, 2007 (proceedings of the 16th International Conference on Control Systems and Computer Science)
6. Guidelines for drinking-water quality. IInd edn. Vol. 3, Surveillance and control of community supplies. World Health Organization(WHO), Geneva
7. Zambonelli JF, Jennings N, and Wooldridge M (2003) ACM T Softw Eng Meth 12(3):317–370

Learn Reversi using Parallel Genetic Algorithms

Daniel Paraschiv and Laurentiu Vasiliu

CIMRU National University of Ireland, Galway
{daniel.paraschiv,laurentiu.vasiliu}@nuigalway.ie

Abstract: This paper presents how parallel genetic algorithms can be used to learn the game Reversi. Parallel Genetic Algorithm can create a distributed environment in which instances of genetic algorithms are executed in parallel. The history of chromosomes is stored in an ontology using OWL language. One computer player takes a decision using the Analytic Hierarchy Process (AHP) method of multiple criteria decision analyses (MCDA) and using game decision tree. To learn the game means to find the optimal weights and rules.[1]

Keywords: Parallel Genetic Algorithms, MCDA, AHP, Decision Trees, OWL

1 Reversi Game

Reversi, also called Othello, is a strategic board game of 8x8 squares for two players [1]. In Figure 1 a is presented the GUI of the application presented in this paper. The application generates (learn) the optimal chromosome and a human can play with this optimal chromosome. The game starts as is seen in Figure 1 a and black has the first move. One player moves by reversing opponent stones, opponent stones have to be between the player stones. All alternatives for black are presented in Figure 1 b and c. When black places in the right dot of Figure 1 b and white above it the game is like in Figure 1 c. The game finish when there are no alternatives for both players. The winner is the one that has more stones on the board.

When choosing an alternative (a move) a player has to consider more criteria. One criterion is to number how many stones are after he or she chooses the respective alternative. Another criterion is to consider the influence of the position of stone on the board. A diagram with the influences of stones on the board is presented in Figure 1 d. Because of the symmetry of the board all position stones are reduced to these 9 positions. The third criteria is presented in Figure 1 e when black places in the dot,

[1] This research has been funded from CFTD/05/312 - M2M Negotiation Enterprise Ireland project, at National University of Ireland, Galway

Daniel Paraschiv and Laurentiu Vasiliu: *Learn Reversi using Parallel Genetic Algorithms*, Studies in Computational Intelligence (SCI) **78**, 295–301 (2008)
www.springerlink.com

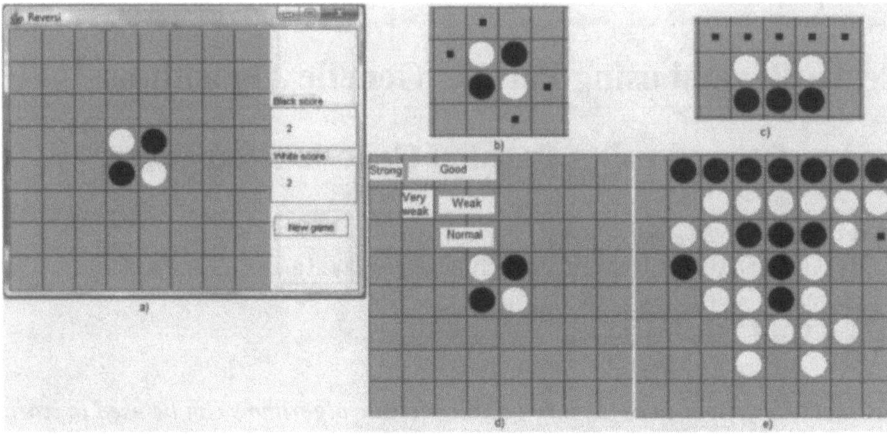

Fig. 1. Reversi game

the stones are connected with the corner and they can not be reversed until the end of the game. Another criterion of taking a decision is by looking ahead meaning that the opponent moves are considered. The application uses these four criteria and it uses AHP for the first three criteria and a decision tree for the last one. Each of the first three criteria has weights and the influence of a stone on the board is implemented with a rule of the form IF stone is at position (x,y) THEN the influence is Ixy. The parallel genetic algorithm is used to learn the optimal weights and rules.

2 Parallel Genetic Algorithms (PGA)

In parallel genetic algorithms distinct groups of chromosomes, called demes, evolve in parallel [2]. The migration process is the process in which chromosomes are copied or moved from one deme to the other. In Figure 2 a is presented a network of demes where the arcs represent the migration between demes. These demes can be in a distributed environment.

In Figure 2 b is presented the PGA. Each deme starts with a random population of chromosomes on which is performed the GA [2, 3]. After a given number of new populations in a deme the best chromosome migrates to other demes as can be seen at "Send" step. The "receive" step checks if new chromosomes came from other demes and if yes these are placed in the new population. When all demes finish its GA the best chromosome in all demes is considered the optimal solution.

One benefit of PGA is that it reduces the crowding problem often encountered in nonparallel GAs in which because of a chromosome that dominate the entire population the system falls into a local optimum [2]. Another benefit is the increase of speed when the computation of the fitness function is polynomial or exponential with respect to the number of chromosomes. In the application of this paper the fitness function is $O(n^2)$ where n is the number of chromosomes in a deme, because each chromosome plays with the others two times once as black and once as white.

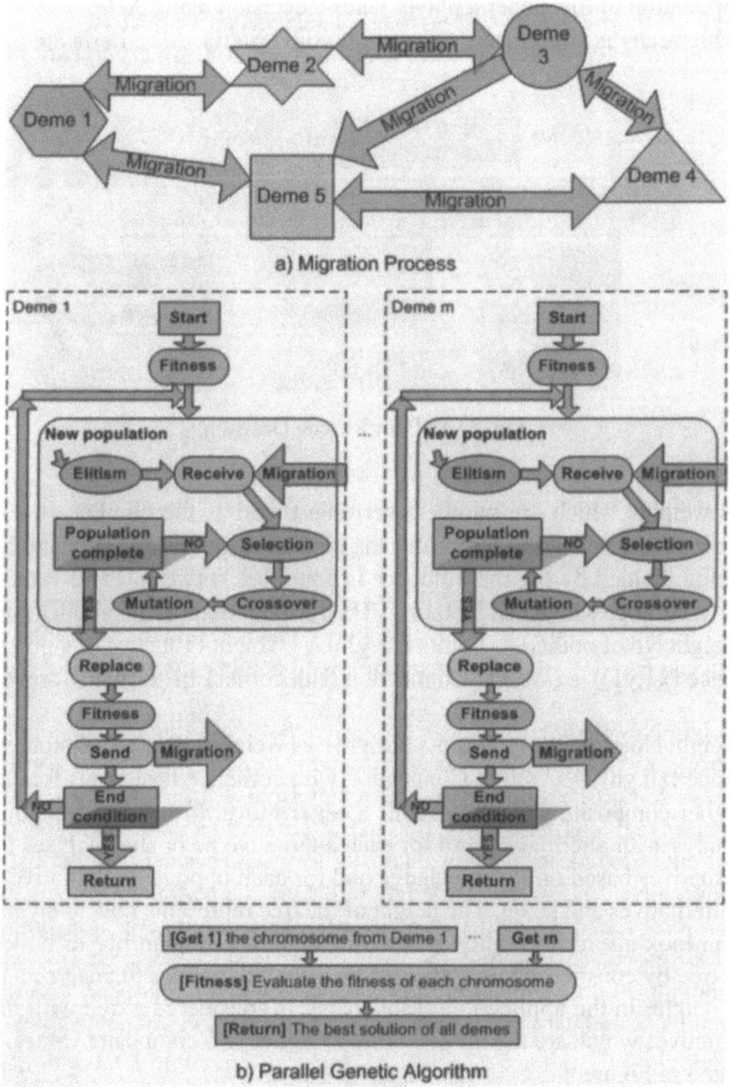

Fig. 2. Parallel Genetic Algorithm

3 Multiple Criteria Decision Analyses and Game Decision Tree

For a player choosing the next move means to choose the best alternative. For this the player builds a game decision tree to consider also the opponent alternatives. All alternatives are characterised by three criteria as presented in introduction section and in Figure 3. To quantify the quality of the move a multiple criteria decision analyses method can be used to create a composite value for each alternative. Some of MCDA methods are SCORING (SAW), TOPSIS, AHP, PROMETHEE or DEA [4].

In the application of this paper a player takes a decision using AHP.

The hierarchy is simple here there are no sub criteria. All criteria are maximise,

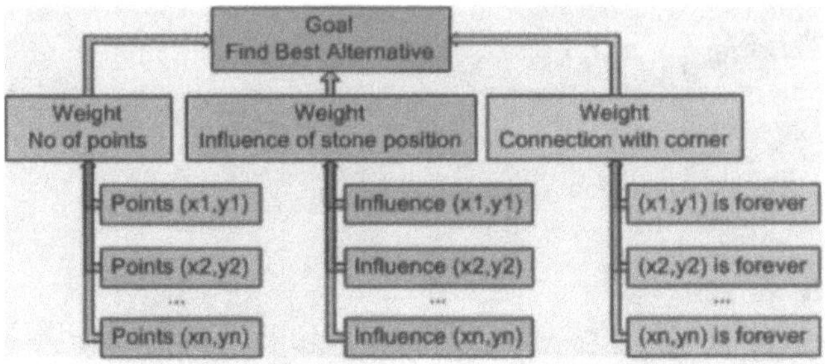

Fig. 3. AHP for Reversi Decision

each has weights which are numbers between 0 and 1, the number of points is a number between 0 and 1 and represents the number of points after placing a stone in this position divided by 64, the influence is a number between 0 and 1 and stone is forever is a number which can be 0 or 1. The formula used in the application is:

C1 = (Weight No of points) * (Points (x1,y1)) + (Weight Influence of stone position) * (Influence (x1,y1)) + (Weight Connection with corner) * ((x1,y1 is forever)

...

Cn = (Weight No of points) * (Points (xn,yn)) + (Weight Influence of stone position) * (Influence (xn,yn)) + (Weight Connection with corner) * ((xn,yn) is forever)

A higher composite value represents a better move. In a game decision tree the player analyses all alternatives and for each alternative he or she analyses the opponent alternatives based on its knowledge and for each opponent alternative analyses its own alternatives and so on. The height of the tree represents how many moves the player can look ahead. In reality in a strategic game a human player build a game decision tree by considering only the two or three best moves to analyse a tree with a bigger height. In the application of this paper is considered a tree with maximum two alternatives which are the moves with the higher two composite values. The tree is presented in Figure 4.

For each alternative in a node is defined a probability which is computed as the

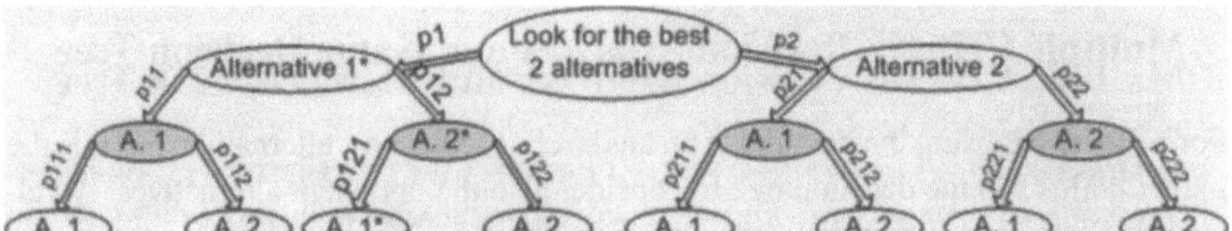

Fig. 4. Decision tree in Reversi game

value of the composite value divided by the sum of the two composite values. If ci, cj are the two highest composite values and p1 and p2 are the two probabilities in the node then p1 = ci/(ci+cj) and p2 = cj/(ci+cj). To decide which alternative to choose between the initial two alternatives the program computes the probabilities to happen for the final alternatives and the highest value is considered. For example for A. 1* alternative, in Figure 4, the probability to happen is p = p1*p12*p121. If the tree is unbalanced the probability to happen is multiplied with 0.5 as many times to create a virtual balanced tree. The player chooses Alternative 1* if the highest probability to happen is in the first half of final alternatives (from the left), otherwise the player chooses Alternative 2.

4 Case study: Learn Reversi using Parallel Genetic Algorithms

The application of this paper is composed of two parts, the part of learning the game and the part in which a human plays with the best chromosome. To analyse the evolution of learning the best chromosome and to create the migration process the application makes use of an ontology [5] build with Protege-OWL [6]. Protege creates automatically java code to query the ontology. The Schema of OWL ontology is presented in Figure 5.

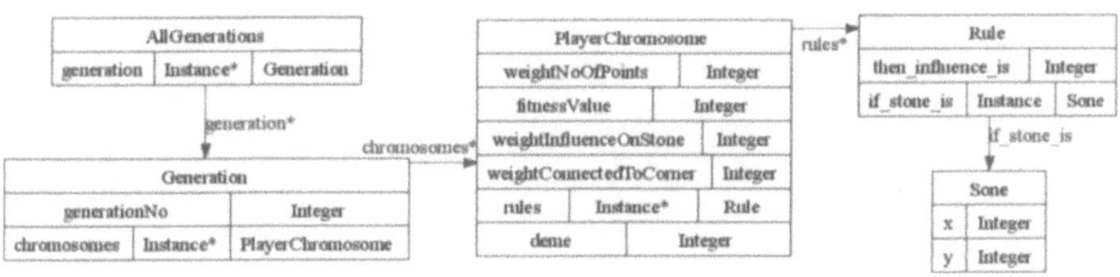

Fig. 5. The meta-data of OWL of Reversi application

The application permits that a human player to play with a computer which has the knowledge encoded in a chromosome using a simple GUI (see Figure 1 a) and the computer to play with computer each computer having its own knowledge encoded in its chromosome. To play a game the computer uses for each move the AHP method and game decision tree presented in section 3 and a chromosome which is encoded as can be seen in Figure 6. The height of the game decision tree is set to three. To learn the game means to find the best knowledge that are encoded in a chromosome. Each weight and influence is a number between 0 and 100 divided by 100 to be scaled between 0 and 1, thus the chromosomes are value encoded. A naive approach would be to compare each chromosome with the others to get the best chromosome. As a chromosome has three weights and nine influences the searching space is $100^{(3+9)} = 10^{24}$. A processor of 1GHz would take at least 10^{15} seconds (32 million years) only to take in its registry these chromosomes and to compute the fitness of the chromosome would take much more. The application uses an evolutionary method to find an optimal chromosome, the used method being the parallel genetic algorithm presented in section 2. Each deme starts with a generation of 20 random chromosomes and then new population evolve. The end condition is true when 12 new populations evolved. All generations are stored in the ontology. To create the migration process new demes are created in sequence and after 6 new generations the best chromosomes from previews demes are loaded from ontology.

The fitness function is computed using a competition in which each player plays with all others two times once as black and once as white. The fitness value represents how many times a chromosome won a game. The computation time of the fitness function is $O((tn)^2)$ where t is the time to play a game and n the number of chromosomes in a deme. The time to play a game is high, but is almost constant because it depends on the size of the board and the height of the decision tree.

The main operations of creating a new population are elitism, selection, crossover and mutation. In elitism the best two chromosomes are copied in the new population from the old one to not lose the best chromosomes. The selection is Roulette Wheel Selection in which two parents chromosomes are selected from the old population with a probability the fitness value divided with the sum of all fitness values. The crossover operation is presented in Figure 6. Two single point crossover masks are

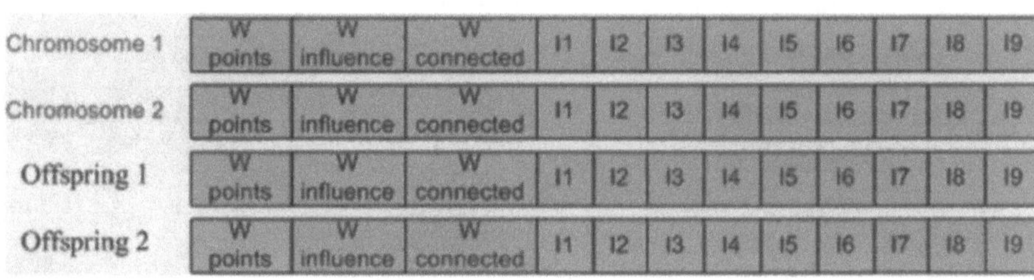

Fig. 6. Crossover operation

used, one for weights and one for rules. In the case from Figure 6 the crossover masks are 100 and 111111000. When the crossover mask bit is 1 the value in offspring one is loaded from chromosome one and when the mask bit is 0 the value is loaded from chromosome two as can be seen with the colours. The others parts of the chromosomes form the offspring two. The number of 1 in each crossover mask is chosen randomly. The mutation operation mutates each value of the chromosome with probability 0.09. Mutating a value means to add the value with a random value from intervals [-0.05,-0.01] or [0.0,0.05]. The mutation operation is useful to avoid the system to fall in a local optimum. Some observations are:

1.Usually one chromosome dominates the populations in a deme after 6 generations.
2.The other chromosomes try to beat the chromosome that dominate, thus if the chromosome that dominate is a good one then all chromosomes become good.
3.When one chromosome migrate from one deme to the other this chromosome that is the best in its original deme can survive in the host deme and may influence the host deme to make all chromosomes better or may die fast if the fitness of the mi-

grated chromosome is low in the host deme.

5 Related work

Reversi is a simple game on which can be proved the efficiency of learning algorithms. Ming-Da Wu, Ying-Hong Liao and Chuen-Tsai Sun [7] present some learning methods for Reversi game using Genetic Algorithm and Game Theory. Kmmerlin [8] did his thesis with Genetic Algorithms with application in game Reversi. Buro [9] presents a paper with evaluation functions for the game Othello in which uses weights like in AHP method. Nejad [10] has a good presentation of Game Decision Tree for Reversi game.

6 Conclusions

In the introduction of this paper are presented the Reversi rules and four criteria that a player uses to move which are the number of points, influences, connected to a corner and look ahead. The weights for the first three criteria and the influences are encoded in a chromosome. This paper presents how this chromosome can evolve to an optimal solution using PGA. In section 2 PGA are presented. In section 3 is presented how a player decide about the next move in Reversi using AHP and game decision tree. In section 4 is presented the ontology and PGA used by the application: the fitness function, selection, crossover and mutation, the migration process and some observations. Some improvements of the application would be to make it work in a distributed environment, because the OWL ontology has proved to be very slow, and the game decision tree can be optimized with game theory as presented by Nejad [10]

References

1. Reversi rules http://www.pressibus.org/reversi/gen/gbdocs.html
2. Mitchell, T.M., Machine Learning. McGraw-Hill series in computer science. 1997, New York: McGraw-Hill.
3. Introduction to Genetic Algorithms http://cs.felk.cvut.cz/~xobitko/ga/
4. Alexander Ablovatski, Multicriteria Decision Aid
5. OWL tutorial http://www.w3schools.com/rdf/rdf_owl.asp
6. Protege http://protege.stanford.edu/
7. Ming-Da Wu, Ying-Hong Liao and Chuen-Tsai Sun, Network Tournament Pedagogical Approach Involving Game Playing in Artificial Intelligence, Journal of Information Science and Engineering, Vol.19 No.4, pp.589-603 (July 2003)
8. Dorothee Kmmerlin, Genetic algorithms for the determination of heuristic evaluation functions in games
9. Michael Buro, From Simple Features to Sophisticated Evaluation Functions, Proceedings of the First International Conference on Computers and Games (CG-98)
10. Hamed Ahmadi Nejad, An Introduction to Game Tree Algorithms

Graph-based semantic description and information extraction in analysis of 3D coronary vessels visualizations

Marek R. Ogiela[1], Ryszard Tadeusiewicz[1], and Mirosław Trzupek[1]

AGH University of Science and Technology
Institute of Automatics, Bio-Cybernetics Laboratory
30 Mickiewicza Ave, 30-059 Krakow, Poland
Tel. +48-12-617-38-54
Fax. +48-12-634-15-68
{mogiela, rtad, mtrzupek}@agh.edu.pl

Abstract. The work presents new possibilities for making semantic description and interpretation of some medical visualization with the use of Computational Intelligence linguistic formalisms. The methodology described in this work makes possible to find, for each considered biological 3D structure, its semantic description whose elements make reference to the medical significance of the entire structure described, while at the same time refrain from the formal differences (i.e. appearance and shape) of individual visualisations. Such methodology may be further used for attaining other goals related to computer-assisted semantic interpretation of selected elements and/or the entire 3D structure of the coronary vascular tree. The obtained semantic information allows to make a description of the structure, focused on the semantics of various heart muscle lesions. Thanks to these, the analysis conducted allows fast and automated interpretation of the semantics of various morphological changes in the coronary vascular tree. Especially it is possible to detect stenoses in the lumen of the vessels that can cause critical decrease of blood supply to extensive or especially important fragments of the heart muscle. The basis for the construction of 3D reconstructions of biological objects used in this work are visualisations obtained from helical CT scans, yet the method itself may be applied also for other methods of medical 3D images acquisition.

1 Introduction

Today's state-of-the-art methods of image diagnostics provide abundant and varied diagnostic and research material, as 3D visualisations are available

Marek R. Ogiela et al.: *Graph-based semantic description and information extraction in analysis of 3D coronary vessels visualizations*, Studies in Computational Intelligence (SCI) **78**, 303–309 (2008)
www.springerlink.com

for practically all human body structures and organs. They can illustrate the pathological changes to a greater degree and more completely, having however assumed that the physicians interpreting the image understand well what they see and what it means. Clinical practice shows that in spite of the frequent use of major computer-assisted diagnostic equipment, the visualisations (including the spatial reconstructions of specific body parts) obtained, later undergo only a rough qualitative assessment by the physician diagnostician, with no in-depth semantic analysis.

Naturally, the practical implementation of the postulate of understanding 3D medical visualization requires resorting to algorithms and information techniques far more advanced than those connected to the traditional automatic understanding and classification. Such, however, is the role of science that it must outpace the potential offered by the methods known and generally applied in pursuit of new approaches that meet practical requirements better: for that reason, in this work, we shall deal with the construction of cognitive systems supporting the semantic analysis of 3D visualizations used in diagnostic systems.

2 The goal of analysis

The processed medical images may be more convenient for visual interpretation or easier for automatic analysis, yet this would not move the question of interpreting its meaning forward. Similarly, image analysis, providing the values of these or those parameters describing objects visible in the image as well as automatic recognition that allows their classification do not in fact answer the most fundamental questions asked by the physician. These are questions concerning the nature of the disease that the patient suffers from, the situation of the potentially dangerous narrowings and possible therapeutic suggestions, which, however, calls for the understanding of the content of the medical visualisation, and not solely its formal analysis. What becomes necessary in such cases is the application of more advanced IT techniques, and particularly reaching for the so-called computational intelligence methods with generalisation attributes. Application of linguistic description, making use of special languages reaching into the semantic field of the patterns registered (also as images) opens the possibility of its understanding with methods derived from parsing of appropriate grammars. The parsers suggested in the article (and their future, updated versions that are the subject of current research work) shall allow the conversion of image description built around the language suggested into a description in the semantic terms and making it possible to understand all the significant medical facts related to the image. The linguistic and cognitive methodology postulated in this article shall allow the creation of hybrid records at further stages of the analysis; such records will contain both the considered 3D images and their automatically generated cognitive descriptions containing the definition of their semantic properties.

3 The manner of developing semantic models for spatial reconstruction of the heart's coronary vessels

To allow the introduction of linguistic formalisms in the form of graph-based image grammar that allows finding the stenoses in main arteries or their ramifications and, at the same time, quantification of the pathological changes discovered, several visualisation sequences were performed for the arteries in the DICOM format. Such sequences were obtained for various patients during diagnostic examinations of the heart conducted with a helical CT scanner. Such visualisations present in a very clear manner all morphologic changes of individual sections of arteries in any plane and they allow performing spatial rendering of coronary vessels. Shown in the Fig. 1.A.B and in the Fig. 2.A.B examples of spatial reconstruction of the heart's coronary vessels, presenting the vessels without the accompanying muscle tissue of the heart. It presents the main arteries together with the minor vessels and stenoses in the initial section and in the lateral branch of the left coronary artery and also in the initial section and in the posterior interventricular branch of the right coronary artery. The proper graph-based grammar describing these structures was defined so that the individual branches of the graph in the description provided identify all start and end points of coronary vessels and all bifurcations or transitions of main vessels into lower-level ones. Thus developed, the graph-based structure will constitute the elements of the language for defining spatial topology and correct vascularisation of the heart muscle together with the potential morphological changes. In practice, the mutual spatial relations

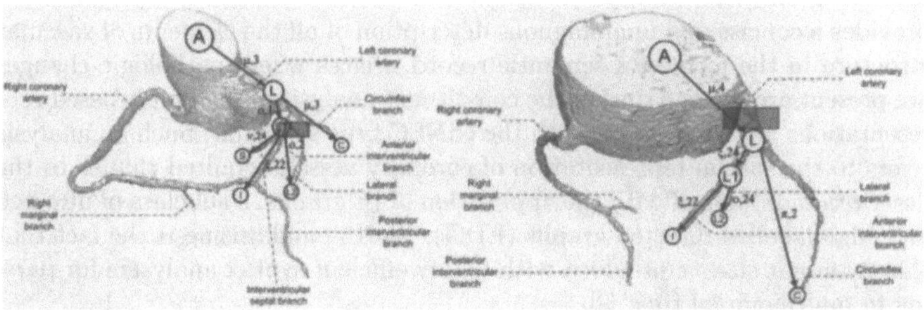

Fig. 1. Spatial labelling of coronary arteries and defining relations occurring between them for the left coronary artery. The grey rectangles mark the places of major stenoses in a coronary artery

that may occur between elements of the vascular structure represented in the form of a graph describing the set of edge labels. Elements of the set are defined by the introduction of appropriate spatial relations: vertical, defined by the $\{\alpha, \beta, \ldots, \lambda\}$ set of labels, and horizontal, defined by the $\{1, 2, \ldots, 24\}$ set of labels on the hypothetical sphere surrounding the heart muscle. These

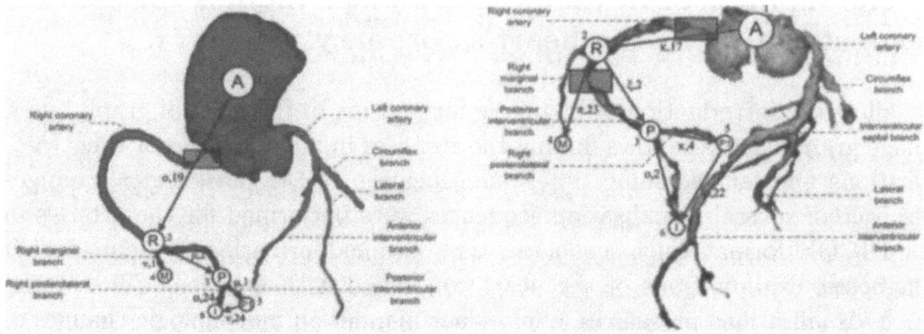

Fig. 2. Spatial labelling of coronary arteries and defining relations occurring between them for the right coronary artery. The grey rectangles mark the places of major stenoses in a coronary artery

labels designate individual sections, each with the angular span of 15 degrees. Next, for labels defined in this manner, edge terminal labels are assigned to each of the bifurcations marked by the start and end points of appropriate coronary artery sections, depending on the situation of sections of the arteries. Once these relations have been taken into account and node labels of the artery-spanning graph have been entered, it is possible to develop a linguistic description for the needs of defining the semantics of the changes sought after and the recognition of pathologic stenoses. Such creating among other things allows for conducting a computer-supported analysis of the rendered structure, oriented on the automatic detection of places of stenoses, the degree of narrowing and also its type (concentric or eccentric). Such a representation provides a concise and unambiguous description of all the elements of vascular structure in the form of a semantic record. Places where pathologic changes are present are tracked during the constituent analysis of the graph-based representations analysed, made with the edNLC type grammar. Such an analysis refers to the spatial representation of coronary vessels acquired thanks to the description developed with the application of IE graphs, a subclass of directed and edge-labelled directed graphs (EDG). Worth emphasising is the fact that this grammar class is provided with highly efficient syntax analysers for parsing in multinominal time [3].

To define the places where vessel pathologies occur the edNLC-class grammar has been proposed with a finite set of productions recorded in paranthesis-based form and with the use of diagrammatic graph-based notation (Fig. 3.A.B) The edNLC-class grammar allows spatial analysis of the structure described with IE graphs. The set of deriving rules in the form of IE graphs, which allow for deriving spatial representation for the right coronary artery described in Fig. 2.A.B is presented in Fig. 4.A.B.

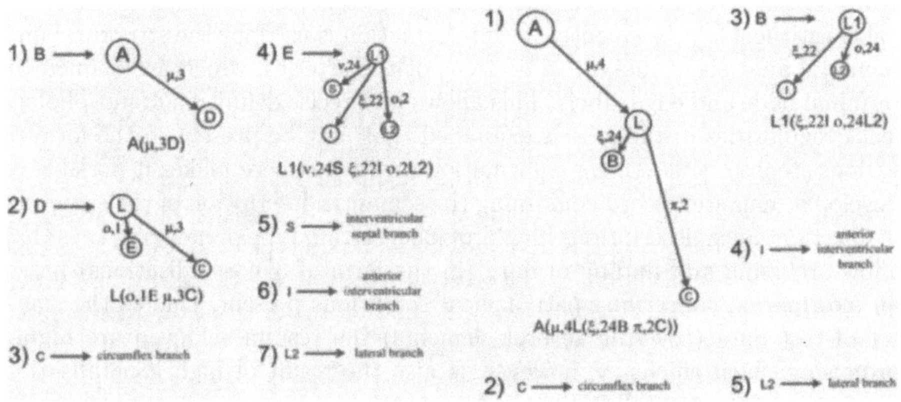

Fig. 3. Production set for the G_{edNLC} grammar deriving the structure of coronary vessels for graphs in the Fig. 1.A.B

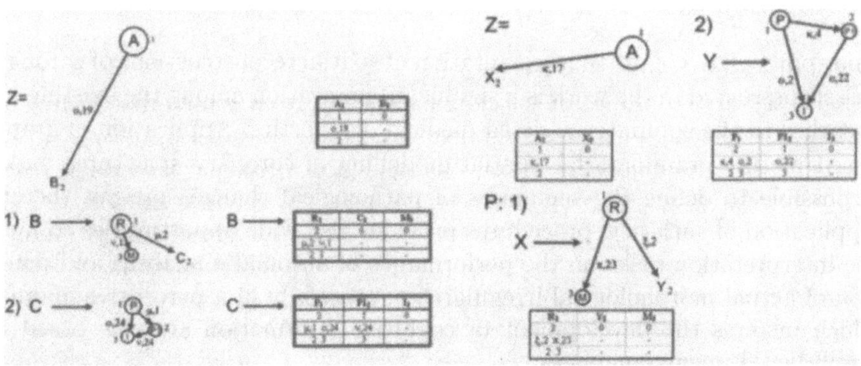

Fig. 4. The set of productions in the form of IE graphs deriving the structure of coronary vessels together with the vertex characteristic description for graphs in the Fig. 2.A.B

4 Results

As the research conducted proves, thanks to the application of the formalisms and graph-based grammars presented in the work, it is possible to develop semantic descriptions and spatially model the morphology of coronary vessels as well as discover in them some lesion symptoms. The discovery of such changes is performed by determining the place of their occurrence i.e. the definition of the position of the vessel and the degree of the stenoses of the artery lumen, and by the designation of morphometric parameters defining its scope and type (eccentric or concentric). The proper element of the syntactic and semantic analysis is parsing that, in the case of structural analysis of the representations developed, provides automatically (thanks to the use of semantic actions and procedures) full information defining the spatial topology of the graph in question, describing the coronary vessels together with the individual constituents. Subsequently, such information is used for the structural and semantic analysis of individual sections of the arteries, which are defined by terminal node and edge labels. This allows for precise definition of morphological irregularities in the vessels examined that may be present in the form of various stenoses [10]. All the information mentioned here makes it possible to develop a semantic record containing the semantic description of the structure in question. Such descriptions may provide a certain type of indexing keys that allow archiving and finding of data (in the form of 3D visualisations) based on information concerning pathological conditions present. Due to the small set of test data (covering several elements) the results achieved are highly promising. Such efficiency, however, is also the result of high generalisation properties of the formalisms applied.

5 Conclusion

The approach to cognitive interpretation of spatial reconstructions of coronary vessels presented in the work is a significant innovation among the existing approaches to the computer-assisted medical diagnostics. Application of graph-based picture grammars for spatial modelling of coronary structures makes it possible to define the semantics of pathological changes present therein. Application of such new procedures provides also wide opportunities to focus the interpretation tasks on the performance of automatic analysis and detection of actual morphological irregularities, especially in a perceptive manner, which inspires the development of cognitive information systems based on graph-based image languages.

Such methods are of profound significance as they allow not only for the discovery of pathologic changes but also for the definition of their semantics, which - in the case of medical diagnostic images - may lead to computer-assisted 'understanding' of their medical significance, and in future also to forecast and plan the optimum therapeutic procedures. Worth noting are the vast development of medical engineering procedures in the field of intelligent information systems, and the deep analogy between the functions of the presented model of analysis and the interpretative cognitive procedures taking place in the human mind [1].

Acknowledgement

This work has been supported by the Ministry of Science and Higher Education, Republic of Poland, under project number N516 025 32/2881 decision No. 2881/T02/2007/32.

References

1. Tadeusiewicz R, Ogiela MR (2004) Medical Image Understanding Technology. Springer Verlag, Berlin-Heidelberg
2. Meyer-Baese A (2003) Pattern Recognition in Medical Imaging. Elsevier-Academic Press
3. Skomorowski M (2000) A Syntactic-Statistical Approach to Recognition of Distorted Patterns. Jagiellonian University, Krakow
4. Khan MG (1996) Heart Disease Diagnosis and Therapy. Williams & Wilkins, Baltimore
5. Netter FH, Colacino S (1998) Atlas of Human Anatomy. Novartis Medical Education
6. Bankman I (ed.) (2002) Handbook of Medical Imaging: Processing and Analysis. Academic Press
7. Higgins WE, Reinhardt J M (2000) Cardiac image processing, 789-804, in Handbook of Video and Image Processing. Ed.A. Bovik, Academic Press
8. Higgins WE, Karwoski RA, Ritman EL, Spyra WJT (1996) System for analyzing true three-dimensional angiograms. IEEE Trans. Med. Imag. 15: 377-385
9. Hoffman KR, Sen A, Lan Li, Chua Kok-Gee, Esthappan J, Mazzucco M (2000) A system for determination of 3D vessel tree centerlines from biplane images. The International Journal of Cardiac Imaging. 16:315-330
10. Ogiela MR, Tadeusiewicz R (2005) Nonlinear Processing and Semantic Content Analysis in Medical Imaging - A Cognitive Approach. IEEE Trans. on Instrumentation and Measurement, 54(6):2149-2155

MIX
Papier aus verantwortungsvollen Quellen
Paper from responsible sources
FSC® C105338

If you have any concerns about our products,
you can contact us on
ProductSafety@springernature.com

In case Publisher is established outside the EU,
the EU authorized representative is:
Springer Nature Customer Service Center GmbH
Europaplatz 3, 69115 Heidelberg, Germany

Printed by Libri Plureos GmbH
in Hamburg, Germany